THE NATURAL AND THE NORMATIVE

Theories of Spatial Perception
from Kant to Helmholtz

Gary Hatfield

"This is a major reinterpretation of Kant's work and its continued relevance to cognitive science as revealed through nineteenth-century attempts to pursue implications suggested by Kant's writings, culminating in the work of perhaps the most important cognitive scientist of the time, Hermann Helmholtz. In a brilliant reinterpretation of Kant's work certain to provoke lively discussion among Kant specialists, Hatfield shows that Kant embraced both the naturalistic and the normative positions in a manner consistent with his philosophical enterprise."
—*Timothy Lenoir*, Stanford University

Gary Hatfield examines theories of spatial perception from the seventeenth to the nineteenth century and provides a detailed analysis of the works of Kant and Helmholtz, who adopted opposing stances on whether central questions about spatial perception were amenable to natural-scientific treatment. At stake were the proper understanding of the relationships among sensation, perception, and experience, and the proper methodological framework for investigating the

(continued on back flap)

The Natural and the Normative

The Natural and the Normative

Theories of Spatial Perception from Kant to
Helmholtz

Gary Hatfield

A Bradford Book

The MIT Press
Cambridge, Massachusetts
London, England

This book was set in Palatino by The MIT Press. It was printed and bound by Halliday Lithograph in the United States of America.

Library of Congress Cataloging in Publication Data

Hatfield, Gary C. (Gary Carl)
    The natural and the normative: theories of spatial perception from Kant to Helmholtz / Gary Hatfield.
        p. cm.
    "A Bradford book."
    Includes bibliographical references.
    ISBN 0-262-08086-9
    1. Space perception—History. I. Title.
BF469.H34 1990
153.7'52'09033—dc20   90-32191
                                                              CIP

Contents

Preface

This is an essay in the history and philosophy of theories of perception beginning with the rise of modern science in the seventeenth century but focusing on the period from Kant to Helmholtz. It examines both philosophical and psychological approaches to perception, particularly to the visual perception of space and spatial properties. Early modern theories of perception typically were embedded in a theory of mind, whether philosophical or psychological. My study looks not only at perceptual theories themselves but also at what they reveal about theories of mind, and especially about the relationship between philosophy and psychology as complementary (or competing) approaches to investigating the mind.

Philosophers from Descartes through Locke, Berkeley, Hume, and Kant—and on to Helmholtz—all exhibited a deep interest in the mind's apprehension of spatial properties. They believed that the investigation of the knowing mind would reveal much about the intellectual foundations of natural science. Space and spatial properties became central to the modern scientific picture of nature during the seventeenth century. Philosophers who wished to evaluate (or to support) the claims of the new science therefore analyzed the cognitive basis for knowledge of spatial properties. Although this analysis was concerned with claims about spatial properties in the "external world," the mentalistic bent of early modern metaphysics and epistemology directed the analysis toward the perception of such properties, diverting the focus from the spatial world to the perceiving mind. As a consequence, theories of spatial perception were at the center of attention in what we would now call early modern philosophy and psychology.

The question arose of how to investigate the knowing mind. The best science of the day suggested that the way to understand any process was in terms of the spatiotemporal interactions of its small parts as governed by laws of motion. Could the same type of account be applied to the mind itself, and particularly to the mind's activity during spatial perception? Some early modern thinkers tried to extend the approach of natural science to the perceiving mind; their work signals the intellectual (as opposed to institutional) birth of naturalistic psychology. Others denied that natural science could provide an explanation of perception considered as an epistemic achievement, not on the grounds that the mind is a spiritual substance distinct from nature, but because thought itself cannot be described adequately in the language of natural science. In more recent terminology, these authors maintained that thought has an inherently normative or evaluative core that cannot be explained, or even adequately described, from within the naturalistic vocabulary of physics, physiology, and mechanistic psychology. The opposition between these contrasting attitudes toward the possibility of a science of the knowing and perceiving mind is my central theme. I discuss both scientific theories of perception and the work of those who saw their approach to perception as logically prior to, or outside, the domain of natural science.

The work of Immanuel Kant provides the primary example of the latter approach. Although Kant did not develop a scientific theory of perception, or even discuss in any great detail the theories developed by others, he was especially concerned to articulate the relationship between philosophical work on perceptual knowledge and actual or possible scientific investigations of the mental. Kant considered philosophical and psychological approaches to perception to be distinct in both problems and methods. He marked off some problems as proper to each, insisting that philosophical questions about perceptual knowledge could not receive natural-scientific answers. Other authors, both before and after Kant, attempted to develop scientific theories that spoke directly to the questions Kant considered peculiarly philosophical. They typically saw themselves as resolving philosophical questions through the power of empirical investigation. Whether they could have been and were successful, whether philosophical problems admit of empirical answers, was and is a contested question. It is a question that immediately arises from contemplating the response to Kant's work by nineteenth-century German physiologists, psychologists, and philosophers. This response culminated in the work of Helmholtz, who saw himself as resolving Kantian questions with research methods and modes of explanation proper to natural science.

In this study I shall not ignore the fact that the question of whether philosophical problems admit psychological solutions remains under dispute. My focus, however, will be on the genesis of the intellectual framework within which the dispute takes place. My purpose is to uncover the conceptual formation of separate philosophical and psychological approaches to perception and to the mind. My concern is with the intellectual issues that underlie the close interaction between philosophy and psychology during their emergence as distinct mental disciplines. My approach to the study of this historical development is that of the historian of philosophy, and of the historian and philosopher of science, rather than of the social and institutional historian, for whom a "discipline" is a subject taught in school. I treat a discipline as a department of knowledge characterized by its own subject-matter and methods: as a "mental discipline" rather than a school discipline. Psychology was conceived as a discipline in this sense long before the institutional and professional matrix of psychology crystallized in the second half of the nineteenth century.

I hope that this book will interest philosophers and historians of philosophy, historians of science, and contemporary perceptual psychologists and cognitive scientists. From long experience of daily or weekly interaction with members of each of these groups, I know that this hope aims high. Indeed, the conjunction of an historical approach with problems of contemporary interest and under contemporary dispute may seem inappropriate to members of all three groups. Historians generally are leery of those who attempt original historical work while keeping contemporary problems in mind, fearing (partly on empirical, if impressionistic, grounds) that such writers will allow their own interest in the subject-matter to distort their view of materials from the past. This is a real danger. But it may be met by keeping one's eye on the methodological ideal of locating and understanding authors and texts in their historical context first, before taking a broader view. In any case, the idea of a philosophically neutral history of thought is in fact naive, and since the aim of this study is to investigate and to evaluate philosophical and psychological arguments and positions, it cannot by its very nature remain neutral.

Philosophical readers raise the question from another standpoint. Although the idea that one way to do philosophy is to do it historically has conspicuously reemerged in contemporary philosophy, there remain those who doubt that the study of past thinkers can serve to advance present thought. The latter doubt is likely to be even more prevalent among contemporary scientists concerned with vision, who sometimes half-jokingly describe anything written more than five years ago as "out of date," and who may conceive the appropriate scholarly activity and attitude toward such materials to be that of the "literature review"—a label connoting that re-viewing something adds nothing to the stock of knowledge but merely condenses a portion of the storage space dedicated to older ideas.

Skepticism about the direct yield of history in solving contemporary philosophical or scientific problems is understandable if it is based on the belief that the historical approach claims to contribute in stepwise fashion to the day-to-day problem solving activities of certain groups of philosophers and scientists. But a different role for history deserves to be acknowledged, in the analysis of the fundamental concepts that frame the problem-space for contemporary theorizing, both psychological and philosophical. This sort of framing investigation examines the very idea of a scientific account of perception, the very idea of a special domain for philosophy, and the particular ideas that have directed the mainstream theoretical tradition in the study of vision since the seventeenth century. It investigates "fundamental" concepts not in the old foundationalist sense of deciding their status once and for all but in the sense of deepening our understanding of concepts that are fundamental to certain ongoing concerns.

Past and contemporary thought about perception connects with everything from neurophysiology and psychology to philosophy of mind, epistemology, and philosophy of science. I have attempted to achieve a concrete focus without overlooking the wide range of connections between perception and other subject-matters by examining this complex of relations in the context of a particular historical episode. The episode centers on Kant, Helmholtz, and their opposing stances on whether central questions about spatial perception are amenable to natural-scientific treatment. In addition to Kant and Helmholtz, who have an acknowledged presence in contemporary thought (Kant among philosophers, Helmholtz among perceptual theorists), the episode includes lesser known figures such as Fries, Herbart, Steinbuch, Tourtual, and Lotze. This array of actors suits my purpose, because it retains a connection with familiar positions while nonetheless providing what history can offer: a laboratory of worked-out positions that are distant enough from present positions to facilitate a certain amount of detachment. Such detachment enables the study of past positions to teach us about possibilities in the problem space that we may not otherwise be familiar with, and it helps us see the contingency of the range of theoretical options that constitute the framework for contemporary thought.

Whether a historical approach can yield the desired understanding of the problems to which it is applied is a question that can be answered only by the product produced. Whatever is good in my results owes much to the institutions and persons who have, over a span of years, provided support for this project. I began research on an ancestor of this book while at the University of Wisconsin/Madison. I gratefully acknowledge the support of a National Science Foundation predoctoral fellowship, a Wisconsin Alumni Foundation Graduate Fellowship, and Vilas Foundation travel funds during that period. This work resulted in a doctoral dissertation completed in 1979. For their expeditious and critical responses to drafts, I am indebted to my advisors, Fred Dretske, William Epstein, Victor Hilts, David Lindberg, and Elliott Sober. They and other readers may recognize revised portions of the earlier work in certain sections of chapters 2 through 5, although the major portion of each chapter has been produced in subsequent years.

During those years, my work has been aided and abetted by a number of colleagues and institutions. During a sojourn at Harvard, Lorraine Daston in History of Science and Burton Dreben in Philosophy read significant portions of the older manuscript; I am indebted to each of them for posing thought-provoking questions. Various groups and associations gave me the benefit of their response to portions of the work in progress during this time: the History of Science Departments at Harvard and Johns Hopkins, to an early version of chapter 5, in 1981; the Department of Psychology at Hopkins and an audience at the biennial meeting of the Philosophy of Science Association, on portions of chapters 3 and 5, in 1984 (a version of the latter talk appeared in *PSA 1984*, vol. 2); faculty and students of the University of Chicago, at a lecture series on Science and the Humanities, on material that has been used in chapter 5, in 1985; and participants in the Edinburgh conference on Science and the Enlightenment, on a portion of chapter 3, in 1986. My colleagues at the University of Pennsylvania have given generously of their time in reading and commenting upon recent drafts; special thanks are due to Alan Kors, James Ross, and Jay Wallace.

Along the way, numerous libraries and rare books collections have made their resources available to me. Between 1977 and 1979 the librarians and curators of the Bayerische Staatsbibliothek, the Rare Books Collection of the Deutsches Museum in Munich, the Rare Books Library of the University of Wisconsin, the Preussischer Kulturbesitz in West Berlin, and the Central Archive of the Akademie der Wissenschaften in East Berlin welcomed me and provided courteous and efficient service. More recently, the staffs of the Rare Books Collection at the University of Pennsylvania, the American Philosophical Society Library, and the historical collection of the College of Physicians in Philadelphia have done the same. In an age when libraries too often are described as "information retrieval systems," it remains a pleasure to visit institutions that specialize in preserving materials in "paper archival format" (as both books and manuscripts have been labeled by librarians I have known) under conditions that allow them not only to be retrieved but also to be read and studied.

Some friends and colleagues have suffered with this project longer than others. David Ring provided many helpful suggestions on the original manuscript and on the early chapters of a recent version. Hannah Ginsborg has long been a stimulating discussant of the issues raised here and has generously provided comments on later drafts of several chapters. Patricia Kitcher read and encouraged the work as it was in 1982 and gave me useful criticisms and suggestions on two subsequent occasions. Harry and Betty Stanton and their staff at Bradford have been quietly patient and encouraging.

The preface of a first book provides an opportunity to offer special thanks to one's teachers. In addition to those already mentioned, I owe a special debt to my father, Chandler B. Hatfield, for teaching me many things, large and small, and to Mira Merriman and Phillip Thomas, for teaching me to think historically while encouraging me to pursue work in both the humanities and natural science.

Finally, I am grateful to Rose Ann Christian for her patience and for her critical green pen.

The Natural and the Normative

Chapter 1
Introduction

Can there be an empirical science of the mind? Opinions are divided. If within the purview of "mind" we include phenomena typically identified as "reasoning," "understanding," and "perceiving," then many philosophers, from the time of Kant until recent days, have answered that there cannot be. Reasoning and understanding, they say, are "normative," are subject to appraisal as true or false, right or wrong, valid or invalid, and so on.[1] As the point is sometimes put, to reason or to understand is to achieve a certain kind of success.[2] The essential features of such mental acts cannot be captured in a naturalistic (nonevaluative) vocabulary that merely describes mental processes and neural mechanisms. Because the question whether someone has judged or reasoned correctly depends upon standards of good judgment and right reasoning, not on the observed performance of candidate judges and reasoners, normative conceptions are indifferent to empirical investigation. Psychologists might empirically investigate the environmental conditions under which right reasoning—reasoning that meets the criteria for success—tends to occur, but the standards for such reasoning are presupposed by such an investigation, not determined by it.

Not everyone has been so pessimistic about the prospects for a science of thought. Many practitioners of psychology and the neurosciences believe they are on the verge of a radically new understanding of the mind and its higher cognitive functions. Empirical psychology has already done much to illuminate some areas of mental life, including many aspects of sensory perception and some aspects of memory and learning. Many investigators are encouraged by the rigorous approach to mental modeling that has arisen from comparing minds with computers as well as by the development of increasingly informative techniques in neuroscience. They foresee a day when the mind will wholly yield its secrets.

The optimistic prediction of a general science of the mind has had philosophical as well as scientific supporters. Philosophers increasingly predict either that future science will capture such notions as "judgment" and "understanding" within a naturalistic vocabulary, or that the findings of science will relegate these and other old mentalistic notions to the same dustbin that received "phlogiston," "vital spirit," and other outdated or falsified concepts.[3] The first prediction foresees the successful reduction of the mental by one or another natural science and so envisions an empirical science of the mind. The second foretells the elimination of mentalistic discourse and therefore denies that there can be a science of the mind, on the ground that there really is nothing to study. In either case a new naturalistic understanding of the mind or of humanity itself is anticipated. It is to have implications for every dimension of our self-description, including the description of ourselves as "knowers."

The mind's place in nature has in one form or another been a topic of philosophical discussion from the earliest philosophical writing. During the modern period, however, investigation of the mind promised an especially large payoff, or so thought a number of influential writers, including early modern philosophers from Descartes to Kant. These authors came to expect that knowledge of the mind's powers or faculties would reveal the content of—or, more modestly, the limitations on—human knowledge in general and the theory of nature (the science of physics) in particular. The most notorious among them was Descartes, who sought to reveal the foundations of physics through a thorough investigation of the knowing mind.[4] Locke, Berkeley, Hume, and Kant also contended that an investigation of what they and their contemporaries sometimes called the "knowing power," or "the faculties of the mind," would reveal how, or whether, the foundations of physics can be known. For authors interested in the foundations of the new science, the cognitive status of geometrical or spatial properties became a principal focus of inquiry, because from the time of Galileo and Descartes, but especially after Newton, spatial or geometrical properties were assigned a fundamental position within the ontology of early modern science. As befit the tendency of early modern philosophers to demand an explanation of how something could be known before accepting claims to knowledge, most such philosophers tested the foundations of modern science by first testing the epistemic standing of our perception or mental apprehension of spatial properties.[5]

Even when early modern authors agreed that investigation of the mind was propaedeutic to securing the foundations of physics and philosophy, they did not agree on the proper conception of the mind itself. Initially, most philosophers, including Descartes and Locke, analyzed the mind into "faculties," such as sense, imagination, and understanding, each with its own characteristic abilities. Subsequently, an increasing number of authors—Hume is a prominent example—spoke less of faculties and more of laws and mechanisms. The first sort of author typically conceived the mind's powers and abilities as fundamentally normative (at least in the case of the faculty of the intellect); the second sort, who spoke of mental laws, often adopted a naturalistic attitude toward the mind.

The project of developing a natural science of the mind first arose during the seventeenth and eighteenth centuries.[6] Most philosophers, however, continued to conceive the mind as a power or agency that was either outside nature or distinct in both substance and mode of operation from the rest of nature. The intellect or faculty of understanding in particular was regarded as a "knowing" or a "truth-discerning" power. The deliverances of this power were conceived to be successful epistemic achievements, that is, essentially normative. The nature of the intellect was to perceive truth, and authors such as Descartes maintained that, left undisturbed, it could do nothing else. Naturalistic explanations were reserved for cases of error. However, as the existence of an intellectual faculty with inherent powers of truth-detection came under ever greater challenge throughout the early modern period, the mind's status as a power or agency distinct from the rest of nature was also called into question. Naturalists such as Hume came to assert that the phenomena of mind are as much a part of nature as are the phenomena of physics, and that they might be charted empirically with equal right. The naturalistic attitude held a strong attraction for nineteenth-century philosopher-scientists such as Helmholtz.

Although interest in the study of the mind has blossomed during the course of the twentieth century, expectations of what such study will reveal about the foundations of science have diminished. From the perspective of physics, it no longer seems likely that the psychology of spatial perception can shed light upon the geometry of space or the fundamental properties of the physical world. From the vantage point of philosophy, findings of psychology have long been considered irrelevant to questions about knowledge and justification. This sharp separation between psychology and epistemology has recently been challenged by those who envision a "naturalized" epistemology. But naturalizing epistemologists generally eschew the project of discovering the foundations of empirical science.[7]

This tendency notwithstanding, some philosophers and scientists continue to urge that scientific investigation of the knower will reveal constraints on and substantive principles of human knowledge.[8] Such hopes may well reveal the continuing influence of the vision expressed by authors such as Descartes, Hume, Kant, and Helmholtz, each of whom posited a strong connection between theories of mind and the substantive doctrines of the sciences. Throughout the twentieth century their works have served as benchmarks that orient discussion in epistemology and philosophy of mind. Because they enjoy a heroic status, their vision lends credence to grand claims for the intellectual benefits expected from the science of the mind. Recent authors who invoke their arguments and texts tend to describe them in the terms of current theory, thereby making the problems and positions found in earlier writings seem conceptually continuous with current work. In particular, terms such as "philosophical," "psychological," and "mental process" are applied as if their meanings were fixed across the modern period and remained the same today.

Redescribing the work of past authors in terms of the locally prevailing philosophy of mind has a double effect. It masks the specific assumptions necessary to make their projects plausible, assumptions which may now seem problematic. More significantly, it conceals the fact that these authors treated the proper understanding of terms such as "philosophy," "psychology," and their ilk as objects of investigation. Descartes, Hume, Kant, and Helmholtz each sought to determine the proper relationship among the disciplines we call psychology, epistemology, physics, and metaphysics. In so doing, they came to different conclusions about the correct conception of both *nature* as an object of knowledge and *mind* as a knowing agent. Their disagreement about how to conceive the mind's activity are of especial interest to me, for their differences exemplify in fascinating detail the confrontation between naturalistic and normative approaches to the mental. Indeed, these authors set the terms of the confrontation. Their differences also served to structure the distinction between psychological and philosophical approaches to perception. Let us see how.

1   Mind and Space from Kant to Helmholtz

It is natural to assign Immanuel Kant and Hermann von Helmholtz leading roles in an historical episode that takes as its narrative theme the interaction among theories of mind, theories of perception, and theories of knowledge. Kant's *Critique of Pure Reason* is a culmination in the development of early modern philosophy. Kant examined and reformulated the problems that had been central to the metaphysical and epistemological writings of Descartes, Locke, Leibniz, Berkeley, and Hume, among others. But his work was not only a culmination: the *Critique* set the problematic for further work in philosophy throughout the nineteenth century, especially in Germany. Similarly, Helmholtz's *Treatise on Physiological Optics* is seminal in the development of modern psychological theories of the senses. His work crowned a tradition of theorizing about spatial perception that included contributions by Berkeley, Reid, the physiologist Johannes Müller, and a number of lesser-known nine-teenth-century German philosophers, physiologists, and psychologists. As with Kant, succeeding generations of sensory physiologists and psychologists looked to writings of Helmholtz for the context from which further work would proceed. Thus, the work of each author represents a closure of previous efforts and a focal point for further work.

The two authors also stand in a substantive relation of their own. Helmholtz took himself to be responding in part to the same problematic that had interested Kant, even to be correcting Kant on certain topics. He brought the techniques of empirical science to bear on questions Kant had regarded as intrinsically philosophical—or, as Kant might have put it, "transcendental"—questions about spatial perception and its relation to the geometrical descriptions found in physics. Kant had distinguished philosophical from psychological approaches to thought in general and to spatial perception in particular, assigning some questions to philoso-phy and some to empirical science. Denying that the techniques and concepts of the empirical sciences could provide a complete understand-ing of human knowledge or of the knowing subject, he maintained that such an understanding must arise from philosophy. Helmholtz, by contrast, sought to bring his natural-scientific theory of the senses to bear directly on questions Kant had set beyond the pale of empirical science. He persistently read—or perhaps misread—Kant's position in such a way that it contained assertions amenable to empirical treatment. Helmholtz thus stood in direct relation to Kant in trying to answer Kantian questions. This relation takes on special interest because Helmholtz tried to answer Kant on grounds that his predecessor had judged to be incapable of yielding such answers.

The disagreement between Kant and Helmholtz includes both content and method. As for content, Kant affirmed and Helmholtz denied that space could be known a priori to conform to Euclid's geometry. Helmholtz characterized the disagreement in terms of the innateness of spatial perception; he maintained that Kant's commitment to the a priori status of Euclid's axioms derived from a mistaken psychological doctrine that spatial perception is an innate ability. Kant, according to Helmholtz, adopted a "nativistic" position; Helmholtz on the contrary held that the visual perception of space is a learned ability. Because he considered Kant's apriorism to stem from the psychological thesis of innatism, he attributed the source of Kant's mistaken position to faulty psychology.

Helmholtz was correct in thinking that he and Kant differed in their conception of the psychology of spatial perception, although I hope to show that the difference was less than he thought. Be that as it may, their disagreement over the epistemic status of geometry reflected a deeper division between the two authors that was both methodological and conceptual. Methodologically, it resulted from different views of the proper way to investigate thought and the mental; conceptually, it reflected differing positions on what is fundamental to the notions of *thought* and *the mental* themselves.

The substantive conceptions of mind held by these two authors represent two framing conceptions of the mental that emerged in the eighteenth century. One may be characterized as *associational*, the other as *judgmental*. Associationists sought to explain all of mental life, including belief formation and judgment, in terms of a few (usually three) laws of association. Typical associationist explanations emulated the explanatory structure of the new physics: mechanisms or processes were analyzed into elemental components (corpuscles or atoms in the physical case, simple or atomic sensations in the mental) and the laws governing their interaction. Since associationists included mental activity within the domain of the natural, they considered it to be a proper object of natural science. Judgmental accounts, by contrast, considered the act of judgment to be primitive and unanalyzable, thus denying that such acts could be decomposed into more primitive acts or processes, associational or otherwise. Adherents of the view that judgment is a primitive act typically described judgments using a special vocabulary, one capable of accommodating notions of sound and unsound judgment, good and bad grounds for judgment, and truth-delivering or error-prone methods of judging. They regarded at least some judgments as inherently normative and therefore inexplicable within a wholly naturalistic vocabulary.

The episode of Helmholtz's scientifically based response to Kant's epistemological doctrines provides a model case of the problems that arise in evaluating the possibility of a general science of the mind. This episode pits an author who denied that thought could be adequately explained naturalistically against an avowed naturalizer who aimed specifically to refute his nonnaturalistic predecessor. For this reason, it may evoke opposite responses from philosophers and psychologists today, depending on their tastes in the theory of mind. Some will immediately suspect that Helmholtz's response to Kant is a version of the "naturalistic fallacy" of psychologism; others may find in Helmholtz an heroic ancestor in the cause of naturalization.

I shall argue that Helmholtz was in fact able to respond effectively to Kant's doctrine on the foundations of geometry, but not by correcting Kant's psychology, or at least not his nativism. Helmholtz's classification notwithstanding, Kant in fact was not a nativist. However, even though Kant's account of the foundations of geometry did not require a psychological thesis of innateness, I maintain that it had empirical and, in a certain sense, psychological implications. One way of characterizing these implications is to say that they pertain to the character of physical space and that because of the peculiar structure of Kant's transcendental program they follow from claims about the conditions on any possible experience (and in this slightly punning sense are "empirical"). Such claims are themselves founded upon Kant's doctrine of sensibility, and for that reason they place certain constraints upon possible findings in sensory psychology. Helmholtz was able to refute these implications of Kant's geometrical doctrine. He did not do so through his appeal to psychology, but through arguments that detailed the conceivability of obtaining a specific set of empirical results through physical measurement. In effect, he was able to show that questions pertaining to the geometry of physical space are empirical in a manner that Kant did not foresee and perhaps could not have foreseen.

The arguments with which Helmholz refuted Kant were not psychological and hence were not naturalistic in the way he had imagined. More broadly, even Helmholtz came to realize that he could not carry through his naturalistic account of scientific inference all the way. Helmholtz's writings reveal that he ultimately recognized the distinction between naturalistic and normative aspects of thought, even though the resulting limits to science were a source of frustration.

2   Historical Constitution of Philosophical and Psychological Approaches

The relationship between philosophy and psychology in general, and between Kant and Helmholtz in particular, has been discussed for more than a century. The distinction between psychological and philosophical approaches to the mind exemplified by these authors and their successors has left its stamp on intellectual life. The distinction has been used to organize departments, journals, research, and graduate education; consequently, it has become embedded in the disciplinary histories philosophers and psychologists produce. Not surprisingly, the two sets of disciplinary histories differ from one another in systematic ways. Although there is no consensus on the question of whether philosophical and psychological approaches to thought and the mental are continuous or distinct, philosophers and psychologists have presupposed or, in some cases, asserted specific answers to this question in writing histories of their respective disciplines. The works of Descartes, Locke, Hume, Kant, and Helmholtz, among others, have been characterized through the disciplinary perspective of the history-teller. Differences in such characterizations have not pertained solely to the content assigned to the work of various authors; indeed, the primary difference has arisen in critical assessments of this content and in the diagnoses offered to explain the defects of these earlier works. I want to consider briefly these differing characterizations and assessments.

Virtually all histories of psychology echo Ebbinghaus' remark that psychology has had a "brief history" but a "long past." The brief history begins in the second half of the nineteenth century when experimental techniques were first applied to the study of mental phenomena, or perhaps with Wundt's founding of what is conventionally described as the first psychological laboratory. The long past consists of both a nonscientific prehistory of scientific psychology in philosophy, which stretches back to the Greeks, and a scientific prehistory in physiology. Within this prehistory, most histories of psychology allot the largest space to the writings of early modern philosophers.[9] Standard histories agree that what marks the beginnings of psychology as a science, distinguishing it from earlier philosophical treatments of mental phenomena, is that, whereas philosophers remained in their armchairs, psychologists ventured into the laboratory. On this view, the absence of the experimental method was what set early modern philosophers apart from their scientific successors; the problems addressed and solutions proposed by philosophers are typically treated as if they were continuous with those of later scientific psychology.[10]

At least two factors underlie this perceived continuity of problems and solutions. First, much of the vocabulary of nineteenth-century psychology is identical with the vocabulary of early modern philosophy: "sensation," "perception," "mind," "faculty," and "association" are but a few of the many terms both disciplines used. This terminological continuity can give the illusion of conceptual continuity where it does not exist. It is explained by the fact that psychology borrowed much from philosophy, even if not all that was borrowed remained the same.[11] Second, historians of psychology have constructed classifications of problems and solutions that they rather promiscuously employ in describing philosophical and psychological thought throughout history; such classification systems can impose terminological continuity even where there was none. The second factor explains how the continuity could be construed as conceptual, for it effectively renders the problem-spaces of philosophy and psychology equivalent. It has been especially prevalent in accounts of the history of theories of spatial perception.

Histories of psychology typically portray the history of spatial perception as a battle between two positions that were polar opposites: nativism and empiricism. Nativism is equated with the position that the ability for spatial perception is innate; empiricism, with the position that this ability is acquired. These opposing positions are then connected with a general nativism or empiricism about psychological abilities, and, even more broadly, with rationalism and empiricism as movements within seventeenth- and eighteenth-century philosophy. According to this story, Descartes and Kant were philosophical precursors to the nineteenth-century nativist physiologist Ewald Hering, while Locke and Hume were the empiricist precursors to Hering's archrival, Helmholtz.[12] Some histories do in fact recognize that the earlier authors were concerned with the theory of knowledge (epistemology) rather than with the search for empirical laws (empirical psychology); but even these histories equate the problems addressed by the earlier authors with the problem of whether knowledge is innate or acquired, and they then treat this latter problem as though it were continuous with recent debates over "nature" and "nurture."[13] Quite generally, histories of psychology have been written as though authors from Descartes and Kant to Hering and Helmholtz all asked similar questions about the existence of "innate ideas" or innate abilities, and as though they chose their answers from those made available by the same pair of polar positions.

Philosophers who recount early modern discussions of the mind and knowledge have tended to ascribe to seventeenth- and eighteenth-century authors aims and problems different from those assigned to them in histories of psychology. It is common to organize the history of early modern philosophy around the theme of skepticism. According to a familiar story, Descartes was responsible for posing the skeptical challenge, and Locke, Berkeley, Hume, and Kant successively attempted to support or to refute the skeptical position. Considerable effort has been spent in isolating and criticizing the aspects of early modern philosophy that allowed skepticism to enjoy more plausibility than it deserved. And so the "theory of ideas" has received special attention, for it allegedly facilitated the "veil of perception" problem—the problem of how knowledge of the external world is possible if one's immediate knowledge is limited to one's own ideas. This problem has kept antiskeptics busy.[14]

Despite these differences histories of philosophy display significant areas of agreement with histories of psychology. For although philosophers in their histories have emphasized the difference between epistemology and psychology, characterizing the aims and problems of past "philosophical" authors as epistemological rather than psychological, many philosophers also believe that Locke, Hume, and Kant were in fact confused about the psychological and epistemological dimensions of their projects. A long line of interpreters, from T. H. Green and James Ward in the nineteenth century, to Sellars and Rorty in the twentieth, have accused Locke, Hume, and Kant of the fallacy of psychologism, that is, of confusedly proposing psychological solutions to philosophical problems.[15] One of the most commonly ascribed confusions is that of attempting to answer questions about the justification or validity of knowledge claims with assertions about the causal origin of ideas. The controversy over "innate ideas" is singled out as a product of such conceptual confusion.[16] While the Sellars-Rorty interpretation differs from its psychological counterpart in insisting that the problems addressed by early modern authors were philosophical, it resembles psychologists' histories in characterizing as psychological the solutions they proposed. But this point of agreement leads to a further disagreement between philosophers and psychologists. Whereas psychologists have charged early modern philosophers with failing to use the experimental method to support their (psychological) answers to their (psychological) questions, philosophers have faulted these same philosophers for giving the wrong kind of answer to the (philosophical) questions they were asking. In the philosophers' view, had early modern philosophers turned to experimentation, they would have only doubled their mistakes by pairing a method unsuited to their problems with the conceptually inappropriate solutions they had already generated.

Although the description of these "confusions" dominates histories of early modern philosophy written in the middle decades of the twentieth century, some philosophers have recently described matters quite differently. This new reading does not question the view that the positions adopted by several early modern philosophers were inherently psychological. Rather, it proposes a different assessment of this finding on the basis of a new understanding of the relationship between philosophy and psychology, both at present and in the early modern period. According to this recent interpretation, Locke, Hume, and Kant were indeed proposing psychological answers to epistemological questions, but this fact provides cause for approbation rather than opprobrium. For, according to the new view, epistemological problems do admit of psychological answers. Early modern philosophers are seen as having been pioneers in the development of a naturalistic approach to the mind, and perhaps to have been early laborers in the field of cognitive science.[17]

Both philosophers and psychologists have found their current problems, and their current construal of the relation between their disciplines, to be directly discernible in works from the past. Their tendency to discover contemporary problems in the "great works" of the past is understandable, given the indoctrinatory purpose of disciplinary history. Problems of current interest had best be discernible in past works, if those works are to be pedagogically effective in introducing new generations of students to the problems of a discipline—or so some may believe.[18] The use of the "classical" status of past authors to legitimate present work also is understandable. Philosophers and psychologists alike may want to establish the importance of recent work by characterizing it as a solution, or as progress toward a solution, to a problem that has remained unsolved since the time of Descartes, or Locke, or Kant.[19]

By contrast with these approaches to past authors, my strategy will be to seek to understand how the authors themselves conceived problems and solutions. This strategy reflects my belief that if we are to learn from reflection on historical developments, we must avoid imposing a ready-made grid of categories on past texts before reading them carefully in their own context. Such a strategy leads to a somewhat more complicated description of the relation between philosophy and psychology in early modern and nineteenth-century authors than usual. More specifically, it leads to a rejection of all three of the above conceptions of the relation between philosophy and psychology. While granting some truth to the contention that psychology grew out of philosophy, I call into question the perception of a general continuity of both problems and solutions between early modern philosophy and the emergent science of psychology. I attempt to show that although some philosophical discussions from the early modern period are conceptually continuous with scientific psychology and with more recent philosophy, most are not. If we are to come to understand the development of both disciplines and to assess the legitimizing claims of philosophers and psychologists both past and present, we must learn to recognize discontinuities as well as continuities between and within philosophical and psychological problematics as they developed over time.

From the perspective of the history of psychology, the notion that, from Descartes, Locke, and Kant through Helmholtz and Hering, nativism and empiricism provide a continuous problematic must be rejected. I shall attempt to show that Descartes, Locke, and Kant were concerned with problems that are conceptually distinct from those addressed by the sensory physiology and psychology of Helmholtz and Hering. I thereby endorse the view that these problems were epistemological rather than psychological. I maintain that Locke's empiricism is quite distinct from the psychological thesis that human perceptual abilities are acquired through experience. Further, I contend that the doctrine of innate ideas endorsed by Descartes and Leibniz, and attacked by Locke, should not be assimilated to the "nativism" of Müller or Hering. In general, I adopt the terminological convention, sometimes used in psychological writing, of distinguishing *empiricism*, the epistemological doctrine that all knowledge is based on experience, from *empirism*, the developmental thesis that we "learn to see." I also distinguish *rationalism*, the epistemological doctrine that certain basic truths are known by reason alone, independently of the senses, from *nativism*, the thesis that certain abilities, such as the ability to perceive distance by sight, are inborn, not acquired by learning. Further, I distinguish between characteristically philosophical and psychological uses of the terms "sensation" and "perception"; although the terms are not homonyms (like "bank" in river bank and savings bank) in their respective philosophical and psychological senses, they have discernibly different implications. These implications bear on the proper understanding of the nativist-empirist dispute about whether there are spatial sensations.[20]

Although I argue for the propriety of categorizing the work of Descartes, Locke, and Kant as philosophical (or epistemological) rather than psychological, I do not take the further step usually made by those who distinguish philosophy from psychology: I shall not contend that their discussions of mental processes were irrelevant to their philosophical aims, or were patently confused. Quite the contrary. I shall contend both that their discussions were directly relevant to their epistemological projects, and that in speaking of mental processes they were not confusedly engaging in the psychologistic fallacy. My conclusion will be that we need to rethink the charge of "psychologism" as applied to early modern authors.[21] If the charge of "psychologism" is to be appropriate in a given instance, it must make sense in that instance to distinguish between logical, or epistemological and psychological, aspects of the mental.[22] But many authors in the seventeenth and eighteenth centuries conceived the mind as a logical and epistemic power. For them, studying the human intellect was not an attempt to bring psychology to bear on logic or epistemology, or even to engage in psychology at all in our sense of that word, but to investigate the logical and epistemic faculty itself. They were committed to substantive claims according to which the eternal truths are ensouled in the intellectual faculty of the individual mind. Because we now consider these substantive claims to have been mistaken we may believe that, in discussing the mind, they could only have been doing psychology (just as, we might say, Priestley was collecting oxygen, not de-phlogisticated air, in his apparatus). However, we must take care not to allow our rejection of their substantive claims to result in our mistakenly attributing to them the psychologistic project of attempting to investigate the logical and the epistemic through the techniques of natural science.

Moreover, we must not assume that the attempt to bring psychology to bear on epistemological questions was always the result of simple confusion. During the eighteenth and nineteenth centuries, "psychological" and "philosophical" projects often stood in the same relation as did those of Kant and Helmholtz: they were attempts to bring two different approaches and conceptual vocabularies to bear on a common set of problems. The identification of natural scientific and philosophical problematics was precisely the thing proposed. In the absence of a conclusive argument that empirical psychology in principle cannot have a bearing on epistemology, the assumption that this identification resulted from confusion begs the question.

None of these remarks is intended to forestall all retrospective distinctions between philosophical and psychological problem domains during the period in question. It is possible to characterize some problems as belonging primarily to the psychology of vision and others as belonging primarily to epistemology. Thus, throughout the modern period the question of how distance is perceived was addressed by virtually all psychological theories of vision. Work on this question led investigators to seek "cues" for distance in optical stimulation, and to speculate about mental processes that might mediate the perception of a three-dimensional visual world on the basis of a two-dimensional retinal image. Is distance directly represented in optical stimulation? Are the mental processes that underlie vision judgmental or associative? By contrast, during the same period nearly all philosophical treatments of spatial perception addressed the question of whether our knowledge of the geometrical properties of material objects is based solely on sensory experience. This question pertains to the means and warrant for putative knowledge of the basic properties of matter. Does the knower come to know the fundamental properties of matter through inferences based upon experience, or are these properties known through the intellect, acting in an a priori manner?

Inasmuch as both sets of questions pertain to mental processes or abilities, they are similar. But they are distinct in that the first set concerns the basic functioning of the senses in the perception of space, while the second pertains to the cognitive grounds for physical or metaphysical knowledge of the fundamental properties of matter. Some authors, such as Berkeley, brought these two problem areas into direct contact, while others focused solely on one or the other. Consequently, we should not seek blanket characterizations of the relation between the two problem areas, or at least not before carefully considering their treatment by individual authors.

Just as no single set of conceptual distinctions captures the difference between philosophical and psychological approaches, no single description captures their historical relations. Empirical psychology may develop by extending the boundaries of natural science to include a new subject-matter, such as the perception of depth and distance, without *ipso facto* altering philosophy. But on some occasions an empirical and naturalistic stance toward the mind was adopted in order to alter philosophy itself. Indeed, during the late nineteenth and early twentieth centuries the experimental approach to the mind was advanced in order to replace the outdated and prescientific approach allegedly pursued by philosophers.

3 *The Natural and the Normative*

The distinction that separates naturalistic from normative approaches to thought and the mental cannot be equated with that between philosophical and psychological approaches, for there have been naturalizers among both philosophers and psychologists. The distinction requires its own analysis.

Although recent writings invoke the distinction with some frequency, it does not bear its implications on its sleeve.[23] It is often presupposed or asserted by those who, in the prefaces of their books, announce a position, pro or con, on the naturalistic program, even while the term "natural" and its cognates are not included in the indexes of their books. Thus, although the charge of "naturalistic fallacy" is often mentioned in contemporary epistemology and philosophy of mind, it is rarely analyzed beyond the bare mention of an analogy with ethics. Similarly, although the project of "naturalizing the mind" is often announced, the notion of what it is to naturalize a given subject-matter is virtually never spelled out. Does the attempt to naturalize the mind amount merely to the use of empirical techniques in its study, or is it rather a matter of admitting only a certain range of concepts—concepts constituting a "naturalistic vocabulary"—into one's explanations? But precisely what is the content and what are the boundaries of a naturalistic vocabulary? Furthermore, what is implied if an author adopts a "normative" attitude toward the mind? Is naturalism precluded? Can there be natural norms?

The terms "nature," "natural," and "naturalistic" have been used in a bewildering variety of senses, some of which contradict others. A partial list of terms contrastive to "natural" would include "artificial," "cultural," and "supernatural"; in the early modern period, freedom was regularly contrasted with nature, and the mental was often considered to lie outside nature. The term "nature" has meant (a) the essence or guiding principle of a thing; (b) the totality of the material world; or even (c) the totality of creation, including both minds and bodies. The terms "norm" and "normative" have a less diverse range of meanings that includes both the sense of the "normal" as the usual or the statistically mean, and the sense of the "guiding norm" as an ideal standard.[24] In the first use, the "normal" is typically identified with the ordinary course of nature. In the second use, "normative" often contrasts with "natural," as when a distinction is made between an aesthetic norm of beauty and the ordinary productions of nature; conceivably, the aesthetic norm might rarely or never be met with in nature. But even in this second use, the normative can be identified with the natural, as when, for specific theoretical purposes, one identifies the beautiful with nature's usual productions ("making nature the norm," so to speak).

In light of the wide variety of meanings attaching to the terms "natural" and "normative," I want to highlight the senses upon which I shall draw. I apply these terms only to mental acts or states, especially to cognitive acts or states, or to the theories that describe such mental acts and states. Even in the restricted domain of theories of mind, it is notoriously difficult to prescribe the range of a "naturalistic" vocabulary. Writing at the turn of the century, Dewey and Baldwin jointly distinguished three definitions of "naturalism" in psychology and philosophy, and, by implication, three applications for "natural," that capture well the range of meaning I intend. According to the first, proposed by Dewey, naturalism is "the theory that the whole of the universe or of experience may be accounted for by a method like that of the physical sciences, and with recourse only to the current conceptions of physical and natural science; more specifically, that mental and moral processes may be reduced to the terms or categories of the the natural sciences."[25] This formulation leaves open what is meant by the "terms and categories of the natural sciences"; Dewey suggested refining the definition negatively, by excluding the "spiritual" or "transcendental." Baldwin added two other definitions. One equated naturalism with materialism (many writers, Baldwin included, have advised against this usage); the other simply contrasted the natural with the supernatural or the mystical. Although Baldwin commended the last meaning for general use, I find it too broad to describe attempts to naturalize the mind, before and after Baldwin wrote. Dewey's definition, which equates naturalism with the application of the methods, terms, and categories of the natural sciences, is closer to the core meaning of "naturalism" as it has been applied to theories of mind, at least in the century before and after he wrote. I shall treat it as the primary sense. Equating naturalism with materialism, as in Baldwin's first definition, may then be seen as a special case of this definition, one that results from accepting the substantive proposal that a strictly naturalistic vocabulary reduces to a materialistic one.

On this usage, to describe the mind "naturalistically" is to describe its processes, mechanisms, or the laws governing its states in a naturalistic vocabulary that excludes terms of epistemic evaluation, such as "true" or "justified" or "valid," from its primitive terms. By contrast, a "normative" vocabulary includes these terms as primitives. To take a normative attitude toward mental states is to treat the applicability of epistemic standards or criteria as essential to their being mental states. Adopting a "normative" stance toward thought does not require denying that some aspects of the thinker *qua* thinker can be understood naturalistically; it does, however, entail a commitment to the view that, within a purely naturalistic approach, the core of thought—what makes thought *thought*—will be missed.

The opposition between naturalistic and normative attitudes toward the mind should not be confused with taking a "descriptive" as opposed to an "evaluative" stance toward mental activity. A purely "descriptive" attitude is equivalent to a "naturalistic" attitude only in the special case in which one stipulates that descriptions must be limited to a naturalistic vocabulary. Some authors who adopted a "normative" attitude toward the mind considered themselves to be *describing* the nature of the mind (or of the intellect); for them, an essential characteristic of the intellect was that the products of its activity set the standards of knowledge, and hence the standards against which claims to knowledge were to be judged. Moreover, although "evaluative" and "normative" are closely related, those who regarded the intellect as setting the norms for knowledge might best be understood not as evaluating the intellect and its products but as using the norms of the intellect as standards of evaluation to be applied to particular claims.

My distinction between the natural and the normative identifies two types of account of human thought and judgment, independently of the familiar distinction between "matters of fact" and "questions of value." Following my distinction, moral judgment, which lies on the "value" side of the fact-value distinction, is susceptible to either a naturalistic or a normative account, the choice depending upon whether one endorses naturalism in ethics. My concern is only with epistemic or cognitive judgments, and particularly with epistemic judgments pertaining to the spatial properties of objects. Such judgments lie on the "fact" side of the fact-value distinction, but are subject to either a naturalistic or normative account.

When used to describe theories of mind that provide accounts of epistemic judgments, the terms "natural" and "normative" indicate two ways of understanding the mind itself. The first suggests that thought (mental activity) can be understood within a naturalistic framework, that human beings *qua* thinkers can be wholly understood by using the techniques and vocabulary of the natural sciences (physics, physiology, and mechanistic psychology). The second suggests that thought is inherently normative, that it cannot be understood in terms of naturalistically described processes occurring in the thinker.

Throughout the modern period and into the twentieth century, the major attempts to "naturalize" the mind can be divided into materialistic and nonmaterialistic variants. Both share the goal of investigating and explaining mental processes with the methods of the natural sciences and, in particular, with methods and modes of explanation modeled after physics. Materialists propose that the naturalization requires that the mind be reduced to a material or physical system; I shall call their position "metaphysical naturalism." Nonmaterialistic versions of naturalism are defined by the attempt to discover "natural" laws of the mind, where "natural" is cashed out through an analogy with the methods and modes of explanation in natural science, instead of by an appeal to ontology. I shall call this version "methodological naturalism." Associative accounts of the mind, which reduce mental processes to simple elements governed by a few laws—where these elements and laws are described as mental phenomena and are not identified with physical processes—have provided the typical example of what I term methodological naturalism. Among modern naturalistic authors, the prevailing mode of naturalism was methodological rather than metaphysical. Indeed, among those authors who substantially contributed to the naturalistic science of the mind, methodological naturalism was far and away the predominant form. Both Hume and Helmholtz pursued methodologically naturalistic projects.

Advocates of both versions of naturalism have held differing opinions about whether naturalistic description is consistent with epistemic normativity. The most stringent of recent naturalists, the "eliminative materialists," would do away with all traditional mental talk, including talk of knowledge and justification. They maintain that all scientific theories must ultimately be reducible to physics, and they predict that mentalistic terms will prove too vague and imprecise to admit of reduction. Others would eliminate neither mental nor normative concepts, but would reduce them to a naturalistic vocabulary, not necessarily that of physics but of physiology or even of mechanistic psychology (a psychology that charts mental laws without recourse to the normative vocabulary of "judgment" or "reason").

By contrast, others have asserted that because thought is fundamentally normative it cannot be reduced to or described within a naturalistic vocabulary. They regard empirical or natural-scientific descriptions of the mind as irrelevant to the discovery and application of standards of epistemic evaluation. Kant's conception of thought is an example. Among authors who conceived thought as irreducibly normative, some hold that the mind is normative in its essence, that it is of the nature of the mind (or of the intellect) to discern the truth, to make valid inferences, to judge correctly. I call this a "metaphysically normative" conception of the mind, recognizing its tendency to explain the mind's special powers by appeal to its immateriality. Descartes is an example. Others who held that thought is essentially normative eschewed metaphysics. This attitude may be denominated a "methodologically normative" approach to thought. These authors concentrated their attention on the concept of *thought* (or *understanding*, or *judgment*) itself, contending that to call something "thought" is to imply that it meets certain standards, not merely that it follows certain natural laws. They maintained that the methods and explanations of natural science are inadequate to describe thought, not because the mind is a special substance with powers that lie outside nature but because thought constitutes a conceptual domain in which it makes sense to criticize instances of thought against standards of success, irrespective of whether those instances are empirically normal. Representatives of this view deny both that a purely naturalistic vocabulary is adequate to describe thought and that natural scientific investigation is suited to the discovery of the norms for thought. They insist that thought must be investigated "philosophically," through the analysis, interpretation, and criticism of our cognitive practices and of the concepts through which we describe such practices.

The question of whether there can be a science of the mind can now be seen as dividing into at least two questions. One asks whether normative mentalistic notions can be eliminated entirely in favor of naturalistic ones. The other asks whether, failing that, the mental is subject to naturalistic interpretation, or must instead be investigated philosophically. To the extent that the new experimental psychology committed itself to methodological naturalism, it came into conflict with those who upheld the normativity of thought.

4 Strategy and Organization

Attempts to use history to attain a firmer philosophical grip on our current conceptual scheme usually involve working at the history rather than simply applying what is already known. After all, if the historical considerations that permit new insights were surface phenomena, they would already be available to contemporary discussion; philosophy as a discipline is noteworthy for the honorific bow it makes in the direction of its own history. However, despite the widespread use of historical texts in philosophical writing and teaching of nearly every stripe—or, more likely, because of it—contemporary philosophy is largely insulated from its past. In order to discover something new through historical investigation, we must put aside familiar categorizations and characterizations of past figures and look at them again with fresh eyes.

For this reason, the conceptual vocabulary I use to analyze the various texts I discuss below will initially be drawn from the texts themselves, or from nearly contemporary works. This strategy follows from the methodological precept that historical texts should be read in their own context. My discussion will not, however, remain wholly within a set of reconstructed past categories, as if the goal were to speak in an eighteenth-century philosophical dialect. Part of my aim is to understand how certain problematics developed in the past in order to gain new insight into their legacy for more recent thought. This aim requires that we speak of past problematics in language that connects them with the present.

The target sighted in pursuit of this aim is the historical episode defined by Kant's theory of mind and spatial intuition and Helmholtz's naturalistic response to it. To understand the conceptions of mental activity available to Kant, I first survey theories of mind and theories of perception that run the gamut from physiology to metaphysics as these theories are found in the works of major and minor authors who were under discussion as Kant wrote. Similarly, before considering Helmholtz's work, I shall examine a variety of writers in the period between Kant and Helmholtz. Immediately following Kant's death, a number of authors sought to respond to his work and particularly to his claims about the status of space and spatial perception. These authors range from philosopher-psychologists such as J. F. Fries, J. F. Herbart, and R. H. Lotze, to the obscure but original physician-psychologist J. G. Steinbuch, to physiologists such as Johannes Müller and C. T. Tourtual. Taken together, their work constitutes the immediate historical context in which Helmholtz wrote; they also reveal one historically influential manner of reading and interpreting Kant. Examination of their writings promises both to reveal various ways of understanding the Kantian problematic and to provide the background of theoretical work on the senses Helmholtz looked to in formulating his own physiology and psychology of vision.

The body of this book consists of four chapters. Chapter 2 surveys the major theoretical positions available within eighteenth-century theories of perception; it examines the physical (optical), physiological, psychological, and epistemological dimensions of major works from both the seventeenth and eighteenth centuries. Chapter 3 examines Kant's distinction between naturalistic and philosophical or "transcendental" accounts of thought and the mental; it applies this distinction to his account of spatial perception, with special attention to the implications of his transcendental account for empirical psychology. While Kant did not develop an explicit theory of spatial vision *per se*, focusing instead on the mind's apprehension of spatial properties in general, his position on the psychology of vision can be reconstructed from his limited statements on the topic. This reconstruction can serve in probing the relation between his transcendental pronouncements and his own beliefs about the empirical psychology of perception. Chapter 4 examines the writings of Fries, Herbart, Steinbuch, Tourtual, and Müller, early nineteenth-century German thinkers who sought to extend or refute what they took to be the psychological implications of Kant's critical philosophy. It also recounts Lotze's survey and synthesis of the work on the senses extant when he wrote, a survey that took into account physiological, psychological, and metaphysical approaches. Lotze attempted to strike a Kantian balance between naturalistic accounts of the mind and normative descriptions of the knower. Chapter 5 analyzes Helmholtz's classic formulation of the empiristic theory of spatial vision and the theory of unconscious inference, in which the physiological, psychological, and epistemological approaches to perception are integrated. The extended discussion of the notion of "unconscious inference" I present is intended to nail down the referent of a phrase frequently invoked in contemporary discussions of perception but rarely analyzed: "Helmholtz's theory of unconscious inference." Helmholtz's theory embodies his attempt to extend methodological naturalism to both perceptual and scientific inference. Chapter 5 also examines Helmholtz's response to Kant. I argue that, by interpreting Kant's position on the status of the geometrical axioms as resting on psychological grounds, Helmholtz made it seem possible to apply natural scientific arguments against that position, arguments Kant would have considered irrelevant on principle. Helmholtz, I contend, was successful in criticizing Kant's theory of geometry, but not on the psychological grounds he imagined.

A summary chapter reviews the material covered. The conclusion examines the status of more recent distinctions between natural and normative conceptions of perceptual judgment. It surveys the distinction between naturalistic and normative conceptions of mind in the light of attempts to naturalize epistemology and to study the normative empirically within cognitive science. While granting that the naturalistic approach to psychology can extend quite far into the domain of the mental, I conclude that the best arguments currently available preclude the complete naturalization of thought.

Chapter 2

Mind, Perception, and Psychology from Descartes to Hume

Theories of mind were presented in, or implied by, a remarkably diverse body of seventeenth- and eighteenth-century writings. At the time, these writings fell under several rubrics, some of which remain familiar and some not. The familiar labels include "psychology" and "mental philosophy"; among the unfamiliar is "pneumatology," which meant the study of the soul. Theories of knowledge, which were more likely to be called "theories of human understanding," included an analysis of the mind's faculties or cognitive powers. One label, "optics," no longer suggests the study of mental phenomena. But from near its inception, optics was a complete theory of vision, including a theory of the mind's activity in perception; it comprised what we would now call physics, physiology, and psychology.

The subjects falling under these labels were all, at one time or another, pursued as part of a recognizably philosophical enterprise by writers who would be recognized as philosophers today. Further, each had an historical connection with the rise of modern psychology, although none of them was equivalent to a science of psychology considered as a branch of natural science. Despite these continuities it is not safe to assume that the problems and solutions found in these writings are continuous with those familiar in present-day philosophy and psychology. In this chapter I shall examine carefully both continuities and discontinuities. I shall work toward an assessment of the positive claims for the theory of mind made by early modern theorists, as well as assessments of the retrospective charge of psychologism leveled at early modern theories by detractors in more recent times and the claims of recent naturalizers to find heroic ancestors among the great modern philosophers. Because both positive and negative assessments of early modern theories of mind have depended upon perceptions of continuity with the goals of naturalistic psychology, I shall be especially attentive to the origin and development of naturalistic tendencies in those theories.

I shall pose two sorts of questions about early modern writings on the mind. One sort asks whether the authors of theories of knowledge and theories of vision explicitly intended to pursue a psychological investigation. In the case of theories of vision, this question pertains to the connection between psychology, naturalistically conceived, and the portion of optics devoted to perception; in the case of theories of knowledge, it pertains to relations between such theories and initial attempts to treat the knower *qua* knower in a naturalistic fashion. The other sort of question abstracts from historical context and asks whether, regardless of their intentions or self-conceptions, various authors did in fact pursue an investigation that we can, with hindsight, see to have been psychological. I start by applying the contextual question to the notion of naturalistic psychology itself, before turning to theories of vision and theories of knowledge.

1 Psychology, Naturalistic and Otherwise

The term "psychology" apparently is a sixteenth-century coinage.[1] Originally it was used to refer to the science of the soul, a science whose origin could be traced back to Aristotle's *De anima*. The term thus referred to a branch of study much older than itself, which comprised a variety of investigations ranging, in our terms, from biology to logic. Because Aristotle conceived the soul as having organic (vegetative, sensitive) as well as cognitive (rational) powers, the study of psychology included the investigation of bodily growth and decay, nutrition, reproduction, animal motion, and the operations of the senses, as well as an examination of the logical and volitional acts of the rational soul. It is doubtful that *any* of these aspects of Aristotelian psychology belongs to psychology conceived as a naturalistic account of the mind. The organic portions may be assigned to biology and physiology. The theory of the senses pertained primarily to the physical and physiological aspects of their operations or to their role in the acquisition of knowledge, but included very little on the topics that became central to modern psychology, such as the perception of depth or distance.[2] Nor did the investigation of the rational soul constitute a properly naturalistic psychology. Of all the soul's capacities, rational intuition was regarded as least dependent upon the body; indeed, *nous* typically was regarded as an immaterial active principle, the operations of which did not require a bodily organ.[3] Although the rational soul constituted the "nature" of the human animal, this nature served to mark the discontinuity of humanity with other natural things: human beings share the power of reason only with the divine intelligences with which Aristotle populated the celestial spheres. Aristotle's psychology is, therefore, not naturalistic, if the latter term is taken to imply an account of the mind that makes human mentality continuous with the rest of the natural order, including the mentality of other animals.

In its Aristotelian sense, it is true that some seventeenth- and eighteenth-century authors were engaged in "psychology"—which is to say that such authors constructed theories of the soul. Their theories were ontological and logical (as they understood the latter term), pertaining to the essence of the soul and its epistemic powers. Such theories did, indeed, speak of various "processes" and "activities" of the soul, as well as various "capacities," "powers," or "faculties." Moreover, such talk was proffered in the spirit of explaining the epistemic capacities of human beings. But this talk of "explanation" via "processes" was not part of an attempt to provide a naturalistic explanation of the mind's cognitive powers, for the vehicle for such faculties and processes was conceived as an immaterial substance removed from physical nature. Several authors, including both Descartes and his Aristotelian opponents, were agreed that by its nature this immaterial substance had an ability to discern the truth; further, both Descartes and his followers held that the processes and abilities of this substance could not be explained by the emergent "new science" or by any foreseeable extension thereof.[4]

Among the multitude of meanings for "natural" indicated in chapter 1, three are especially germane for understanding "naturalism" in the seventeenth and eighteenth centuries; all were in use at the time, and to avoid confusion they must be kept distinct. Two of these meanings are suggested by considering two contrastives for "natural": "accidental," and "supernatural." When compared with "accidental," "natural" has its root philosophical meaning in the School Philosophy of the seventeenth century: to study the nature of something was to study its essence. Accidental properties, such as the skin color of human beings, were regarded as nonessential; to ask whether a property or an activity is "natural" to something was to ask whether it flows from the thing's nature (as in "the soul's nature is to think"). When compared with "supernatural," the "natural" pertains to the entire created world. According to Aristotelian doctrine, the term "nature" in this sense applied to the region below the sphere of the moon, including the four elements, but also including, by many but not all accounts, the individual souls of human beings.[5] Descartes' use of the term "natural light" to distinguish the ordinary use of the intellect from its illumination by the "light of grace" provides another example of this usage; the term "natural" marks a distinction between the use of natural reason (found in all human minds) as opposed to the receipt of divine revelation.[6] These first two meanings allow an immaterial soul to be classified as "natural": the soul may be identified with an instantiation of the essence or "nature" of human beings,[7] and it may be regarded as "natural" by contrast with the supernatural. Some post-Aristotelian authors, including the eighteenth-century Leibnizian, Christian Wolff, maintained a dualistic conception of mind and matter but considered the natural world to include all creation, including the soul. When "natural" is equated with "created," thought counts as a natural power, even if thinking substance and its operations are not subject to the laws of physics.

According to the third meaning of "natural," "nature" is equated with the system of physical causes, excluding the immaterial intellect, or other immaterial agencies (such as angels, or a deity). This meaning, which provides the first sharp contrast between nature and mind, is the most difficult to define non-circularly. Some examples may serve to make this meaning sufficiently clear. In Aristotelian philosophy, the world of nature in this third sense was the sublunary world, restricted to include only those "natures" that arise from the four elements. In Cartesian and Newtonian physics, the world of nature included the entire system of the physical world, earthly and heavenly. To be "natural" was to be a part of this system, which meant to be governed by its laws. These laws were conceived in such a way that they excluded the attributes traditionally ascribed to the mind: they were described as deterministic, nonteleological, and mechanical.

In this third sense of "natural" paradigm dualists such as Descartes separated the mind, or at least the faculty of the intellect, from nature. Although Descartes treated mind and matter equally as created substances, he excluded the mind from the purview of physics, the science of nature. The physical part of his *Principles of Philosophy* is devoted to "The Visible World," thereby excluding the invisible, immaterial mind.[8] Furthermore, the mind is distinguished from passive matter and compared with God and angels in its activity and capacity for thought. The mind does come into contact with material nature, and this contact plays an explanatory role in theories of the sensations and the passions.[9] It would even be correct to say that Descartes developed a psychology of vision in the present-day sense of that word. Nonetheless, his conception of what he considered to be the essential faculty of the mind—the intellect—was of something outside nature. And so while dualists such as Wolff may have dubbed "thought" a natural activity, Descartes, the paradigm dualist, did not.

Attempts to "naturalize" the mind during the early modern period often began by denying that the mind or soul stands outside nature and is free from nature's laws. There were two species of such naturalism in the seventeenth and eighteenth centuries. The first and most familiar contends that human minds are part of nature because materialism is true. On this view, which I have dubbed "metaphysical naturalism," the mind is identified with the activity of organized matter; the implication of this identification is that the properties of the mind can be fully explained by the laws that govern matter. This form of naturalism denies that human beings possess a special rational faculty, equates thought with the activity of the senses and the imagination, and suggests that this activity can be reduced to matter in motion. It is as old as ancient atomism. Lucretius and later revivals of his thought provide examples; Hobbes, La Mettrie, d'Holbach and Priestley are modern adherents.[10] Yet, perhaps because their main interest was in establishing a broad and thoroughgoing naturalistic attitude, rather than in developing a detailed theory of the mind's activities, such authors provided little by way of a naturalistic psychology.[11]

The second species of naturalism in the early modern period is a version of what I term "methodological naturalism." Its early modern embodiment consisted in attempts to provide the phenomena of the mind with the type of explanation already achieved in mechanics and astronomy. Such "naturalism" need not be committed to materialism; it earns its title through its attempt to develop explanations of mental life that exhibit the same type of appeal to general laws found in Newtonian mechanics. In the eighteenth century, this effort took the form of a competition to become the "Newton of the mind"; David Hume and David Hartley were principal contestants.[12] Their theorizing recognized a domain of mental phenomena, which they attempted to explain naturalistically. In this context "naturalistic explanation" requires more than simply speaking of "laws of thought" as any theorist of the soul might; the naturalist must reduce the activities of the mind to processes that do not include irreducibly rational or judgmental operations. Reason and judgment must themselves be explained by appeal to laws that do not include bare cognitive powers; otherwise, the would-be naturalist has taken the Wolffian tack of simply classifying thought as natural. Methodological naturalists typically attempted to naturalize the intellectual powers of the soul by reducing them to the imagination and the senses, as did the materialists; they differed from the materialists in not equating sensations (and other mental states) with matter in motion: neither Hume nor Hartley was committed to materialism. Nearly everyone was agreed that sense and imagination were possessed by the beasts. If humanity's apparently singular possession of "reason" could be shown to reduce to imagination and sense, human mental abilities could be made continuous with those of the animals;[13] the human mind would be as much a natural object as the mind of a horse or a monkey. As the century proceeded, sophisticated attempts at naturalization were less likely to be expressed through simply equating the intellect with the senses and imagination. The predominant form of naturalism sought to explain all mental phenomena by appeal to "mechanistic" laws of the mind, modeled after (but not necessarily reducible to) Newton's laws and devoid of explanatory appeal to mental powers such as judgment or reasoning.

Hume's naturalism provides a paradigmatic instance of the project of producing a "science of man," that is, a science of the mental and moral life of human beings.[14] As the subtitle to his *Treatise of Human Nature* attests, this was Hume's stated project: "An Attempt to introduce the experimental Method of Reasoning into moral Subjects"; his Introduction contains a thinly veiled comparison of his work with that of Newton. And, in fact, both the conception of science and the explanatory structure embodied in Hume's "science of man" are properly Newtonian. Just as Newton in his *Principia* disavowed any claims to discover the nature or essence of gravity or matter, so Hume disavowed any claims to uncover the essence or nature of the mind or the understanding. Hume characterized his endeavor as that of rendering "all our principles as universal as possible, by tracing up our experiments to the utmost, and explaining all effects from the simplest and fewest causes."[15] In practice, this amounted to engaging in a "mental geography," or a cataloguing of mental phenomena, in order to uncover the elements of the mind, together with an attempt to discover universal principles, or general laws, that account for the combinations and dynamic relations among the mental elements. Hume's picture is of simple ideas—e.g., for vision, punctiform sensations—combined with one another or entering into chains of succession with one another in accordance with three so-called "laws of association": the laws of resemblance, contiguity, and cause and effect.[16] By means of these basic laws Hume promised to analyze the customs and habits that underlie our beliefs regarding matters of fact, particularly those based on "causal" reasoning (now reduced to association).

Hume conceived these laws as laws of imagination, as opposed to intellect or reason. They were couched not in the language of clear and distinct intellectual perception of the truth, as in Descartes, but in the epistemically neutral language of connections established through custom and habit. In terms of Hume's "mental geography," these connections are tokened by an experiential characteristic of impressions and ideas, viz., their vivacity. In certain broad categories of mental life, Hume reduced belief to vivacity of perception.[17] At the same time, he freely admitted—nay, he forcefully argued—that such vivacity is neutral with respect to truth. I want to examine this point more fully.

Hume's reduction of belief to vivacity reveals a characteristic feature of early naturalized accounts of the mind: they posited belief-producing processes that were neutral with respect to whether the beliefs produced are true or false. In Hume's case, the notion of truth itself was not called into question: he endorsed the standard definition of truth as "an agreement or disagreement either to the *real* relations of ideas, or to *real* existence and matter of fact."[18] Of course, Hume restricted the "real existences" that may be known to present and remembered sensory impressions;[19] but the standard of truth remained correspondence, even if the correspondence was between ideas, or between ideas and sensory impressions (as when one comes to expect an impression and then finds the expectation confirmed). The "truth-neutral" aspect of Hume's theory of belief derives from the fact that the vivacity of ideas, arising from custom or habit in accordance with the laws of association, provides no mark of the truth. Hume expressed this point by insisting that beliefs in matters of fact are formed by the imagination, not the understanding; they have no rational backing. So far as we can rationally determine, the processes that yield beliefs in matters of fact (that go beyond current and remembered sensory impressions) are equally capable of producing true or false beliefs. Hume of course did not deny that some ways of inculcating customs or habits have been more success-promoting than others in the past, that is, have proven to be better at predicting the course of subsequent impressions; he spoke of a "harmony," established over time, between the regularities of nature (as manifest in patterns of received impressions) and the associative regularities of ideas. But, famously, it was one of Hume's central tenets that the formation of associative connections in the previously most successful manner is not inconsistent with future falsehood.[20] Unlike Aristotle and Descartes, Hume posited no faculty of the mind whose proper operation suffices for the production of (or sets the standard for) true beliefs in matters of fact; the same faculties produce both true and false beliefs, and it is the luck of the draw whether one's habit-producing experiences are a reliable guide to further experience.

Hume did not apply the "truth-neutral" account of beliefs to all judgments. He acknowledged that in some judgments the grounds for belief are such that an attentive reasoner cannot be in error. He termed these judgments perceptions of "relations" between ideas, the prime instances of which are the "intuitive" knowledge of the relations between ideas found in arithmetic, algebra, and geometry.[21] As he explained in the *Inquiry*, judgments pertaining to geometrical and arithmetic propositions depend wholly upon applying the principle of contradiction to the "ideas" implicated in the propositions; that "three times five is equal to half of thirty" can be seen to be necessarily true, because its negation expresses an evident contradiction.[22] Hume maintained, however, that matters of fact cannot be known through perception of relations between ideas. With the exception of mathematical knowledge, then, he claimed to have reduced thinking to a series of (as it were) mechanical interactions among mental elements.

The perception of relations between ideas was a last vestige of "rationalism," or of the operation of intellect, in Hume's system. Hartley and other more radical methodological naturalists went beyond Hume, attempting to reduce mathematical judgments, and indeed all judgments of truth, to association. Hartley posited that the brain resonates similarly in each case in which, say, pairs of pairs are present. The subject may learn through association to say "two and two is four" in such situations—and, if truth is in question, to add "is true." In this way, judgment allegedly is reduced to nonmentalistic, "stupid" laws of resemblance, contiguity, etc., mediated by harmonic vibrations in the head.[23] Hartley carried the "truth-neutral" account to its naturalistic extreme.

The "Newtonian" naturalism of Hume and Hartley must be distinguished from another eighteenth-century sense of "Newtonian," which denotes nothing more than a commitment to approaching nature empirically. Although naturalistic approaches typically have been committed to the possibility of applying empirical techniques to the mind, the pursuit of an "experimental" or observational approach to human nature is not enough to qualify a particular approach as "naturalistic." In the century prior to Hume and Hartley, "experimental" might mean simply "experiential."[24] Numerous authors in this period set out to investigate human nature by "observing" and reflecting upon their own actions and the actions of others—by "experimentation," in this broad sense. But such observation does not by itself make one's descriptions and explanations "naturalistic" in any of the senses noted above. Moreover, "observing" the mind's activities was conceived by some authors as the observation of the products of the mind's logical and noetic operations. An empirical investigation of the mind would count as naturalistic in the present sense of the term only if the concepts it used to describe human nature, or the human mind, were restricted in the manner required by either metaphysical or methodological naturalism.

The contrast between "experiential" and "naturalistic" may be clarified by comparing Hume with Locke. From the eighteenth century onward, there have been those who perceived Locke's project as strongly kindred with the naturalism of Hume. Locke, however, was not a naturalizer of either the metaphysical or methodological variety. For although he claimed to base his *Essay* on observation and in this way is similar to Hume, he did not attempt to provide a naturalistic explanation of thought. Moreover, although he was, as Hume would later be, suspicious of the Cartesian intellect and eager to disprove many of the purely intellectual powers attributed to it, he did not attempt to reduce the understanding to the senses and the imagination, nor did he give a naturalistic account of belief formation of the type produced by Hume.

Locke's "plain, historical method" does not amount to an attempt to use observation to develop a naturalistic psychology of the knower, because the fundamental concepts with which he describes the mind's knowing power are not naturalistic in the required sense. Locke's project was fundamentally epistemological. In repeated pronouncements he distinguishes his project from that of providing a physiological or mechanistic understanding of thought. Near the beginning of the *Essay*, Locke announces that "I shall not at present meddle with the physical consideration of the mind; or trouble myself to examine, wherein its essence consists, or by what motions of our spirits, or alterations of our bodies, we come to have any sensations by our organs, or any ideas in our understandings." His purpose is "to enquire into the original, certainty, and extent of humane knowledge; together with the grounds and degrees of belief, opinion, and assent." He uses the "historical, plain method" to give an "account of the ways, whereby our understandings come to attain those notions of things we have, and can set down any measures of the certainty of our knowledge, or the grounds of those perswasions, which are to be found among men."[25] His "observation" of the human understanding amounts to reflections on human cognitive practices with an eye toward setting proper standards for assent.

The initial and abiding focus of Locke's investigation is the grounds of belief, a fact that connects his project with the normative investigation of justification rather than the naturalistic search for causal mechanisms. But despite Locke's clear indications of his intent, numerous authors have attributed to him the project of—confusedly and incorrectly—attempting to answer questions of justification through a kind of "mechanics of the mind."[26] It is easy to locate the basis for such claims. Locke describes his investigation as an inquiry into the "original" of human knowledge and the "ways" in which human beings attain notions of things. When these turns of phrase are considered in light of his analysis of the contents of experience into simple and complex ideas and his long discussion of innateness, it may look as though Locke avowedly pursued questions that we would today regard as psychological, despite what he says elsewhere. For we are likely to see the distinction between simple and complex ideas as the product of analyzing the mind into the basic elements of psychological processes, and to regard questions of origins or innateness as psychological or at least as causal, and therefore as distinct from a properly epistemological concern with justification. For such reasons Locke has been described as an early victim of the psychologistic fallacy.

The tension between the stated epistemological aims of Locke's project and the seeming psychological character of the questions he asks should give us pause and should lead us to reflect on our own retrospective characterization of these questions as "psychological." It may be granted that others, including Hume and a host of nineteenth-century psychologists, attempted to analyze the mind into simple constituent elements as the first step in a naturalistic account of the mind's operations. It is also true that the question of the causal origin of various abilities—the question of whether they are innate or learned—has been a subject of controversy within psychology for much of the twentieth century. It may nevertheless be misleading to read Locke's concern with these issues as continuous with properly psychological projects.

According to the reading I propose, Locke's analysis of the contents of the mind into simple and complex ideas may be seen not as an attempt to discover the psychological primitives from which to construct a mechanics of the mind but as part of an empiricist investigation into the logical origins of our concepts. Some concepts, Locke argues, are primitive, while others are derived through the composition of such primitives.[27] This version of conceptual atomism is not driven by an interest in psychogenesis but by a desire to investigate the epistemological standing of simple and complex notions. Their differing standings take center stage in Locke's arguments pertaining to substance and real essences; his argument against knowledge of real essences comes down to a denial that we could ever attain knowledge of the necessary co-occurrences of simple ideas.[28] Locke's distinction between simple and complex ideas is a distinction between what is epistemically primary and what is not. The associated doctrine that complex ideas must have their origin in simple ideas is designed to provide the proper analysis of the content or meaning of such concepts, not to give an account of their etiology.

As for Locke's extended discussion of innateness, it is important to ask what was at stake in attributions or denials of innateness. This question may be sharpened by comparing the Lockean dispute with the most recent discussions of innateness.[29] Today, to ask whether a concept or an ability is innate or acquired is to ask about its etiology during the lifetime of an individual; that is, to ask whether it is "programmed in" or is produced as a result of interaction with the environment. Typically, such questions are considered to be distinct from the epistemological concern with the justification of knowledge claims. That a certain belief is innate, for example, does not guarantee that it is correct: it is not incoherent to postulate an innate tendency toward error. Accordingly, to link innateness with the justification of knowledge is to fall prey to a conceptual confusion—it is to confuse a question about causal origin with a question about evidential basis.

Although this conceptual point is clear enough and has a certain plausibility, it does not in itself establish that any linking of what is "innate" with what constitutes a "sound basis for knowledge" is necessarily mistaken. Rather, it shows that such a linking would require justification. Such justification has been offered in recent discussions of "evolutionary epistemology." Appeal has been made, for instance, to the possibility that the "ecological validity" of an innate disposition might be established by showing its origin through natural selection; not innateness alone but innateness with a certain etiology provides epistemic credentials for innate tendencies. The extent to which such credentials should be given epistemic weight is rightly contested; but the point remains that an evolutionary story, and hence an etiological story, arguably could be relevant to an assessment of the epistemic status of an innate mechanism or belief structure.[30]

In the seventeenth century, there were much clearer grounds than are accepted today for supposing that the origin of an innate idea or ability could provide it with epistemic credentials. It was sometimes claimed that innate ideas or cognitive abilities have been bestowed on humankind by the deity, and that they therefore carry a divine warranty to be truth-generating (on the assumption that God is no deceiver). Either particular innate ideas, such as the idea of God, or reason itself to the extent that it is conceived as having an ability to recognize truths about the world, are held to be trustworthy because of their divine etiology. Although we are likely today to reject these substantive claims about the divine origin of reason, such rejection does not render the doctrines in question psychological. There is a large difference between (a) debating whether God has implanted truths in the minds of humankind, and (b) disputing over competing psychological explanations for the origin of a trait. On my reading, Locke was doing the former, not the latter.

The comparison of Hume with Locke also suggests that the line between naturalistic and nonnaturalistic attitudes toward the mental, as well as the distinction between naturalistic psychology and epistemology, will not always be easy to draw. It is perhaps easiest to draw in the work of Hume, Hobbes, or even Lucretius, who avowedly conceived their projects as naturalistic. With Locke, where the attribution of naturalism is made retrospectively, the classification is made with greater peril.

Nonetheless, the extent to which figures such as Locke, or even Descartes, who explicitly set the mind apart from nature, are subject to the charge of psychologism has by no means been settled. Even if they did not conceive of their projects in psychologistic terms, and even if talk of innate ideas, or of simple and complex ideas, need not be psychologistic on the face of it, it may still be the case that their liberal invocation of mental faculties and mental processes *eo ipso* convicts them of psychologism. This charge of a deep and implicit psychologism can be evaluated only by considering whether the processes and activities posited in these instances can be interpreted in a manner that does not assimilate them to the causal processes described in physics, a question to which I shall return.

2 Theories of Visual Perception

The problem of perception stood at the center of early modern epistemology and metaphysics, and the sense of vision was the focus of theories of perception. The heavy emphasis on perception within epistemology and metaphysics was understandable, given the integral relation posited between the investigation of nature and the investigation of the mind that would know nature. However, a variety of intellectual concerns led thinkers to an interest in visual perception. Not all these concerns were linked to topics that we should now characterize as metaphysical and epistemological. In addition to epistemological interests in evaluating perception as an instrument of knowledge, an independent tradition of investigating vision traced its lineage from antiquity through the middle ages and into the early modern period. These writings went under the title of "optics," where this term was understood to denote a complete theory of vision.

Metaphysical writers who were investigating the possibility and extent of human knowledge virtually always took the optical writings into account as providing a basic description of the conditions for sensory knowledge by means of vision. The authors of optical works might or might not have taken an interest in developing a general picture of the knowing faculties and the mind's cognitive operations. Many optical treatises from the seventeenth and eighteenth centuries were extensions of the optical tradition embodied in the works of Ptolemy, Alhazen, Witelo, and Kepler.[31] The writings of some authors—Alhazen is the most noteworthy example—contained a discussion of the role of vision in the acquisition of knowledge, but such discussions were secondary and derivative.[32] Other optical writings, including those of Pecham and Kepler, barely mentioned such topics. Moreover, a writer such as Descartes, for whom the theory of the senses constituted an important component of a revisionist metaphysics, produced in his *Optics* a work that observed the boundary between the theory of vision proper and the metaphysical concerns of the *Meditations* and the *Principles of Philosophy*.

Whatever the interests of particular authors, the theory of vision itself provided a framework of fact and theory that conditioned metaphysical and epistemological discussions of perception. My discussion of such theories is restricted to the problem of visual space perception, which accords with the emphasis placed on spatial perception by the authors themselves. The central problem of spatial perception was to explain the perception of the size, shape, and distance of various objects in the field of view.

From the mid-seventeenth century until well into the nineteenth century sensory perception typically was conceived to be the result of a causal chain starting outside the body and ending in the mind. The earlier stages of the perceptual process were viewed as mechanical or physical: an object, perhaps acting through a medium such as light, makes an impression upon the nerves in a sense organ. This impression is transmitted mechanically to an organ in the brain variously known as the "common sense" or "sensorium." Here the mechanical impression initiated at the sense organ has its effect upon the mind, causing a sensation or idea the character of which is determined both by the type of nerve carrying the impression (auditory, optical, etc.) and by the mechanical character of the impression (one sort of mechanical effect causing a sensation of red, another of blue, etc.).[33] I say "has its effect upon the mind," implying a difference between states of the nerves and mental states, because such a distinction was accepted by most writers on sensory physiology from the mid-seventeenth until the latter part of the nineteenth centuries (independently of whether they expressed a commitment to the mind as a separate substance). As applied to vision, this picture of the sensory process led to a division of the visual process into three stages: a *physical* stage, which includes those processes by which light is reflected by objects into the eye and is formed into a retinal image; a *physiological* stage, which includes the neural events beginning with the effect of the retinal image upon the optic nerve and proceeding into the brain to the sensorium; and a *mental* or *psychological* stage, which includes those events occurring when the mind is affected by the physiological processes of the sensorium, as well as any subsequent operations upon or interactions among these properly mental events.[34] My primary interest is in accounts of the third or mental stage of vision, which for ease of exposition I label "psychological" (a designation that will require justification). However, since conceptions of this third portion of the process of sense-perception must to some extent be conditioned by beliefs about the physiological and physical events that precede it in the causal chain, it will be necessary to consider these earlier stages as well.

Within early modern theories of the psychology of vision, two types of explanation for the phenomenal content of a sensory idea may be discerned. The first type, which may be termed "psychophysical," appeals only to the immediate effects of physiological processes upon the mind. The second type, which may be termed "psychological," includes whatever operations are believed to mediate between the sensations produced by physiological processes and the final perceptual experience. If one thinks in terms of a causal chain, the psychophysical link is prior to the psychological one: brain events cause sensory ideas that may then be modified by psychological operations. Within the sort of dualistic framework commonly accepted in the eighteenth century, the role attributed to psychological processes during vision generally varied with two factors: (1) what one took to be the character of the "sensory given," the immediate effect of brain processes upon the mind,[35] and (2) what one took to be the phenomenal character of the end product of the process of vision, the experience of a visual world. As the gap between these two factors widened, a greater role was assigned to psychological processes, for there was more work to be done in mediating between the sensory given and experience of the visual world. Those who believed there was no gap *ipso facto* proffered a purely psychophysical account of visual experience. The difference (if any) postulated between the sensory given and the visual world turned upon theoretical considerations, and opinions were varied.

The early modern natural philosopher had access to two links in the physiopsychological portion of the causal chain of vision: he knew that the physiological portion of the chain was initiated by the retinal image, and he had access to the end product of the visual process through his own visual experience and from the published reports of other natural philosophers. The retinal image belongs to the domain of physics, and visual experience to that of psychology. Between the two there were thought to be the following stages: the effect of the retinal image upon the retina, the transmission of this effect along the optic nerve into the brain, the immediate effect of the neural processes upon the mind, and the operation of brain-independent mental processes (if any) to yield the final product.[36] The evidence available regarding these various stages was severely limited. Anatomically, the path of the optic nerves had been traced to the chiasma and beyond to the thalamus. But the location of the sensorium, the end point of the process of neural transmission, was disputed, even if most authors agreed that it was not the pineal gland, as Descartes had proposed. There was similar lack of direct evidence about the immediate mental effect of the neural process in the sensorium. The structure and site of the sensorium were unknown, hence there was no firm evidence on the physiological side. Because of the mental status of the so-called sensory given, it might seem that introspective evidence would be available regarding its character; however, those who posited a difference between the sensory given and ordinary visual experience generally maintained that introspective evidence regarding the former is unattainable. As optical writers from Alhazen to Descartes to Berkeley and his eighteenth-century successors lamented, or at least grudgingly admitted, the mental operations intervening between the sensory given and the perceptual outcome occur so rapidly that they are unnoticed and hence inscrutable.[37] The task of the natural philosopher was to give an account of these inscrutable operations, based upon his beliefs about the starting point (the sensory given) and the outcome (experience of the visual world), together with his views about the relation between brain events and the sensory given and about the nature of mental operations in general.

Even though phenomenal access to the sensory given seemed precluded, there was nearly universal agreement regarding its characteristics. The predominant view was that it is a mental representation of the retinal image. This view was a key feature of the influential theory of vision put forth by Berkeley in 1709; it had been widely accepted before Berkeley wrote, and it was repeated as a matter of course by a variety of authors throughout the eighteenth century, including Hume, his fellow countryman Reid, and the German philosophers Lambert and Tetens.[38] When Berkeley built his theory of vision on the view that the proper object of sight is a picture-like correlate of the bidimensional retinal image, he offered no justification for his position beyond an appeal to previous visual theory. In particular, he was precluded by his immaterialist metaphysics from offering a psychophysical explanation based upon the character of the neural events at the sensorium and their subsequent mental effect.[39] Yet, notwithstanding his criticism of other authors for their postulation of "unperceived" ideas and intellectual operations, he did not attempt to provide an observational justification based upon introspection.[40]

Other writers, however, saw fit to offer an account of why the immediate object of vision should correspond to the retinal image. The most typical account was cast in terms of assumptions about the structure of the nervous system and the relationship between neural activity and the resultant visual sensations. Johann Gehler,[41] writing near the end of the eighteenth century and drawing upon the most important writings on vision of that century, captured the spirit of this explanation in a manner that recommends itself for its explicitness and clarity:

> The retina may be a delicate web of uncountable nerves, which come out of the brain, together form the optic nerves, and themselves end at the surface of the retina in extremely fine papilla and endings. These endings are the true seat of vision, just as tactual papilla are the seat of touch. Through the nerve to which it belongs, each ending brings, into the sensorium, the sensation of the light ray that strikes it. These nerves proceed with the same order among themselves in which their endings lie. Thus the sensorium receives the impressions in the very order in which the light rays fall upon the retina, and therefore in which the parts of the image lie. The order of the

incoming light rays determines therefore not only the ordering of the parts of the image, but also simultaneously the sensation that the soul receives of the ordering of objects and their parts.[42]

The correspondence between the retinal image and sensory given is explained on two assumptions: (1) the retinal image is topographically isomorphic with the neural events at the sensorium (owing to an hypothesized ordering of the fibers in the optic nerve), and (2) the spatial ordering of the brain events in the sensorium directly determines the (two-dimensional) spatial ordering represented in the sensory given. These two assumptions, which had been implicit in Descartes' theory of brain function, were later accepted by Johannes Müller and presented in his *Handbuch der Physiologie* of 1838-40.[43]

Gehler's postulation of a bidimensional sensory given was common doctrine in the eighteenth century. Given this account of the spatial structure of the sensory given, a question naturally arises concerning the relation of this sensory given to the final product of the perceptual process. This question can be answered only by determining the phenomenal character of that end product. If it were supposed that the visual experience of the typical perceiver is a copy of the retinal image, then the explanation of vision could end with the account quoted from Gehler. Yet, as Gehler himself noted, vision acquaints us with the size, shape, and location of objects as they are distributed in three dimensions, and not simply with these properties as they are represented in the two-dimensional retinal image (and therefore in the sensory given).

The example of the perception of size makes clear the disparity between the spatial properties of objects as experienced and as represented in the retinal image. Gehler remarked that there were two senses in which the term "apparent size" was used by optical writers. The first equated apparent size with visual angle (the angle subtended by an object's retinal image). This first sense is acceptable, remarked Gehler, if one is seeking a geometrically precise definition of apparent size, such as would be useful for recording astronomical observations of the apparent distance between two stars or for defining apparent size in a treatise on perspective. It does not, however, reflect psychological reality. Following those writers from Alhazen onward who had concerned themselves with accounting for actual perceptual experience in addition to providing geometrically precise constructions in the service of astronomy or perspective, Gehler maintained that perceivers typically do apprehend the objective size of objects, at least at moderate distances. According to Gehler and numerous previous writers, the end product of the perceptual process represents the spatially elaborated world of objects distributed in three dimensions, rather than the bidimensional retinal image.[44]

The problem of explaining our perception of a three-dimensional visual world, given that the retinal image is two-dimensional, was shared by all optical writers in the period under discussion, no matter what their beliefs about the transmission of the image to consciousness. The problem is, as Berkeley succinctly put it, that "distance, of itself and immediately, cannot be seen. For, distance being a line directed endwise to the eye, it projects only one point in the fund of the eye, which point remains invariably the same, whether the distance be longer or shorter."[45] The retinal image, as a perspective projection of objects in the field of vision, is ambiguous with respect to the third dimension. The perception of distance is requisite for perception of the objective sizes and shapes of the objects in the field of view, but distance itself is not immediately perceived, since it is not directly represented in the retinal image (and not, therefore, in the sensory given). Nonetheless, humans do perceive distance, size, and shape. The explanation for the ability to perceive distance, and hence to experience the world in three dimensions, was found in the fact that there exist "signs" or "cues" for distance in the optical stimulus itself and in the motions of the ocular muscles. In general, there were two requisites for something to serve as a sign for the perception of distance (or anything else): first, a connection must exist between the sign and the thing signified (an optical sign for distance must vary as distance varies); second, the perceiver must be able to interpret or make use of the sign.[46] The first requirement calls for a lawful relation between the physical properties of the object and the properties of its optical representation; the second specifies that the perceiver have the ability to exploit that lawful relation.

Visual theorists writing in the second half of the eighteenth century typically listed the following six signs for distance:[47]

1. Accommodation of the eye for vision at various distances; the degree of accommodation varies with the distance of the focal objects. While the precise nature of the accommodative mechanism was not settled until the nineteenth century, most writers supposed that there must be some mechanism that functions through intraocular muscles.

2. Convergence of the eyes so as to direct the optic axes on a single point of an object; the angles formed by the intersection of the optic axes and a baseline running between the centers of the two eyes, together with the length of that line, carry determinate information with respect to distance.

3. Apparent size (visual angle); since the visual angle subtended by an object of constant size varies in direct proportion with its distance

from the eye, the distance of objects of known size could be determined from a given visual angle.

4. Intermediate objects or imagined intervals along the continuous ground space between the observer and a more or less distant object; counting the number of equal intervals yields the distance.

5. Lightness and liveliness of the colors, also known as "atmospheric perspective"; due to the effect of the intervening atmosphere, the colors of distant objects are softened, and may take on a bluish cast.

6. Clarity of the small parts of objects; the farther off an object, the less distinctly its small parts are represented in the retinal projection.

The first two of these signs depend upon the action of the intra- and extraocular musculature; they must make their mark in the mind separately from the visual image, perhaps through the effort of will exerted in acts of accommodation and convergence, or through muscle sensations that signify the degree of accommodation and of convergence.[48] The other four signs depend upon variations in an object's retinal representation and hence are included in the sensory given (they are imagistic or "pictorial" signs). In either case, the mere representation of a sign within a sensory state does not guarantee its use in distance perception. Various explanations were given of how the perceiver's mind (or nervous system) uses each sign in the perception of distance.

Early modern accounts of the mediation between sign and distance-perception fall into two types. The first type postulates that the various physical manifestations of the signs—including both the retinal image (with its four pictorial cues for distance) and the state of the ocular musculature—causally influence brain events at the sensorium in such a way that the idea of distance is directly elicited in the mind without the intervention of further psychological processes. This type of account was constructed by Descartes, who hinted at it in his *Optics* and fully explicated it in his *Treatise of Man*. According to this view, the physiological events responsible for the innervation of the ocular muscles have their causal origin in the sensorium or "common sense," which Descartes located in the pineal gland. He supposed that there is a range of continuously variable states of the pineal gland that are causally responsible for a range of continuously variable states of the ocular musculature, varying with near and far vision. The same physical state that initiates the physical process responsible for the contraction of the ocular muscles also directly elicits in the soul an idea of distance appropriate for the degree of accommodation and convergence determined in the ocular muscles. Thus Descartes proposed a direct psychophysical account of distance perception.[49] He did not, however, spell out the process by which this psychophysically-induced idea of distance is utilized in the perception of size and shape. Nor did he rely exclusively on psychophysical explanations; he also formulated a mentalistic explanation of distance perception that invoked a process of judgment.[50]

To my knowledge, Descartes was the only major writer on optics in the seventeenth and eighteenth centuries to advance a direct psychophysical account of distance perception. By far the predominant account of vision in the eighteenth century conceived the operations underlying distance perception as mental, involving operations upon elements already in the mental realm (the sensory given) to produce an end product in the mental realm (experience of the visual world). The mind is given a correlate of the retinal image that contains the four pictorial signs for distance, and it is given sensations corresponding to the state of convergence and accommodation. From these sensory elements, none of which is itself an idea of distance but each of which is a sign for distance, the mind is left to construct a perceptual image or representation of the spatial layout of the objects present in the field of vision. There was virtually universal agreement that the sensory given is altered into experience of the visual world on the basis of previously learned information supplied by the sense of touch: touch educates vision by supplying the missing information about the third dimension, thus allowing interpretation of the signs or cues in the sensory given. However, there was a lack of agreement on the character of mental processes that use this tactual information.

In the eighteenth century there were two theoretical conceptions of the mental operations mediating between the sensory given and the end product of the perceptual process. According to the first view, judgment is the mental operation that combines the sensory elements to produce a perceptual image of the three-dimensional visual world. This view treats judgment and other cognitive acts as primitive powers of the soul: they cannot be reduced to or explained by other operations.[51] Judging is primitive in the psychology of vision in the way that many authors took gravity as primitive in physics. According to the second conception, the processes in vision are associational. The transition from sensory given to visual world is effected by the blind operation of laws of association. The distinction between judgmental and associative theories remained central in the subsequent development of perceptual theory.

Gehler, writing in the 1790s, described the judgmental account. Having remarked upon the tendency of children to touch everything they see, he continued:

> Through such exercise there is finally acquired the ability to judge, quickly and correctly, those things which daily have come before the eyes and have often been compared with touch. Eventually this

ability becomes routine, and interweaves itself so completely with vision itself, that ultimately we no longer see without simultaneously making a rapid judgment about the distance, size, and other qualities of the visible object. This judgment becomes so habitual that the common man confuses it for vision itself, and it requires serious thought in order once again to abstract and distinguish the pure optical presentation (which agrees with the image in the eye) from the judgment of the soul about the visible object.[52]

The child learns to interpret the signs given in immediate sensation by comparing his or her visual ideas with those of the other senses, especially touch; as this learning proceeds, the intellectual operations performed upon the sensory given drop out of consciousness, leaving the product of these operations as the percipient's phenomenally immediate perceptual experience. The observer does not first notice how things look and then make a judgment of their distance and size; rather, the distance and size that he judges an object to have determine the distance and size that the object appears to have. In other words, despite the fact that judged distance and judged size are psychologically mediate, they are phenomenally immediate, and thus seem to be direct products of sensation.

As an example of the degree to which judgment is confused with the activity of the sense organ itself, Gehler referred to the so-called visual illusions. These occur, he remarked, "if, in abnormal circumstances, we nevertheless judge the sensation aroused in the eye according to the normal rules." The result of the application of the normal rules under abnormal conditions is that we judge falsely.

Since, however, we are not clearly conscious of this judgment, and confuse it with vision itself, when the error is discovered we believe ourselves to have falsely seen, and to have been deceived by our eye. For this reason these errors have been named visual illusions, and disputes have occurred over the deceptiveness of the senses, even though the representation is always right, i.e., appropriate to the laws of light and the construction of the eye, so that the mistake lies only in the judgment that we render about the representation.[53]

This analysis of perceptual error, which attributes falsehood or error to judgment rather than to the operation of the senses, is as old as Aristotle. Gehler's account is remarkable for the explicit manner in which it describes the involvement of adjudication in the production of visual experience; visual judgments become so enmeshed with pure sensation as to be indistinguishable from it (except, of course, in theory). A false judgment results in an illusionistic appearance.

In the account of learning to judge distance that Gehler described, the mental activities of comparing and judging are primitive operations; the soul is simply ascribed the power of comparing tactual distance with its visual sign, and of discerning a rule or law relating the two, which it then applies in subsequent perceptual encounters. In the associational accounts, the allegedly primitive powers of comparing and inferring were reduced to laws of association: sensations or mental elements interact with one another in accordance with their qualitative character, unmediated by conceptual or judgmental apprehension of that character. The associational account received its first extensive statement (though without use of the term "association") in Berkeley's *Essay toward a New Theory of Vision* (1709).

The primary thesis in Berkeley's account of the psychology of vision was that "distance is suggested to the mind by the mediation of some other idea which is itself perceived in the act of seeing."[54] The ideas serving to mediate the perception of distance include the familiar six signs for distance, which become paired with tactual ideas regarding distance. The key operation in Berkeley's theory is that by which the "pairing" occurs. Berkeley called this operation "suggestion." According to his account, visual ideas come to suggest ideas of distance and size as perceived by touch as a result of repeated experience of their conjunction; in other words, repeated experience establishes a connection between specific visual ideas and particular constellations of tactually perceived spatial properties, so that when the former occur in the sensory given, the latter are conjoined with them.

Berkeley's account—with an emphasis on the role of touch—may seem to be scarcely different, when stated in this brief form, from the one described by Gehler. Indeed, adherence to the notion that touch educates vision is sometimes seen as sufficient for identifying a theory as Berkeleyan. But it is not sufficient, because Berkeley had his own specific conception of the process of education. Berkeley's theory differs from theories such as Gehler's by its distinction between the process of "suggestion" on the one hand and that of judgment or inference on the other:

> To perceive is one thing; to judge is another. So likewise to be suggested is one thing, and to be inferred another. Things are suggested and perceived by sense. We make judgements and inferences by the understanding. What we immediately and properly perceive by sight is its primary object, light and colors. What is suggested or perceived by mediation thereof are tangible ideas, which may be considered as secondary and improper objects of sight. We infer causes from effects, effects from causes, and properties one from another, where the connection is necessary. But how

comes it to pass that we apprehend by the ideas of sight certain other ideas, which neither resemble them, nor cause them, nor are caused by them, nor have any necessary connection with them? The solution of this problem, in its full extent, does comprehend the whole theory of vision.[55]

It is clear from this quotation that Berkeley considered the process of "suggestion" to differ from judgment or inference. However, despite the crucial role played by suggestion in Berkeley's theory, he provided precious little direct analysis of it. Visual ideas come to suggest tactual ideas because the ideas are conjoined in experience, but the connection is not mediated by judgment or inference. Berkeley compares the relationship between the proper and secondary objects of vision to that between words and their referents: just as words come to be connected with their referents through an often experienced conjunction, so that eventually perception of the word "suggests" the idea it has been made to stand for, so too the proper object of sight comes to suggest the secondary object.[56] Perhaps most importantly, he attributes this act of suggestion (in both the visual and linguistic cases) to the faculty of imagination, as opposed to the understanding. Thus, Berkeley held that arbitrary connections, mediated by the imagination, are established between ideas simply from conjunction in experience; this amounts to a statement of the theory of the association of ideas. Berkeley's use of the process of suggestion in his highly articulated theory of vision confirms this impression.[57]

Two associational principles were implicitly used by Berkeley in his visual theory and were made explicit in the theory's proper, associational, successors. The first is the familiar principle of association through repeated conjunction. If ideas a and b (say, a sensation of the degree of ocular muscle strain involved in fixating an object, and a tactual sensation of the distance to the object) are often found conjoined in experience, a disposition will develop in the mind of the beholder so that idea a acquires the tendency to call forth idea b from memory, in the absence of any current sensory cause for idea b. Association through repeated conjunction was assumed by Berkeley to be a primitive principle of mental operations. (Later, Hartley attempted to give this associative law a physiological foundation in the doctrine of neural vibrations.)[58] An example of the use of this rule in Berkeley's theory is to explain the associative connection between (a) the signs for distance available in the sensory given, including sensations of the ocular musculature, and (b) ideas of distance received from touch. These visual signs are often found in conjunction with a tactual report of an object at a certain distance, so that eventually the sign itself becomes sufficient for "calling up" the associated idea of distance. At this point the individual is able to "see" objects at a distance without having to rely on touch.

The principle of the association of ideas by repeated conjunction is not by itself a sufficient explanation for the associative processes assumed by Berkeley and others to be at work in vision. If it were simply the case that certain visual ideas ("signs" for distance) became associated with tactual ideas in such a way that the tactual ideas entered consciousness whenever the visual ideas were present, when viewing distant objects we should seem to *feel* the distance, rather than to see it. An associational account of distance perception must explain why the association of the bidimensional visual given with tactual sensations causes us to seem to see a three-dimensional visual world. To this end, a second principle of the association of ideas was employed: when simple ideas become associated together, they form a complex idea that need not preserve the phenomenal character of the simple ideas.[59] Thus when the visual signs become associated with tactual reports of distance, the two form a compound in which the tactual ideas serve to give three-dimensional form to the bidimensional array of light and color. The immediate object of vision gives to the complex idea its visual character—the phenomenal qualities of light and color; the tactual ideas, which do not preserve their phenomenally tactual character, serve to give the visual array its phenomenal three-dimensionality. Just as two substances may combine chemically to form some third substance that differs in quality from the other two, our ideas may become conjoined through association in such a way that the product of the association does not preserve the phenomenal character of the ingredients.[60]

Later writers contrasted judgmental and associational explanations, explicitly placing them on either side of a distinction between intellectual or rational processes on the one hand and nonintellectual sensory processes on the other. Earlier, I suggested that the distinction between judgmental and associative approaches parallels that between normative and naturalistic accounts (chapter 1). This parallelism is nearly correct when applied to comprehensive theories of the mind, but it does not work when restricted to theories of the senses. The reason is simple: both judgmental and associative theories of the senses are just that, theories of the senses, and nearly all investigators agreed that sensory functions are connected with the "animal" side of human nature. The theory of the senses was by its subject-matter a theory of a "natural" capacity of the mind.

Although it is accurate to describe associationist accounts of vision as "naturalistic" and as constituting an early chapter in the development of naturalistic psychology, not everyone who adopted an associative account of vision extended that type of account to all of mental life. Berkeley

drew a sharp distinction between the operation of associative suggestion, and the judgments and inferences of the intellect; he separated
"natural habit" from normative inference. Later authors drew a similar
distinction: Reid distinguished processes that link signs to their referents
by "custom" from those that do so by "reason";[61] a number of Kant's near-
contemporaries attributed a significant role to association in some aspects of mental life but refused to extend association to "higher" cognitive
judgments. Others, Hume and Hartley among them, did extend the
associative account to all or most cognitive operations, including the
formation of beliefs. One such author was Erasmus Darwin, whose
Zoonomia was popular in German translation, going through three
editions between 1795 and 1805. Similar accounts were developed in
France by Condillac and Bonnet.[62]

Conversely, the fact that a particular author adopted a judgmental
account of vision did not preclude him from treating sensory judgments
as "natural." Descartes himself, who serves as the paradigm of someone
taking a normative attitude toward the mind, illustrates the possibility of
distinguishing an appeal to "natural" judgment from an appeal to
judgment as a normative act. The distinction is implicit in a passage from
the sixth set of *Objections and Replies*, in which Descartes distinguished
among three grades of sensory act: a physical and physiological stage,
involving the transmission of a pattern of light into the eyes and a
corresponding transmission into the brain; a pure sensation, corresponding to the brain pattern and hence to the retinal image; and "all the
judgments about things outside us which we have been accustomed to
make from our earliest years."[63] The latter included both the judgment
that an external object is colored (as opposed to the pure sensation of
color in grade two), and rational calculation of its size, shape, and
distance. Now, according to Descartes, such judgments and calculations
are made by the very same intellect that renders conscious judgments,
although, he observed, such sensory judgments are rapid and habitual,
and for that reason are not distinguished from simple sense-perception
(cf. Gehler). But despite the fact that they stemmed from the same faculty
of judgment, he distinguished the unnoticed judgments of the third
grade of sense from the mature judgments of the intellect. These mature
judgments of the intellect, when carried out by a mind trained in
Cartesian meditation, embody the clear and distinct perceptions that
serve as the foundation of Cartesian metaphysics. They provide a standard by which sensory judgments may be corrected.[64]

Elsewhere, Descartes explicitly characterized the habitual judgments
formed in childhood as "teachings of nature," and therefore as belonging
to the mind "naturally," and distinguished them from the deliverances
of the natural light, regarded as an instrument of truth bestowed upon

the knower by a nondeceiving deity.[65] In so doing, he distinguished habitual judgment as a natural process from the perceptions of an unclouded intellect. The former, just as Hume's associative processes, operates similarly whether it is forming true or false beliefs; it is thus "naturalistic" by the criterion of the previous section. The intellect, however, sets the standard for knowledge. It is essentially an instrument for discerning truth and falsity; it does not set the standard for truth itself (which is correspondence to an object), but its normal operation allegedly is adequate to ensure the perception of truth. The intellect can therefore set the normative standard by which all rational inquiry is assessed.

Descartes' distinction between the "teachings of nature" and the pure perceptions of the intellect illustrates the distinction between metaphysical and methodological normativity. If we retrospectively apply the notion of methodological normativity to the judgments instantiated by the "teachings of nature," the judgments themselves—but not the processes by which they are formed—will be conceived normatively: they are subject to correction by appeal to standards of truth and falsity, standards that cannot be reduced to statements about how perceivers do in fact judge.[66] Indeed, Descartes' own treatment of such judgments implies that they are normative in this sense, for his complaint that such judgments are typically false implies that they are subject to standards of truth and error. (By contrast, the states of the body, even those implicated in the production of illusory sensory states, are not considered to manifest error.)[67] Nonetheless, he considered the process by which these habitual judgments occur to be "natural," or part of our animal nature, in a way that the judgments of pure intellect are not. The deciding factor for Descartes in this case was not the standards by which an act is judged, but whether the act itself can serve to constitute the standard. His distinction between the "natural light" and the "teachings of nature" was a distinction between judgments (or a faculty of judgment) that set the standard for knowledge, and judgments (or a natural inclination to judge) that serve to preserve the economy of the body; the latter judgments are not metaphysically normative because they do not constitute a standard for knowledge.

I underscore the extent to which methodological normativity differs from Descartes' own stronger conception of metaphysical normativity. It was precisely the standard of metaphysical normativity allegedly provided by the deliverances of the Cartesian intellect that came under attack in the eighteenth century, not only by prospective naturalizers, but also by the metaphysically modest proponents of methodological normativity.

3 Mind, Perception, and Knowledge

The epistemological writings of the great philosophers from Descartes to Kant have served for many years as a kind of training ground in philosophical instruction, and to some extent as a proving ground for contemporary theories of knowledge. Descartes' First Meditation, for instance, has been used to introduce the problems of epistemology at various points in the curriculum, and also as a standard reference for motivating the "skeptical threat" in contemporary epistemological discussions. Berkeley's *Principles* and portions of Hume's *Treatise* and first *Inquiry* have been used in a similar manner. Selected paragraphs from Locke's *Essay* have been used as a foil for Berkeley's and Hume's skepticism, while Kant's first *Critique* has been viewed as an attempt to escape the skeptical conclusions of the latter authors.

In connection with these contemporary uses of past texts, a standard account of the problematic of early modern epistemology has arisen. A central element in the traditional story is that the problem of knowledge arose from acceptance of the theory of ideas and the attendant problem of the "veil of perception." According to this story, the acceptance of the "veil of perception" arose from a mistaken model of knowledge and a mistaken ontology. The mistaken model of knowledge was the "visual metaphor," according to which knowing is like seeing: to know something is to see it clearly with the eye of the mind. But the "something" seen and known was not an ordinary object; an indubitably known object of perception was needed in order to respond to the challenge posed by skepticism. Thus a special mental object, the idea, was invented. In accordance with the visual metaphor, the things known best are those that can be taken in by the understanding "at a single glance," so to speak. The same model was applied to sensory perception; "sensory ideas" became the immediate object of perception. But once ideas had been spirited into existence as third things interposed between perceiver and external world, the knower was stuck behind a veil of ideas. Hence, as in Thomas Reid's criticism of early modern epistemology, the acceptance of the theory of ideas becomes a first step along an inevitable path to Humean skepticism.[68]

According to a second strand of this story, the theory of ideas facilitated the "invention of the mind" by Descartes, after which the mental was taken as a special domain having its own laws and principles. Subsequently, authors such as Locke and Kant confusedly attempted to treat the laws of the mind as a set of explanatory principles that could serve as the foundation for a general theory of knowledge. According to this diagnosis—which, as applied to Locke, dates from the work of T. H. Green and may be found more recently in Sellars and Rorty—early modern epistemology was thick with psychologism. As Rorty has put it, modern epistemology got its mistaken start when Locke confused "a mechanistic account of the operations of our mind" with the properly epistemological concern of "'grounding' our claims to knowledge."[69] Support for this charge has seemed easy to come by, since early modern theories of knowledge make heavy use of various divisions among the faculties of knowledge (such as sense, imagination, and understanding), and make liberal appeals to internal mental acts or processes (such as intuition, comparison, and abstraction).

For the purpose of exploring the genesis of naturalistic, and hence of psychologistic, accounts of the mind, we should suspend judgment on this entire picture, and start over by considering more closely the problem-space of early modern epistemology as it was actually constituted.[70]

The problem of knowledge as conceived in the seventeenth and eighteenth centuries was not the abstract, disengaged "problem," or unmotivated puzzle, that it has seemed to a number of twentieth-century critics. The "theory of knowledge" derived its importance from the significance of the intended objects of knowledge: God, the soul, and nature. The tremendous intellectual energy that went into investigations of the possibility and limits of human knowledge during this period was motivated by a desire to decide whether substantive bodies of doctrine could be established regarding these objects. The desire to know the deity and to establish the immortality of the soul on rational grounds, or to know whether such knowledge is possible, was of central importance in the intellectual and moral landscape of early modern Europe. The desire to know nature, or to know the extent to which nature can be known, was no less central. Conclusions (and implied motives) differed from author to author, ranging from Berkeley's claim to establish the world as a whole as a text of ideas provided to individual minds by God, to Kant's plea for tolerance in matters of theological doctrine on the grounds that the existence and attributes of a deity are beyond human knowledge. Similarly, conclusions regarding the investigation of nature differed, ranging from Bacon's and Descartes' optimistic vision of a mastery of nature, to Locke's cautious defense of a mechanistic approach to nature, to Berkeley's attempt to radically undermine the claims of natural philosophy to describe a mind-independent material world. The stakes were high.

The aims of early modern epistemology contrast with those of more recent theories of knowledge, which typically aim at providing a general analysis of the conditions under which an agent may be said to know a proposition. The project is to discover the necessary and sufficient conditions for affirming that "S knows that P," where S is any rational agent, and P is any factual proposition. Two conditions are widely accepted as necessary: that the person believe the proposition, and that the proposition be true. Further discussion has sought to determine what else is required; for example, whether it is sufficient that, in some sense to be specified, the person is *justified* in believing the proposition to be true (and believes the proposition on the basis of this justification), or whether a more stringent standard than "justified true belief" is required for knowledge.[71]

The primary difference between this activity and the classical early modern theories is not to be discovered in the basic conception of knowledge itself. In early modern theories it went without saying that if someone knows a proposition, he or she must believe it (or hold some suitable mental relation to it) and it must be true. Further, it was assumed that theories of knowledge would be general, in the sense that they would apply to any human cognizer, and could therefore serve as a guide to rational assent. Throughout both early and recent discussions, the aim has been to give an analysis of knowledge that could also explain the normative force accorded to claims to knowledge—to explain how it is that if person X has conclusive reasons for believing proposition P, person Y, if presented with the reasons, should also grant the proposition.[72]

The chief difference between early and recent epistemology pertains to the object of epistemic evaluation. Early modern epistemology did not have as its aim analyzing the conditions for asserting particular factual claims. Rather, the theoretical enterprise was aimed at describing in a quite general way what can be known and how it can be known, where candidates for the "what" are God, nature, and the soul and candidates for the "how" are the various cognitive faculties: the senses, the imagination, and the intellect or understanding. Moreover, because the faculty of the understanding was regarded by many authors as an instrument for grasping the "nature" or "real essence" of mind-independent reality, investigation into the possibility of such knowledge became propaedeutic to investigation into the nature of reality. Because authors such as Hobbes, Locke, Berkeley, and Hume raised serious questions about the possibility of such knowledge, license for the legitimate pursuit of metaphysics came to require the successful answering of epistemological questions.

The characteristic breadth of early modern epistemology is revealed in its treatment of spatial perception. In theories of vision the spatial properties of things were simply taken for granted in describing the causal process by which sight occurs. This attitude contrasts sharply with that taken in metaphysical and epistemological investigations, in which the ontological status of spatial properties was disputed. In the course of evaluating the claims of Cartesian and Newtonian physics, theories of the mental apprehension of spatial properties were thrust to center stage. Metaphysical and epistemological investigations led beyond the spatial perceptions of vision to the apprehension of spatial properties by other mental faculties such as the imagination and the intellect or understanding. "Spatial perception" meant not only the perception of size, distance, and shape by means of sight, but also, and often more fundamentally, the intellectual apprehension of the putative fact that the spatial properties are the fundamental properties of matter.

I discuss the place of spatial perception in early epistemology in three following sections, which examine the contribution of theories of the senses to the "veil of ideas" problematic, the place of spatial perception in theories of the intellect, and arguments about the ability of the intellect to know an external reality.

3.1 Causal Chains and the Veil of Perception

The initial formulation of the "veil of perception" problem drew upon theories of vision. A typical theory of sensory perception regarded the perceptual process as a causal chain starting with physical objects outside the body and ending with a sensation or idea in the mind; the mind was therefore viewed as an entity locked away inside the sensorium, able to gain knowledge of the external world only inasmuch as it received impressions from the nervous system. But if the mind is sequestered inside the brain, how could it be aware of objects at some distance from it? The theory of the senses seems to have suggested, or at least reinforced, the "veil of perception" doctrine by suggesting that we are only indirectly acquainted with the external world, through the mediation of a causal chain.[73]

However, the skeptical problematic defined by the "veil of perception" should not be regarded as a necessary result of a causal-chain analysis of perception. For although such an analysis can lead to "veil of perception" skepticism, it need not. A causal-chain analysis can facilitate the veil-of-perception position by lending credence to the notion that perceivers stand in a direct perceptual relation only to something that is in each of their heads, viz., an image or picture in the sensorium or "common sense." This view of the perceptual process is sometimes characterized as the "homunculus view," on the assumption that it is implicitly committed to positing a little person inside the head who views the images or pictures. The obvious objection to positing homunculi is that it merely moves the relation between perceiver and perceived into the head while leaving the process of perception unexplained (indeed, it may even be seen as the initial step in a regress of homunculi).

It is true that some writers were guilty of adopting something like the homuncular position. Newton, in Query 28 of the *Optics*, wrote that the images of external things, "carried through the Organs of Sense into our little Sensoriums, are there seen and beheld by that which in us perceives and thinks."[74] But other writers were more careful, explicitly warning against the dangers of homuncularism. For instance, in a well-known passage Descartes remarked that although (according to his conception) a material "image" or "picture" bearing some resemblance to external objects is transmitted into the head via the optic nerves, "we must not hold that it is by means of this resemblance that the pictures cause us to perceive the objects, as if there were yet other eyes in our brain with which we could apprehend it"; rather, "it is the movements of which the picture is composed which, acting immediately on our mind inasmuch as it is united to our body, are so established by nature as to make it have such perceptions."[75] We are not to understand the relation between the mind and the "neural image" as one in which the mind perceives or sees the image, but rather as one in which the character of the motions constituting neural events causes the mind to have a specific type of sensation. A hundred years later Porterfield put down the "vulgar error" that we see our own retinal image, arguing that it would also be an error to contend that we sense images in the brain, even though such images are causally implicated in perception.[76] Thus, according to Descartes and Porterfield, the mind does not stand in a perceptual relation to events in the brain; rather, these events give rise to sensations of external objects (including sensations of the states of the sense organs themselves). We do not perceive our neural states, but we have sensations and perceptions of external objects and bodily states by means of neural activity.

Thus, the causal-chain analysis need not give rise to homuncularism, but this may seem cold comfort to those who are worried about the wider problem of the veil of perception. For Descartes, Porterfield, and nearly everyone else could agree that in any case it is the sensations or ideas caused in the mind by the brain events in the sensorium that constitute the immediate objects of consciousness. If these sensations or ideas are the immediate objects of awareness, doesn't the mind then stand in a perceptual relation to them? And isn't this introduction of ideas as "third things" mediating between perceiver and perceived the real source of the veil-of-perception problem? Isn't the veil of perception really a veil of ideas? This diagnostic claim is better founded and potentially more damaging than the preceding one.

3.2 Ideas as Objects of Perceptual Awareness
Since it was agreed by nearly everybody that the objects of direct awareness are mental states, the central problem for theories of knowledge was to determine the extent to which such mental states, or ideas, inform the knower about an independently existing material world (or about the mind

itself, or about a deity). There were numerous conflicting opinions with respect to this question, ranging from belief that our sensory ideas do inform us of a material world (Descartes and Locke), to the denial that there is such a world (Berkeley), to the view that the question of the relationship between our ideas and a world of things in themselves cannot be answered (Hume and Kant). However, of equal or greater interest than these diverse epistemological positions themselves is the fact that they constitute distinct formulations of the "problem of knowledge." These distinct formulations depend upon variations in the conception of "ideas" as objects of awareness, in the conception of the external world, in the conception of possible sources of knowledge of that world, and in the standards that putative knowledge must meet in order to qualify as knowledge. As a way of scouting this diversity, I shall examine three generic positions, using the familiar label "representational realism" for the position of Descartes and Locke, "idealism" for Berkeley's position that only minds and ideas exist, and calling Hume's and Kant's denial of the possibility of knowledge of transcendent reality "metaphysical agnosticism."

Basic to the representational realist position is the belief that ideas give us knowledge of an external, material world of mind-independent objects possessing spatial properties. This position is "realist" in positing such a world; the "representations" come into play as part of the project of evaluating perception as a means of knowing that world. Representational realists ask whether perceptual states are good or poor, or accurate or inaccurate, representations of the external world. Such questions may be as general as the question of whether our individual sensory ideas provide us with any knowledge at all, or they may be more narrowly focused on differences in the informativeness of the various kinds of sensory ideas (e.g., of ideas of shape as opposed to color).

It has seemed that, in posing questions about representational accuracy, the representational realist opens a sufficient fissure to allow insertion of the skeptic's wedge. This familiar objection may be stated briefly. Any judgment about the faithfulness of our sensory ideas as pictures or images of the external world must depend upon a comparison between the properties that objects have in themselves and the images that allegedly constitute our sensory ideas. The problem may be compared to that of deciding whether a particular photograph or portrait is a good picture, or portrays a good likeness, or "resembles" the thing of which it is a picture.[77] With a recent portrait of an individual, we can compare the portrait with the person to determine whether it is a good likeness.[78] Matters stand differently when we wish to compare our sensory ideas with material objects, because a direct comparison is impossible. Since our direct experience is limited to our own sensory ideas, we are never in a position to make the needed comparison between our perceptual images of objects and the objects themselves. We are given only one side of the desired comparison.

Recent scholarship has gone far to remove the burden of skepticism from representational realism and the theory of ideas by showing either that Descartes and Locke did not hold the objectionable positions commonly attributed to them, or that, in holding certain positions, they did not provide any special aid to the skeptic. What positions did they actually hold? Did they have ontological commitment to ideas as third things—that is, to ideas as entities distinct from both minds and external objects standing in a relation to the mind analogous to that in which external objects are ordinarily considered to stand to the senses?[79] It is notoriously difficult to determine precisely what doctrine of ideas *should* be attributed to Descartes and Locke, despite the fact that each was aware of the doctrinal importance and the comparative novelty of his own use of the term.[80] But careful attention to the writings of Descartes and Locke suggests that neither of them considered ideas to be third things, distinct from the minds that have them.

Within Descartes' ontology, everything is either a substance or a modification of a substance. The two substances, mind and matter, are each characterized by a chief attribute—thought in the one case, extension in the other. The "modes" or modifications of substance are specific determinations of these chief attributes. In the case of extension, its modes are size, shape, position, and motion; these should be conceived as determinations of otherwise undifferentiated extended quantity. The modes of thought include various mental actions, such as willing or fearing, and ideas. Ideas are modes of thought; as such, they inhere in minds.[81] Minds do not stand in relation to ideas as separate existents; rather, minds have ideas.

Although equating ideas with modes of mind precludes the notorious third thing from having separate existence, it may seem that it does so at the high cost of raising a new difficulty about a central aspect of ideas. This putative difficulty pertains to the content of ideas, that which makes them the idea of one thing rather than another. If ideas are regarded as third things, then ideas themselves can have the properties (such as color and shape) that constitute their content. But if minds have ideas of particular colors and shapes, the ideas themselves surely are not colored and shaped; as conceived by Descartes, the states of an immaterial mind cannot themselves literally *be* colored and shaped, even if they have colors and shapes as their phenomenal content.[82] Where does that leave the color and shape found in the content of the idea? Descartes foreclosed the options of denying their existence, or moving them to the external world. In the case of color, the content found in the idea obviously cannot be equated with color as a real quality of external objects, because Descartes denied that color is a real quality; yet he affirmed that phenomenal color is present to the mind.[83] Similarly, Descartes was committed to there being ideas of shape in which shape is phenomenally present in the content of the idea. Again, it won't do to equate the shape in the idea with some external shape—say, the shape of a brain state—because Descartes explicitly denied that the shape we experience during vision is the same as the shape in the brain.[84]

Descartes attempted to solve the problem of content by adopting a scholastic distinction between the "formal" and the "objective" reality of ideas.[85] The mind has ideas *formally* in a way similar to that in which matter has shape; ideas and shapes are each the modification of the chief attribute of a substance. But their formal status as modes of mind does not exhaust the description of ideas, for ideas have content: objects exist "in" them *objectively*, or by way of representation. Thus, a sun or a stone exists "in" the idea as object of the idea, where "object" is to be understood as intentional object, or object of thought. To have an idea is to have a perception of a particular sort, which may be typed by its intentional object. Thus, to have an idea of a red thing is for the mind to be formally modified such that it has a particular idea. Neither the mind nor this idea, however, are formally (actually) red. It is, on this view, a peculiar and amazing feature of states of mind that besides simply being states of mind—or perhaps in virtue of being such—they also can represent shaped and colored objects, and represent them in such a way that color and shape are phenomenally available to consciousness, even though the states of mind themselves are not colored and shaped.

Considered in accordance with the doctrine of objective and formal reality, Descartes' ideas are best regarded as perceivings, not as objects that are themselves perceived.[86] An idea of red is a perceiving of red; a perceiving of red is not the perception of a distinct entity, of an idea that is itself red, but is the possession of a mental state with a particular phenomenal content. A similar reading has been provided for Locke's theory of ideas, notwithstanding his tendency to use language that suggests reification; it has been argued that Locke's talk of "perceiving ideas" should be understood to mean that one has a certain perception, rather than that one stands in a perceptual relation to an idea considered as a thing.[87] And indeed, when speaking carefully, Locke describes ideas as states of mind; thus, in the case of sensory ideas, he cautions "when I say the senses convey [sensible qualities] into the mind, I mean, they from external objects convey into the mind what produces there those *perceptions*."[88]

When "ideas" are construed as perceivings, the assertion that we are directly aware only of our own ideas may be given one of two interpretations. Either it means that our thought is directed toward the intentional object of our idea (we are aware of the object of our perception), or it means that we are aware that we are having a particular idea. (These might be called "first-" and "second-order" awarenesses.) The first kind of "awareness" may be glossed as perception per se, the second as awareness that we are having a particular perception. Ideas then become "that by which" we perceive external objects (in those cases in which the intentional object successfully represents a physical object), not that which is itself perceived.

When ideas are treated as perceivings, their representational content is equated with an intentional object. It must be admitted that to interpret "perceiving an idea" as "having a modification of the mind with a certain intentional object" does not make the notion of an idea any less problematic than the notion of an intentional object. It must further be admitted that in a certain sense intentional objects are "third things," although of a queer sort. They are distinct from external objects but not distinct from states of mind; and yet they are not to be equated *simpliciter* with states of mind, because intentional objects possess color, for example, but states of mind do not. Thus, although ideas are not separate existences, we still have the following three things: external objects, states of mind (considered as acts of perception), and intentional objects of those acts. Nonetheless, ascribing content to ideas by way of intentional objects removes the impetus toward skepticism that was alleged to result from the theory of ideas itself. Even the "anti-idealist" Reid was committed to a distinction between perceptions as acts of mind and the objects of those perceptions, and even he admitted that those objects need not coincide with an external object.[89] The positing of objects of thought distinct from external objects is more widespread than is adherence to the theory of ideas.[90]

Even if this interpretation undercuts the familiar picture that the "way of ideas" inevitably leads to skepticism by interposing a "third thing" between perceiver and perceived, that in itself does not make proponents of the theory of ideas immune to skeptical challenges of the sort posed by Berkeley or Hume. Which is all to the good, for adherents of the theory of ideas are not immune to skeptical challenge. But are such challenges peculiar to the theory of ideas? Reflection suggests they are not. Skepticism with respect to the senses can be posed within any position that distinguishes between perceivers and things perceived; it is not a matter of having *three* things (perceiver, perceived, and *tertium quid*); two will suffice. Other things being equal, the skeptic's wedge can find an opening as long as one is willing to allow the possibility that a particular state of a perceiver, of the type that ordinarily constitutes the perception of an external object, can occur in the absence of its standard external cause.[91] This possibility does not, however, depend upon the theory of ideas. The fact that skepticism toward the senses can be converted to "brain in the vat" skepticism suffices to establish this point. And so the theory of ideas is not a necessary condition for skepticism regarding the senses; whether it is a sufficient ground is a question that could be answered only by evaluating the substantive claims of Descartes, Locke, Berkeley, and others, each of whom adopted some version of the theory of ideas but nonetheless claimed knowledge of objects lying beyond the veil of ideas.[92]

Be that as it may, we have yet to find a reason for taking seriously the representative realist's concern with accuracy of representation. According to some recent accounts, the concern with accuracy was nothing more than an unfortunate consequence of the fact that early modern philosophy had been captured by the "mirror metaphor" for the mind. According to this analysis, the image of the mind as a "mirror of nature" encouraged a concern with accuracy of representation, and did so for no good reason.[93]

In the cases of both Descartes and Locke, the question of representational accuracy is couched in terms of the "resemblance" (or lack thereof) between the contents of ideas and external objects. Their talk of resemblance, which has been much maligned, becomes less objectionable (or more understandable) when seen in context. For both Descartes and Locke, the question of resemblance was posed to them by their common adversaries, the scholastic Aristotelians. As Descartes and Locke understood their opponents, the cause of our experience of color is the quality of color per se, considered as a primitive property of objects. According to this theory, the quality of color as we experience it is identical with the quality of color in objects; explanation of the physical basis of color appeals not to particles or waves, but to color as a real property. No closer resemblance between experienced color and physical color could be desired.[94]

Descartes and Locke wanted to allow that some types of sensory idea might present "resemblances" of properties of material objects, but to insist that others do not. In particular, they wished to deny that ideas of color and other secondary qualities provide an accurate imagistic representation of the basic properties of material objects, where "accuracy" is spelled out in terms of resemblance. This denial resulted from a particular conception of the basic properties that should be admitted into physical explanations. To say that color as experienced does not "resemble" anything in objects is to say that the best scientific account of color perception will explain the property of objects that causes us to perceive color in terms of a set of basic properties that does not include color as a primitive but only size, shape, and motion. To say that ideas of shape are (or could be) accurate representations is to say that shape is a basic property of material objects that is (or can be) imagistically portrayed in ideas in such a way that the shape in the idea is a good likeness of the shape in the object. More generally, if the so-called primary qualities exhaust the basic properties of bodies, or of material substance, then only ideas of primary qualities stand a chance of counting as likenesses of the basic properties of material objects;[95] ideas of secondary qualities cannot accurately represent the basic properties of objects, because color, for example, is not a basic property of such objects. The question of representational accuracy becomes the question of deciding which of the experienced qualities of objects should be considered basic properties within physics. According to this reading, Descartes' and Locke's discussions of the "resemblance" or "accuracy" of sensory ideas grew out of one of the central intellectual projects of the seventeenth century—that of developing an adequate science of nature.

Descartes' and Locke's respective arguments for the representational accuracy of the ideas of primary qualities were somewhat different. Both authors wished to ascertain whether the faculties of the human mind are adequate to discover the fundamental properties of matter, the properties that will be basic in physical explanations. They differed in their assessments of those faculties, and these differences resulted in differing conceptions of the status that should be accorded the primary qualities in the ontology of physics. Descartes maintained that the mind is capable of discovering the essence of matter, and he was willing to assert that extension constitutes that essence; Locke denied that the mind could discern real essences, and he adopted an epistemically (and metaphysically) more modest attitude toward the primary qualities.

Descartes provided a criterion for determining the basic properties of matter that was alleged to be independent of sensory ideas. The independent source of knowledge is the intellect itself, conceived as a faculty capable of intuiting essences without sensory input. Thus, when Descartes claims to perceive clearly and distinctly that bodies are possessed of size, shape, and motion, the clear and distinct perception is not, in the first instance, sense perception; it is intellectual perception. Clear and distinct perception pertains to the intellect operating independently of the senses and imagination. In this mode of operation, the intellect has no grasp on phenomenal color as a possible property of material substance, but it does have a perception, or a "pure understanding," or an intellectual intuition, of geometrical quantity or extension.[96] Descartes builds upon his previous argument that the clear and distinct perceptions of the intellect are true by asserting that he clearly and distinctly perceives that extension is the essence of matter. If the argument and the ensuing assertion about the essence of matter are accepted, then it follows, in accordance with the ontology of attributes and modes, that all aspects of matter must be modifications of extension, and modifications that can be apprehended by the pure intellect. Thus, any sensory property that is not included among the clear and distinct perceptions of the intellect operating independently of the senses, as color is not, cannot be regarded as a basic property of matter; if color is ascribed to matter as a property, it is a secondary one, and so is derivative from the basic properties.[97]

Descartes' justification of the distinction between primary and secondary qualities was not open to Locke. Although there are passages in the *Essay* that sound very much like Descartes' appeal to the mind's perception of the essential properties of matter,[98] Locke clearly was not in a position to appeal to perceptions of the "pure understanding" as a source of knowledge about the external world. According to Locke, the intellect by itself is unable to provide *any* knowledge of the material world. He rejected the notion that the intellect can grasp essences independently of the senses; this rejection is one implication of the doctrine that all knowledge of the external world comes through the senses.[99] The passages on perception of the "essential" properties of matter are best read as reflections on what is conceivable in accordance with the most plausible scientific account of the operation of the senses. And indeed, immediately subsequent to these passages, Locke invokes the mechanistic account of sensory stimulation.[100] According to this account as Locke develops it, color in objects is just a particular surface texture that causes light to be reflected in such a way that it produces a certain effect in the nervous system, which in turn regularly produces a sensation, or sensory idea, of a particular character. This last part, the production of the sensation, Locke did not attempt to explain; he regarded it as a given fact that sensory ideas have their standard causes in the nervous system.[101] The doctrine of primary qualities does not explain the efficacy of these neural events to produce sensations; it applies to explanations of how external objects affect one another and act upon the sense organs.

Thus, as Descartes and Locke conceived the "problem of knowledge," their concern with representational accuracy did not provide special aid to the skeptic. Nor were their discussions motivated by a perceived skeptical challenge to knowledge in general. Their investigations were largely directed at assessing the possibility of metaphysical and scientific knowledge, and at justifying the primacy of a certain set of properties— Descartes' modes of extension, Locke's primary qualities—in physical explanations. Because spatial properties were central to the new science that Descartes hoped to establish and Locke wished to support, spatial properties and spatial perception were at the center of their investigations of the knowing mind. Skepticism was either a mere tool (for Descartes), or a position to be brushed aside (by Locke).[102]

Berkeley is the chief representative of idealism, the second general epistemological position named above. He used the theory of ideas and his own theory of spatial perception in order to deny the existence of mind-independent material substance. Berkeley began his attack on material substance by remarking that those who defend its existence must do so by appeal to sense or to reason.[103] But, he maintained, the first ground will not work because all we can be certain of by means of sense alone is that we have sensory ideas.[104] Reason must thus justify the postulation of material objects as a cause for sensory ideas. Berkeley had two arguments against such a posit, both metaphysical. The first, which questioned the intelligibility of the interaction between mind and matter, does not concern us here.[105] The second argument drew upon his account of spatial perception and led to the conclusion that we are unable to form any coherent conception of matter itself. The leading component in this argument was an attack on abstract ideas, particularly on the abstract idea of extension.[106] It is important to see that Berkeley's attack on the abstract idea of extension is different from his attack, say, on the abstract idea of a triangle. In the latter case, he was questioning the notion that we have a single idea that stands for all the particular triangles, or even types of triangle (whether right, scalene, etc.). When Berkeley denied that we have an abstract idea of extension, he meant to deny that the same set of spatial properties is represented in both visual and tactual ideas and is described by geometry; he meant to deny that "extension," and particular modifications of extension, such as a particular right triangle, are common to sight, touch, and geometrical description. In both instances, he denied that there exists a single idea representing what is "in common" among a range of other ideas. The second instance is the more radical, inasmuch as it denies that the properly visual idea of a right triangle presents the same species of triangle as the properly tactual idea of a right triangle.

It is important to understand the kind of obstacle such a conclusion would represent for Descartes and Locke. For Locke, it amounts, first, to a denial that the simple ideas of extension, size, and shape are common to sight and touch, and, second, to the denial that one can even understand what would be meant by positing mind-independent bodies having these properties. For Descartes, it amounts to these two denials, plus a denial of the doctrine that the intellect can grasp extension independently of sensory ideas; according to Berkeley, there is no intellectual content, "extension in length, breadth, and depth, independent of color or tactual sensation," to grasp. Berkeley's conclusion would undercut the attempts of both authors to justify claims about the basic physical properties of matter, and would thus put them in the skeptical position envisioned in the problem of the veil of ideas. Whether they are in this position does not, however, depend upon whether they actually adopted the theory of ideas as traditionally understood but upon the success of arguments pertaining to the primacy of spatial properties. Berkeley's second argument does not turn upon perceptual relativity or other considerations limiting the epistemic value of sensory ideas;[107] rather, it questions the very intelligibility of positing a world of bodies possessed of primary qualities.

The third generic position, metaphysical agnosticism, was shared by Hume and Kant. Despite their many differences, these authors each held the view that matter is unknowable if regarded as a mind-independent substratum; each took an agnostic attitude toward the traditional metaphysics of material substance. Moreover, each justified this agnosticism by attacking what Kant called a "real use" of the intellect—the alleged use of the intellect, as in Descartes, to grasp mind-independent reality without the aid of the senses.[108] A portion of Hume's famous analysis of causation was an attack on the ability of reason to know matters of fact or to establish causal connections a priori.[109] In a similar vein, Kant, in his well-known restriction of the a priori to a condition on possible experience, retreated from his precritical position that the intellect can grasp an intelligible world in a manner distinct from the presentation of the world through the senses. Each author in effect denied Descartes' claims for the power of clear and distinct intellectual perception. If they each arrived at a position that restricted human knowledge to sensory impressions or to "appearances,"[110] this restriction arose as the conclusion of various arguments about the power of the intellectual faculty; it was not a starting point determined by a commitment to "ideas" as objects of awareness.

By highlighting several central aspects of the history of epistemology from Descartes to Kant, I have endeavored to reveal the shortcomings of placing the "veil of ideas" problematic at the center of that history. In my own version of this history the guiding thread of the portion devoted to epistemology and the theory of mind would not be the theory of ideas but the real use of the intellect. Modern philosophy after Descartes should be seen as a set of reflections on the prospects for or failures of intellectual intuition to provide direct access to what Kant would call the noumenal world. Skepticism assumes a variety of roles in this story: Descartes used skepticism in the context of a set of "cognitive exercises" intended to better acquaint the reader with his or her intellectual faculties, Berkeley employed skeptical arguments in questioning the intelligibility of material substance, Hume used skepticism to attack the suggestion that the intellect exists as a faculty capable of grasping mind-independent reality.[111] But these roles are parts for supporting players, and the center of action is elsewhere.

Much of modern epistemology and metaphysics after Descartes may be seen as reactions against his bold claims for the metaphysical power of the intellect. The differences between Descartes and Locke on primary and secondary qualities can be attributed to their contrasting attitudes toward the power of intellectual intuition. One of Berkeley's central arguments against the claim that material substance exists was directed against the claim that extension is an object of the pure intellect. The power of the intellect or understanding also was central to Hume's and Kant's investigations of the limits to human knowledge. Within early modern philosophy the status of the faculty of the intellect was of focal interest.

3.3 Intellectual Intuition

Every major modern philosopher accepted the notion of intellectual intuition. The knower's ability to grasp self-evident (or "intuitively certain") propositions was generally attributed to the intuitive use of the intellect or understanding. Such propositions extended from the law of contradiction, to the axioms of geometry, to, in some cases, substantive metaphysical principles. Explanations of how these propositions are known commonly spoke of "mental inspection" (Descartes) or of the "intuitive perception" of agreement and disagreement between ideas (Locke and Hume).[112] Yet, in their conception of the objects of intuition and of the scope to be ascribed to intuitive knowledge, these philosophers differed greatly. Whereas Descartes conceived the intellect as providing, in paradigm cases, its own objects of contemplation (ideas of God, extension, and thought), Locke and Hume considered the intellect to be powerless to act except when provided with materials that originated in the senses.[113]

Whatever their disagreements about the source of the ideas of intuition, all major early modern philosophers agreed that intuitive knowledge possesses the highest degree of certainty. If a proposition or judgment could be established "intuitively," one might rightfully expect intersubjective agreement or assent. Mathematics provided the primary example of a domain in which agreement among individuals could be expected solely on the assumption that they are possessed of common intuitive faculties. Whether in terms of "intuitive evidence," of the "perception of relations between ideas," or of the "figural synthesis of imagination," early modern philosophers ascribed the certainty of mathematics and the possibility of mathematical demonstration to intellectual intuition.[114] It would be simpleminded to explain such ascriptions in terms of an unwary seduction by mirror metaphors. There was a need to understand the possibility of mathematical demonstration, when the actual constructions used in such demonstrations—e.g., lines on paper—clearly failed to realize the properties of the objects of such demonstrations—e.g., breadthless lines. One can understand the temptation to posit a special intellectual object in order to explain at once the "perfection" of mathematical objects and our ability to grasp the force of mathematical propositions; or—in cases where the temptation to posit such objects was resisted—to posit a faculty of abstraction and the ability to perceive the logical and demonstrative relations among mathematical objects "abstracted" from their sensory context.[115] The appeal to such objects and faculties need not be regarded as empty metaphysics; there was no other framework within which to understand the possibility of mathematical knowledge.[116]

Although philosophers agreed that mathematical knowledge is based on the "perceptions of the understanding," they disagreed about the extent to which this mathematical knowledge could be transferred to the material world. More generally, they disagreed about the extent to which intellectual intuition provides any type of sense-independent knowledge. Rationalist metaphysicians such as Descartes and Spinoza agreed that intellectual intuition provides access to a mind-independent reality, and they emphasized mathematical (or geometrical) properties in their conception of the material portion of that reality. They thereby disagreed with their Aristotelian opponents, who required that the intellect have sensory material upon which to operate and who did not regard the mathematical properties of natural things as essential to their substantial natures. However, Cartesians and Aristotelians alike agreed in conceiving the world as constituted by a substance or substances possessed of an essence or essences, and in attributing to the intellect the ability to grasp such essences—the Cartesian intellect through pure intuition, and the Aristotelian *nous* by means of abstraction.[117]

Authors in the British tradition tended to revise downward the number of objects regarded as subject to intellectual apprehension. Locke denied that real essences could be grasped either through experience or by means of innate ideas, principles, or powers. Even though he allowed the intellect a power of abstraction, and granted that it could perceive necessary connections among the nominal, mathematical, and moral essences of its own creation, he nonetheless contended that the intellect is unable to grasp the essential connections among the properties and powers of substances or complex things.[118] Berkeley denied that the intellect grasps pure extension. But, contrary to Locke, he did not consider the failure of the intellect to grasp pure extension as revealing a limitation on the power of that faculty but as providing a premise toward his argument against the existence of material substance. He affirmed the power of the intellect to grasp active, thinking beings, and, in particular, to understand that God's agency is implicated in the production of ideas.[119]

It is clear that the question of the scope of the intellectual faculty was of central importance in early modern epistemology. This question pertained to the existence and the activity of a particular faculty of the mind. It was not, however, a question to be decided by naturalistic psychology. Positing the intellect as an epistemic power was part of "psychology" understood as the doctrine of the rational soul; so conceived, "psychology" took *nous*, and its descendent, the Cartesian intellect, as its object. But the Cartesian intellect was not conceived as a natural object; its activities were not considered to be part of the natural world, and so were not seen as "natural" processes. Investigation of the intellect was not the province of natural philosophy or natural science. Consequently, my claim that the investigation of "mental faculties" and "mental activity" need not be conceived as part of naturalistic psychology has been substantiated. The belief that human beings do not possess a Cartesian intellect might incline one toward a psychologistic account of human rationality if one took the further step of asserting that in the absence of such an intellect the human mind must be regarded as a natural object. However, the affirmation that the mind does possess such an intellect is not a claim in naturalistic psychology, but a metaphysical assertion about the power and scope of a noetic faculty.

We can now see how Hume's skeptical attack on the power of the understanding was of a piece with his naturalism. If Hume could successfully show that the faculty of reason is powerless for most of the purposes to which it had previously been regarded as fundamental, he could achieve two results. First, he could restrict epistemic access to such objects as God and the human soul, objects previously thought to be known only through the abstractive power of *nous* or the pure intuition of Cartesian intellect. That would support his a-theistic naturalism. Second, if he could show that the operations accorded to reason in judging matters of fact actually belong to imagination, he could then assert that human mental abilities are continuous with those of the animals; for Aristotelians and Cartesians agreed that animals are possessed of senses and imaginations.[120] That would support his naturalism with respect to the human mind.

Hume's "naturalism" regarding belief formation has implications for the possibility of appealing to common norms of objectivity regarding matters of fact. In denying that human beings are possessed of a truth-guaranteeing intellectual faculty, Hume at the same time denied that the mind possesses its own constitutive norms of knowledge and truth. If norms are now to be given for the truth seeker, they cannot consist, as they did for Descartes, in deliverances that are sought within the confines of reason itself. On the Humean account, the ascription of a common set of faculties to all human beings cannot provide the basis for the expectation of intersubjective agreement on matters of fact, as it could for Descartes. The faculties of sense and imagination as conceived by Hume are truth-neutral with respect to judgments of matter of fact that go beyond occurrent experience. Concomitance of faculties yields no expectation of objective agreement, because the grounds for that expectation, the belief that proper use of the faculties guarantees the perception of truth, have been undermined.

The implications of Hume's naturalism may be placed in perspective by considering what would follow from accepting Descartes' or Hume's theories of knowledge. If Descartes' theory of knowledge is correct, it provides a criterion of truth; it proffers methodological advice which, if followed, is guaranteed to yield true judgments.[121] If Hume's theory of knowledge is correct, objective truth about significant matters of fact cannot be attained; the best we can do is to aim to cultivate habits of imagination that will prove useful in the course of experience. We shall never be justified in claiming the security that would follow from knowledge of the true principles of the natural world.

The deflated status of epistemic norms in Hume's naturalized science of the mind is brought into relief by considering what sort of method-ological advice he could legitimately offer. Such advice can only be a suggestion for using the truth-neutral processes of the mind to good effect in the project of charting empirical regularities, and he offers it in the section of the *Inquiry* in which he explicitly compares human beings with animals.[122] Hume insinuates that "custom has the same influence on all," including animals and human beings; nonetheless, he is willing to grant that (some) human beings surpass all animals. In order to explain how this is possible, he charts the differences among human beings, ranging them under several heads. His main point is that human beings differ in their ability to observe and follow up the constant conjunctions that underlie the discovery of useful generalizations regarding nature because they differ in attention, memory, accuracy of observation, attention to detail, and so on. Although the listing of such differences clearly constitutes a recommendation toward attention, accuracy of observation, and so on in the formation of one's habits, it is equally clear that Hume could not and did not offer a rational justification of this recommendation. In principle this recommendation could be based on no more than observed success in discovering salient constant conjunctions up to the present. But, notoriously, Hume himself had already argued that previously observed success is no guarantee of future success. From a Humean perspective, normative considerations—recommendations for how one should guide the development of one's cognitive habits—have the status of recommendations for training based on what has been successful. Such norms do not, as in the case of the Cartesian intellect, constitute instructions for the proper use of a faculty that is inherently truth-finding. Thus, in the Humean context, the normative aspect of the theory of knowledge is deflated from setting the criteria that mark sure knowledge to offering recommendations for cultivating useful habits.

4 The Mind in Epistemology as Opposed to Psychology

From my examination of the terms in which early modern arguments were actually framed, I draw two conclusions of special interest. First, questions pertaining to the status of mental faculties—including not only the senses, but, more crucially, the intellect or understanding—formed the primary theme in early modern epistemology. Second, the mental faculties so conceived are on the whole best regarded as postulated noetic powers whose boundaries were being charted in the service of evaluating the foundations of metaphysics. The first of these findings confirms the widespread perception, emphasized by Rorty and others, that the theory of mind played a special role in early modern epistemology. The second finding repudiates the charge that the mental faculties and powers posited in such theories typically were conceived as objects of a "mental mechanics" analogous to Newtonian mechanics.

Clearly the "unmasking" of seventeenth- and eighteenth-century authors as confusedly pursuing psychologistic projects is a perilous task. Questions of innateness, when taken in context, may be either psychological or epistemological. The contemporary tendency to treat such questions as inherently psychological perhaps reflects a rejection of the substantive claims of past authors about the character of mental processes, followed by an assimilation of all discussion of mental processes to psychology (by default). Thus, if we judge Locke to have been confusing psychology with epistemology in posing questions about innateness, our judgment will most likely reflect a refusal to take seriously the substantive doctrines that he was opposing, such as the notion that the innate has a divine warranty, for such doctrines contained not psychological but metaphysical claims. In the case of Descartes, if we judge that his discussion of pure intuition was actually a confused foray into introspective psychology, this most likely amounts to a veiled rejection of his substantive claims about the power and divine origin of the pure intellect. The moral is that recent attributions of psychologism to Descartes and Locke may be what need unmasking, rather than the naturalistic psychologies fictionally attributed to them. For inasmuch as they pursued what they would call psychology, it was not naturalistic, and inasmuch as they pursued what we would call psychology, it did not determine their theories of knowledge.

Yet we should not stop with having "unmasked" the charge that early modern authors promiscuously engaged in psychologism. We need to recall and acknowledge that some authors, such as Hume, openly attempted to provide a naturalistic account of belief-formation. It would thus be a mistake to accuse him of unknowingly confusing psychology with epistemology; rather, those who would criticize his project would do better to reject the attempt to treat belief in a naturalistic manner. Indeed, Kant can be seen as providing precisely that sort of response to Hume.

One particular relation between epistemology and the psychology of perception will become increasingly important in later chapters. This relation has its basis in the fact that Descartes and other thinkers recognized that the senses play a functional role in the guidance of action. Various early modern explanations of the existence of this functional endowment were given. Berkeley, for example, held that any regularity in our sensory ideas—and thus any possibility of using one sensory idea (say, a visual perception) to guide action that is predictably accompanied by yet other sensory ideas (say, various gustatory sensations and the feeling of sated hunger)—is owing to God's benevolence (since he is, on Berkeley's view, the author of all ideas).

For representational realists such as Descartes and Locke (and for most sensory psychologists), a feature of the problem is missing in its Berkeleyan form. Unlike Berkeley, the representational realist is faced with the problem of explaining his supposition that some of our sensory ideas are not merely the standard effects of a given cause (even in Berkeley's account our ideas have their standard causes, inasmuch as a causal relation exists between God and the mind), but are reasonably accurate images of their causes. Descartes and Locke solved this problem by maintaining that our primary-quality ideas represent to us, in their phenomenal spatiality, the properties of objects that actually are spatially extended. Why should there be this match between properties as represented in certain kinds of human sensory ideas (ideas of primary qualities) and properties that objects have? Especially when mind-body dualism is accepted, along with the standard proviso that any relation might obtain between brain events and the phenomenal character of their standard mental effect (e.g., vibrations of one character giving rise to ideas of red, other vibrations to ideas of green), why should some brain events give rise to representations that are "good semblances" of their causes?[123] Descartes explained the match between human forms of perception and properties of matter by appeal to God. In Descartes' view, God created both our minds and the world and ensured a match between some kinds of our ideas—the ideas of geometrical properties—and the real properties of material things.[124]

Nineteenth-century writers did not rely upon a supernatural explanation of the match between our forms of representation and the properties of matter. Instead they offered a psychological explanation, seeking to show how this match might be derived through learning that involves interaction with actual material objects. As they saw it, any match between a form of representation, such as spatial representation, and the properties of material objects is acquired through interactions between a perceiver and some actual spatially extended objects. Still other writers gave an evolutionary account of any such match. In all these cases, we shall be interested in the extent to which such naturalistic theories were intended to speak to the epistemological problems posed by Descartes or Locke, and the extent to which they could be successful if that was their goal.

Chapter 3

Mind, Space, and Geometry in Kant: Transcendental and Naturalistic Conceptions of Thought and the Mental

In comparison with other seventeenth- and eighteenth-century philosophers, Kant drew a singularly sharp distinction between his philosophical investigation of the mind and naturalistic approaches. His aim in framing an explicit distinction between naturalistic and what he termed "transcendental" approaches to thought and the mental was not to deny the possibility of a naturalistic account of mind: he endorsed the legitimacy of the naturalistic approach, affirming that everything within the purview of human experience is subject to natural law, including the mental. But he also asserted that knowledge of the natural laws of the mind would not provide an understanding of thought—where "thought" is taken to include paradigmatic cognitive achievements, including geometrical and natural scientific knowledge. In order to understand the latter, Kant contended, a philosophical, or critical, or transcendental investigation is required.

In sections 1 through 3 I shall examine the place of psychology within Kant's (and his contemporaries') conception of philosophy and natural science, emphasizing his distinction between naturalistic psychology and transcendental philosophy. Section 4 applies this distinction to Kant's treatment of spatial perception, as manifested especially in his doctrine of outer sense from the *Critique of Pure Reason*,[1] but also as revealed in his scattered remarks on the psychology of the senses. The final sections seek to determine the constraints that the transcendental philosophy might place on naturalistic conceptions of the mental, and vice versa.

1 Kant and Psychology

Within the history of psychology, Kant is notorious for his proclamation that a science of psychology is impossible. This negative assessment was, in fact, Kant's considered view: he barred psychology "from the rank of what may be called natural science proper," because he believed that mathematics cannot be applied to psychological phenomena, that is, to the succession of representations that follow one another in an individual mind. Moreover, he did not think that empirical psychology would ever rate as a "systematic art of analysis," or even as an "experimental doctrine" (successive steps down from natural science proper). Kant was convinced that psychological experimentation based upon introspection would fail, because unlike experimentalists in physics, introspective observers are unable to separate and recombine at will various numerically distinct psychological states. He further believed that the observation of one's fellow thinking beings is even less suited as a method of experimentation, since "the observation itself alters and distorts the state of the object observed." According to Kant, empirical psychology can "never become anything more than a historical (and as such, as much as possible) systematic natural doctrine of the internal sense, i.e., a natural description of the soul, but not a science of the soul."[2]

Kant also rejected the possibility of "rational psychology" as conceived by his predecessors.[3] Rational psychology as traditionally conceived pertained to the nature of the soul itself. It was concerned with such questions as the essential properties of the soul as a spiritual substance, its immortality, and its relation to body. Although Kant held rational psychology—considered as the attempt to determine the characteristics of the soul as it is in itself through a priori means alone—to be a legitimate enterprise as late as 1770,[4] he firmly rejected its possibility in the first *Critique*, concluding that "the whole of rational psychology, as a science surpassing all powers of human reason, proves abortive, and nothing is left for us but to study our soul under the guidance of experience" (A382).[5] In light of his remarks on the status of empirical psychology as a science, the prospects for such study are limited.

Kant's misgivings about the possibility of a science of psychology (whether rational or empirical) could easily be misinterpreted. In particular, emphasizing his denial of rational psychology could give the impression that he was an opponent of immaterialism, and emphasizing his denial of empirical psychology could lead to the impression that he exempted mental phenomena from natural law. Neither impression would be correct. Although Kant rejected traditional rational psychology, he by no means negated the fundamental tenet of such psychology, that the soul is immaterial. Before his so-called "critical" period, he asserted the immateriality of the soul; during his critical period, he contended that both positive and negative answers to the theoretical question of the soul's immateriality were beyond the reach of metaphysics, and he continued to show antipathy toward materialism.[6] Furthermore, even though he denied the possibility of a science of psychology, he expressed a steadfast commitment to the proposition that mental states are subject to natural law.

In the Doctrine of Method from the first *Critique*, Kant included both rational and empirical psychology as branches of "physiology," or the doctrine of nature in general (A846-7/B874-5). Empirical psychology is defined as the empirical study of the phenomena of inner sense. If, owing to methodological limitations, it is unable to establish its own laws, the existence of such laws is guaranteed by rational psychology (now understood as a Kantian discipline), which along with rational physics constitutes rational physiology, or the rational doctrine of nature in general. These disciplines are rational and a priori not in the sense of providing insight into mental and corporeal nature as these are considered in themselves, but only insofar as mental and corporeal states constitute objects of possible experience. According to Kant, just as all knowledge of corporeal phenomena must begin with experience but need not all arise out of experience, so too with knowledge of psychological phenomena. The very same principles that yield a priori knowledge in physics when applied to the phenomena of outer sense also apply to the psychological phenomena of inner sense. The principles of the rational doctrine of nature are delivered by transcendental philosophy in the form of the Analytic of Principles as found in the *Critique*, and thus include the law of cause, among others. And so, having denied the legitimacy of traditional rational psychology, Kant reintroduced the term to denote the critical discipline that contains the a priori laws that apply to the mind considered as a part of nature. Presumably, then, methodological considerations alone prevent empirical psychology from attaining results similar to those of empirical physics.

Kant's acceptance of the idea that the mental can be regarded as a part of nature and as subject to laws was not limited to his architectonic remarks on the division of the sciences. Additional statements that the phenomena of the mind—or the phenomena of "inner sense"—are subject to universal laws of nature may be found in both the *Critique* and the *Prolegomena*. Thus, in the first edition of the *Critique*, Kant describes the laws of association as "empirical laws" and as "laws of nature" (A100, 112). Later on, in passages appearing in both editions, he speaks of the possibility of an empirical psychology based upon "observations concerning the play of our thoughts and [upon] the natural laws of the thinking self to be derived from these observations" (A347/B405). And in the opening sections of the Second Part of the *Prolegomena*, he defines "universal natural science in the strict sense" as a science that "must bring nature in general, whether it regards the object of the external senses or that of the internal sense (the object of physics as well as psychology), under universal laws" (295). He then gives the causal law as one of the few principles that can attain the generality appropriate to a universal science of nature, thus indicating its applicability to the play of inner representations.[7]

Kant even allowed a version of mind-body interaction to fall under the law of cause. Within the framework of empirically manifest sequences of outer and inner intuitions, there is an apparent "association" between outer phenomena and the sequence of representations of inner sense (A386-7; see also Ak 29:907-9). Such empirical sequences must, presumably, fall under the law of cause. But this "interaction" must not be conceived as a relation between a material body, considered as a thing in itself, and an immaterial soul, considered as a thing in itself. Because knowledge of a real *commercium* between mind and body so conceived would amount to knowledge of the mind and body regarded as things in themselves, in his critical writings Kant rejected the possibility of knowledge of such a *commercium*. It would be fair to say that he attempted to dissolve the problem of mind-body interaction, so conceived.[8]

The term "psychology" as found in Kant's writings manifests a manageable variety of uses and changes of use. But in the course of interpretation and critical response to Kant, the term has been applied to his texts with still further meanings. In particular, some critics have contended that central portions of Kant's own philosophical enterprise, portions he explicitly set off from what he termed "psychology," should nonetheless be regarded as psychological with the clear vision of hindsight. Some of Kant's followers, such as Fries and Beneke, believed that Kant's project should rest upon introspective psychology; they sought to provide such empirical foundations.[9] Other interpreters of Kant have charged that he mistakenly allowed psychology to enter into his transcendental or critical investigation.[10] Still others have maintained that Kant's use of psychological terminology in the *Critique* constitutes a "transcendental psychology," which they distinguish from empirical psychology, perhaps in part to ward off the charge of psychologism.[11]

In order to untangle the multiplicity of senses of "psychology" applied to and used within Kant's writings, I explore the various conceptions of psychology prevalent in Germany as Kant wrote. I can then approach the question of whether Kant's transcendental doctrines were psychological despite his intentions.

2 German Psychology as Kant Wrote

The notion of psychology was in flux as Kant wrote. Numerous authors were seeking to develop psychology as a science, often with differing aims and methodologies. In discussing the status of psychology as a rational or empirical doctrine, Kant was stepping onto ground well-trodden by Wolff, Crusius, Baumgarten, and Tetens, to name only a few.[12] Varying conceptions had been advanced about nearly every aspect of the idea of a science of the mind, from the notion of "science" itself, to the method that should be followed in investigating the mind, to the place that psychology should hold within the system of philosophical disciplines, to the relation of the mental to the rest of nature (and especially its relation to matter).

In the eighteenth century the Newtonian model of science informed a surprising variety of attempts to develop a systematic account of the mind. Materialistic theories of mind (my "metaphysical naturalism") were discussed by Kant and his contemporaries, but I have found no articulated expression of materialism in any German writer of the period who also developed a body of psychological doctrine. By contrast, methodological naturalism, the attempt to explain thought and the mental within a naturalistic vocabulary—most often through a reduction of thought to association—was well represented in Germany. The writings of Hartley and Bonnet were translated during the 1760s and early 1770s, and Hume's first *Inquiry* had been available since the mid-1750s in an annotated translation by J. G. Sulzer (an acquaintance of Kant's).[13] These and other works were widely discussed by German writers. Moreover, a particularly thorough statement of methodological naturalism was produced by the German writer Johann Christian Lossius, in his *Physische Ursachen des Wahren* of 1775.

Lossius' work was being discussed as Kant was working on his first *Critique*.[14] Although not committed to materialism, Lossius argued that in cognition the mind or soul must always have an instrument that is not itself immaterial. Each idea, and the operations by which ideas are conjoined, must have a basis in the brain. In accordance with previous physiological thought, he regarded the fibers of the brain as providing this basis.[15] He considered the laws that govern the creation and conjoining of ideas to be explicable at two levels of discourse: psychological and physiological. Each level of explanation is "naturalistic" in a robust sense: the psychological explanations appeal only to the qualitative character of ideas and the associative laws by which such ideas are combined, and the physiological explanations are couched in terms of the vibrations of fibers. Normative mentalistic notions such as "judgment" or "assent" are to be reduced to one or the other of these vocabularies.[16]

Perhaps most revealing of Lossius' naturalism is his account of the human tendency to seek and to recognize the truth. Lossius considered this tendency to be a *"Grundfactum"* of psychology, comparable to the principle of attraction in physics. Truth itself he defined relative to human faculties; it is "the pleasant feeling resulting from the satisfaction of the inclination of the soul to broaden itself with respect to the understanding; or, if one will, the pleasant feeling resulting from the harmony of the vibrations of fibers in the brain." When a new idea fits harmoniously with previous ideas, it receives assent or is judged to be true. The tendency to assent (or dissent, if the idea, or its underlying vibrations, conflicts with previous brain organization) again is compared to a basic principle of physics, in this instance, to the law of fall.[17] The dependence of truth upon the constitution of human faculties is developed further in an analogy with aesthetics: just as the beautiful is to be understood relative to the faculty of taste, so too the true is to be understood relative to a healthy human faculty of reason. Finally, the fact that human beings come to agree about the truth is explained by conjecturing that our brains are all built in the same way.[18]

Lossius' naturalism seems familiar to the twentieth-century reader. In mid-eighteenth-century Germany, however, the dominant conception of psychology *qua* natural science was continuous with the traditional conception of psychology as the science of the soul, where "science" means simply a systematic body of doctrine, whether empirically based or not, and "soul" is taken with its full metaphysical and theological connotations. In the mainstream tradition of Wolff and later writers such as Tetens, a Newtonian conception of science as an empirically-based body of demonstrative knowledge was applied to the traditional conception of psychology, taking as its object a particular type of simple substance, viz., one imbued with various powers of thought as the fundamental or characteristic principles of its being. The result was an attempt to produce a science of *thought* in the fullest sense of that word. The object of psychological study in this approach is a Cartesian mind, or a Leibnizian apperceptive monad. Because this generally foreign conception of psychology as a natural science formed the backdrop against which Kant worked, I want to sketch its outlines as manifested in the mainstream Wolffian tradition.

A proper understanding of the Wolffian psychology requires that it be placed more broadly within the Wolffian conception of philosophy and metaphysics. Wolff divided philosophy into logic, metaphysics, physics, and practical philosophy; metaphysics itself was divided into ontology, cosmology, psychology, and theology.[19] Within the architectonic of Wolffian metaphysics, the soul (*Seele, anima*) was a species of spirit (*Geist, spiritus*).[20] Ontology, which gave the principles of being in general and was organized around a systematic survey of the predicates of being, was too abstract to speak of particular types of being such as spirits or souls, but it provided a background framework for discussing the properties of simple substances. Cosmology pertained to created beings; although psychology could have been included within the scope of cosmology, Wolff gave it a separate status and restricted cosmology proper to corporeal nature.[21]

Wolff is responsible for introducing the division between "rational" and "empirical" psychology. As conceived by Wolff, rational psychology was "rational" in two senses. First, it claimed to achieve its results a priori: it applied ontology to the idea of the soul as a simple substance, hoping thereby to resolve questions pertaining to the immortality of the soul. Second, it sought to develop a body of doctrine in a demonstrative manner, through long chains of reasoning from concepts or principles, whether the concepts and principles were themselves a priori or empirical. Empirical psychology, by contrast, charted the specific powers and faculties of the soul, taking the inner play of representations as its data. This might suggest that empirical psychology was the poor relation of rational psychology, given the latter's grounding in ontology. Wolff, however, portrayed the relationship between the two the other way around. He considered empirical psychology to be more trustworthy than rational psychology, and indeed to be capable of providing the principles from which rational psychology would derive its demonstrative arguments. In addition, he considered empirical psychology as a source of principles for logic; he maintained that "generally you see that throughout logic, principles are borrowed from empirical psychology. The more deeply you look into the human mind in psychology, the more light you will see spread on logic."[22] The contrast with Kant on both of these points is striking, for Kant portrayed (traditional) rational psychology as ostensibly the highest and most sure science of the soul,[23] and he considered the prospects to be nil of advancing logic through any type of psychology.

Wolff's conceptions of empirical and rational psychology and their relations with logic make sense within their own context. Empirical psychology was more trustworthy than rational psychology because it was subject to constant test by experience; he even wrote that it serves "to examine and confirm discoveries made a priori concerning the human soul."[24] Wolff compared empirical psychology to experimental physics; he conceived empirical psychology as providing the principles for rational psychology in a manner analogous to the way in which experimental physics provides the principles for rational or dogmatic physics. And although his further claim that the empirical should be conceived as confirming the a priori may seem odd to us, that is perhaps because we are used to considering the a priori, following Kant, as that which conditions all possible experience. For a transcendent metaphysician such as Wolff, "things in themselves" or metaphysical entities such as the soul reveal their properties to us in experience, and thus experience could be regarded as a concrete test for principles derived rationally from the concept of the soul as a simple substance.

It is important to emphasize that Wolff regarded empirical psychology as uncovering the characteristic operations of the soul, among which of course were to be numbered the logical operations of the intellect or understanding. This seems problematic to our eyes, for we are likely to think of an empirical study as being unable in principle to yield inherently normative data such as the rules of logic, just as we do not expect to be able to extract the rules of grammar by induction from randomly collected snatches of conversation. For Wolff, it made sense to think that empirical data could reveal logical rules because of the special status of the particular object of empirical investigation, the human soul. The soul was regarded as the natural vehicle of human knowledge of the eternal truths of logic and mathematics (where "natural" is understood by contrast with "supernatural," and includes all created beings, corporeal or spiritual).

But shouldn't Wolff then be subjected to the standard criticism leveled against such "psychologistic" attitudes during the nineteenth century: that in studying thought empirically, the investigator will be placed in danger of mistaking what is particular and accidental for the universal and necessary laws of logic? For Wolff and others had to admit that the logical activities of the soul are only "confusedly" (we might say "noisily") manifested in the content of actual thoughts, so that even if the mental is considered by definition to be guided by logical laws, the manifest imperfection of individual minds at carrying out chains of logical reasoning raises a problem for empirical psychology. How can one sort the logical wheat from the chaff of confused ideas and bad habits of thought without relying on the very logical faculties that one is allegedly investigating empirically? And if one is willing to call them into play here, then why not base logic directly on the insights of the understanding rather than seeking to extract these rules by sneaking up on them as they are revealed in empirically actual sequences of representations?

Wolff's answer to the last question would have been to agree that logic should be based on "experience" of the use of the logical faculties regarded as special faculties of the understanding. The equation of logical intuitions with empirical evidence may again strike us as problematic. But such an answer is consistent with Wolff's own conception of the proper method in philosophy. Wolff regarded the use of empirically based propositions as proper and indeed essential to philosophy, and he regarded all mental acts, including logical acts, as objects of experience.[25] This attitude was carried through by later Wolffians, such as Tetens and

Eberhard, who took an avowedly empirical attitude toward the soul, but as essentially a thinking and willing thing.

We are now in a position to examine more closely the sense in which Wolffian psychology might be regarded as taking a naturalistic attitude toward the mind. The program in empirical psychology is certainly naturalistic in the sense of adopting an empirical attitude by explicit analogy with experimental physics. If we now ask whether it is naturalistic in the fundamental sense of adopting a naturalistic account of the operations of the mind, the answer becomes complicated. Wolffian psychology does not seek to naturalize the mind by equating it with a material system, nor does it attempt to banish mentalistic notions from its conception of fundamental operations or laws of thought— these remain the laws of understanding and reason. But although the fundamental powers of the mind are not reduced to a neutral set of laws (which themselves do not invoke notions such as *thought, judgment*, or *truth*), might not thought itself (*Denkkraft*) be considered a natural power of the mind, so that Wolffian psychology is naturalistic in the sense of investigating the mind's natural powers?[26]

This type of "naturalism" is found in both Eberhard and Tetens. Eberhard characterized the project of his book on the *Allgemeine Theorie des Denkens und Empfindens* as that of developing a theory of these two "fundamental powers" (*Grundkräfte*) based upon observation. The model of the enterprise is nominally Newtonian, in that it is constituted by the attempt to find basic laws stated in terms of simple elements and their combination; it even states some laws as inverse proportions.[27] Among the fundamental powers of the mind that he posited were powers of thought, *simpliciter*; he posited what Descartes had called "having a clear idea" as a fundamental ability of the soul, along with the ability to perceive the truth.[28] Similarly, Tetens indicated that he would base his investigation of human nature on the method of natural philosophy (*Naturlehre*); the investigation took as its object "human understanding, its laws of thought and its basic faculties, as well as the active power of willing, the nature of the soul, and its development." It attempted a "psychological analysis of the soul, based on experience"; among the things it sought to determine were the "mode of origin and the laws of the operation of the forces [*Kräfte*]" that govern the modifications of the state of the soul.[29] But he made clear that this investigation would neither reduce mental activity to material brain processes, nor seek a close connection between the two.[30] Neither would it allow the reduction of thought and will to principles or laws that do not contain these notions as primitives, such as the laws of association. Although Tetens gave wide scope to the laws of association in explaining the regularities found in mental life, he did not believe that they could provide a complete explanation of either sensibility or understanding. With respect to the use of these faculties in the acquisition of knowledge, Tetens allowed that the laws of association might help explain errors that resulted from habit or accident.[31] But he further maintained that *Denkkraft* and spontaneity (*Selbsttätigkeit*) must be posited as distinct powers of the soul to account for the creative and rational achievements of the human mind.[32] If the "power of thought" and the spontaneity of the will are regarded as natural powers, they are natural powers shared with Cartesian minds and Leibnizian apperceptive monads. They cannot be described in the nonintellectual language of sense and imagination, the faculties that human beings share with animals and that traditionally were explained without appeal to a rational soul. Tetens essentially dubbed the intellect a "natural" power.

Finally, it must be remembered that not all writers in mid-eighteenth-century Germany adopted even the nominally naturalistic stance of Wolff and his followers. During this period, the conception of rational psychology as an a priori or nonempirical doctrine of the soul was ably represented by the work of Crusius. Unlike Wolff, Crusius banished empirically-based statements from metaphysics, on the grounds that metaphysics should take as its object only a priori necessary truths. He therefore excluded empirical psychology from metaphysics proper.[33] The metaphysical science of the soul was, in Crusius' terms, "metaphysical pneumatology," or the metaphysical doctrine of spirits. It consisted in the a priori analysis of the concepts of thought, will, and so forth. Crusius admitted that our conceptions of thought and the will must have their basis in the perception of our own thinking and willing. Nonetheless, once the concepts of thought and will are formed, metaphysical pneumatology proceeds to analyze these concepts in a purely a priori manner, for example, arguing on conceptual grounds to the simplicity and immortality of the soul.[34] As would Kant (in connection with the metaphysical foundations of physics), Crusius allowed that empirically based concepts could provide the object for an a priori analysis that yielded necessary truths.

At the time Kant formulated his conception of psychology, then, what I have termed "methodological naturalism" was by no means the dominant conception of psychology in Germany. The conception of psychology as an a priori investigation of the soul was still strong, but perhaps the dominant conception was that put forward by Wolff, in which the soul (traditionally conceived) was treated as an object of empirical investigation. In effect, Wolffian empirical psychology constituted an attempt to determine, through empirical means, the properties of the soul as it is in itself.[35] Kant, of course, rejected both the rational psychology of

Crusius and the empirical approach to the thing in itself. His systematic attitude to empirical psychology was closest to that of Lossius. But he differed from Lossius in rejecting all attempts to extend the domain of naturalistic psychology to encompass the normative aspect of thought itself, including judgment and assertion.

3 Naturalistic Psychology and Transcendental Philosophy

Although Kant was willing to countenance naturalistic psychology, he maintained that the naturalistic point of view is insufficient for understanding thought. There are two related claims here that Kant did not always distinguish. The first is conceptual. Kant maintained that the products of mere associative processes—the processes of the mind conceived as a part of nature—are in a different conceptual domain from acts of judgment, acts that might claim the title of knowledge. Kant expressed this distinction in his famous contrast between questions of fact and questions of right: as applied to cognition, questions of fact pertain to the psychogenesis of cognitive states, whereas questions of right pertain to the legitimacy of cognitive states, considered as claims to knowledge (A84-5/B116-7). This conceptual distinction prepared the way for his second claim about the insufficiency of naturalistic psychology, which was methodological. Kant contended that questions about the mind considered as part of nature and as the object of empirical psychology are amenable to empirical study (under methodological limitations), whereas questions about the mind considered as the subject of cognitive claims (questions of right) require a different methodology, viz., his so-called transcendental method.

When stated abstractly, Kant's conceptual point about the insufficiency of psychology as an account of thought is similar to the point made by Tetens: psychology is insufficient to account for our conception of ourselves as beings that have objective knowledge and experience. Kant maintained that in describing the knower naturalistically in terms of laws of association, the essential ingredient that makes a knower a knower—the possession of knowledge—is necessarily omitted from the account. Accounts of belief-formation that appeal to mere laws of habit, as did Hume's, fail to explain the essential objectivity of knowledge. Kant explicated the objective validity of knowledge in terms of universality and necessity, which he in turn explicated in terms of intersubjective validity. If a person claims to have made an objectively valid cognitive judgment, he or she expects that others should accept it.[36] A central purpose of Kant's first Critique is to explain how such judgments are possible.

Kant's response to Hume's theory of causal judgments is an example of his contention that the natural psychology of judgment is insufficient for explicating judgments considered as acts of thought. Having denied the "real use" of the intellect in discerning necessary (causal) connections between events, and having argued that induction cannot provide adequate justification for the assertion of necessary connections, Hume contended that causal judgment reduces to a psychological process of association: through experience of the constant conjunction of events of types A and B, we come to expect B, given A. Although Hume himself continued to label such habit-based judgments "causal," Kant described Hume's conclusion in different terms. According to Kant, Hume had concluded that the notion of cause is itself illegitimate.

As understood by Kant, the law of cause states that every event has a cause.[37] It is an assertion that events are generally subsumable under a general rule of necessary succession. Kant's reason for rejecting Hume's associative account of causal judgments was not that it was empirically false as a description of the phenomena of inner sense; Kant admitted its truth.[38] Further, Kant's objection was not to the fact that Hume's positive account of causal judgment made the causal law depend on experience, which is an insufficient basis for the causal law as understood by Kant (but not by Hume). Rather, he objected to Hume's account because it denied the existence of the necessity that he found essential to the notion of cause itself. Kant complained that Hume's account of causal judgment wasn't really an account of causation at all:

> The very concept of cause so manifestly contains the concept of a necessary connection with an effect and of the strict universality of the rule, that the concept would be altogether lost if we attempted to derive it, as Hume has done, from a repeated association of that which happens with that which precedes, and from a custom of connecting representations, a custom originating in this repeated association, and constituting therefore a merely subjective necessity. [B5][39]

In adopting his associative account Hume gave up the notions of strict universality and necessity as applied to causal sequences. That amounted, on Kant's construal of causation, to giving up the notion of cause itself.

But despite his rejection of Hume's own positive account of causal judgment, Kant agreed with his predecessor's objections to previous attempts to justify the causal law. There were two sorts of such attempts. The first appealed to the "real use" of the Cartesian intellect to establish the causal law through direct intellectual intuition. Although this means of establishing the law of cause would be adequate to Kant's conception of causal necessity, it was not open to Kant, for he agreed with Hume that a "real use" of the intellect must be rejected. In rejecting the "real" application of the intellect to causal sequences, Kant added to Hume's "intuitive" arguments (arguments that appeal to the failure of the "real use" in particular cases) his own contention that in principle we cannot justify an explanation of the possibility of the real use of the intellect. Kant maintained that the only possible grounds for such an explanation would be an appeal to a "preestablished harmony" between the intellect and mind-independent reality, a harmony explained by appeal to a benevolent deity; but, Kant remarked, knowledge of such a deity and its benevolence would already presuppose the very capacity it was supposed to legitimate, for the existence and benevolence of the deity could only be established through the real use of the intellect.[40]

The second type of attempt to establish the causal law was through the empirical observation of actual causal sequences. Especially within the Aristotelian tradition, the power of discerning necessary or essential connections as a consequence of sensory experience had been ascribed to *nous*, or the rational intellect. Among such necessary connections were those governed by principles of motion or change, including laws of efficient causality.[41] Here again, Kant agreed with Hume's negative assessment; he virtually never tired of repeating the refrain that "appearances do indeed present cases from which a rule can be obtained according to which something usually happens, but they never prove the sequence to be *necessary*" (A91/B124).[42] Although experience might guide us to causal sequences, empirical evidence could never by itself establish any particular causal law, much less establish the validity of the causal law in general. Hume and Kant could agree that the problem of induction precludes the attainment of universal and necessary knowledge on the basis of experience alone.

Although Kant found Hume's arguments effective against previous validations of the necessity attributed to the causal law, he nonetheless was determined to establish the necessity of that law as governing all events that we might experience. Those familiar with the story know that he turned to transcendental arguments in order to break out of the forced choice that Hume seemed to offer between acceptance of the real use of the intellect (whether Cartesian or Aristotelian) and denial of the necessity of the causal law. Transcendental argument starts from some given body of knowledge, or some given cognitive achievement, and asks how it is possible. If it can be shown that the cognitive achievement in question is possible in only one way, then, given that the achievement is actual, the only possible means for its possibility must be actual, too. Kant's candidates for the starting point of transcendental arguments included: Euclid's geometry, Newton's physics, and the fact that we have "experience," where experience is regarded as objective and demanding of intersubjective agreement. His transcendental arguments concluded by positing his famous "categories" and "forms of intuition." Thus, in the case of the category of causality and the law of cause, he argued that our actual knowledge of Newton's physics, and our actual ability to make intersubjectively valid assertions, can be explained only by positing the a priori validity of the law of cause. Kant contended that the necessity of causal connections is an aspect of human cognition that cannot be argued away.

As might be predicted from even this brief statement of transcendental methodology and of the conclusions Kant drew from it, a host of questions have arisen about both the structure and the soundness of transcendental argumentation. Many of these questions focus on the "given" starting point of transcendental arguments. What Kant takes as "given" is a question of some importance and sensitivity, since he might be accused of begging the question if he started by taking the achievements of Euclid and Newton—or even the possibility of intersubjective agreement—as given.[43] The evidence is, I think, quite strong that he believed the existence of Euclid's geometry and Newton's general science of nature sufficient to show the actuality of objective knowledge in opposition to Hume.[44] It is, I think, also beyond question that he believed he had a more general argument based on the possibility of experience itself, where the notion of experience includes objectivity or intersubjective validity, and instances of such experience includes perception of the spatiotemporal properties of middle-sized objects.[45]

In taking these starting points as given, Kant was remarkably oblivious to the charge of begging the question. We may begin to understand his attitude by considering Kant's objective in the *Critique* with respect to these bodies of knowledge. Kant's aim was not to secure the objective validity of geometry and physics for the sake of these sciences—as if the practitioners of these sciences needed a philosophical imprimatur in order to proceed in the face of an alleged skeptical challenge—for neither he nor Hume was skeptical about the legitimacy of these bodies of knowledge. Rather, Kant was interested in understanding the grounds for the possibility of knowledge he considered to be actual, in order to determine whether transcendent metaphysics could expect similar success in adjudicating the existence and character of God, of an immaterial and immortal soul, and of the world as a thing in itself.[46] Kant was quite explicit in contrasting the aim of his transcendental philosophy with the aims of special sciences such as geometry and physics. The geometrician establishes the truth of the propositions of geometry by direct appeal to spatial constructions produced in accordance with the axioms and postulates of Euclid's geometry; she is satisfied if her proofs are recognized as intersubjectively valid, without inquiring into the cognitive grounds for such recognition. Similarly, according to Kant, the physicist asks only that his laws meet the test of universal confirmation in experience; he does inquire how or whether such confirmation establishes his laws as valid.[47] By contrast, transcendental philosophy investigates the grounds and extent of all human knowledge; it, therefore, attempts both an explication of the possibility of geometry and physics, and a determination of whether the same type of knowledge can be attained with respect to God, the soul, and the system of natural causes in themselves.

For my purpose, it is not as important to attend to questions about the validity or soundness of transcendental arguments as it is to understand how Kant conceived the method by which he sought to establish the premises for such arguments. It is here that Kant departed methodologically from Hume, as well as from Crusius, Wolff, and Tetens. Having given up the "real use" of the intellect, he could not, like Crusius, claim to have insight into the mind and intellect "in themselves." Nor could he claim to approach the intellect "in itself" empirically, as had Wolff and Tetens. Hume's approach was out because Kant needed not to establish the empirical laws of the mind but to discover how acts of cognition with intersubjective validity are possible from the point of view of their justification. It is precisely at this point that Kant's special "transcendental" methodology was supposed to come into play, to back his transcendental arguments.

Transcendental methodology is notoriously difficult to pin down,[48] in part because Kant provided little direct discussion of it. Despite the fact that he described the first *Critique* as a "treatise on method," rather than a systematic statement of the conclusions of transcendental philosophy (Bxxii), he provided surprisingly little explication of his transcendental methodology. In his official discussion of methodology he spent considerable time distinguishing ordinary philosophical method from the methods of mathematics, but he stopped short of giving an account of the method of transcendental philosophy (A738/B766). He provided some additional discussion in various scattered remarks,[49] but the best guide to his transcendental method is the actual arguments he presented. From his practice, it is clear that Kant believed transcendental philosophy should pursue its task of explaining the possibility of objective knowledge by beginning with a list of the possible grounds for knowledge and proceeding by a process of elimination. His list included, from his predecessors, empirical grounds, and the intuitive grasp of transcendental reality, to which he added transcendental characterization of the subject's contribution to knowledge. He eliminated the first two by the arguments described above, which left only the third remaining in the field.

While it is evident that Kant used a process of elimination, it is less clear, and has been the subject of much discussion, what he considered to be the grounds of his knowledge of the possibilities to be eliminated. On what basis did he claim to know that experience cannot yield necessity, or that claims for the real use of the intellect could not, in principle, be given an adequate foundation? Although the answers to such questions are not obvious, one thing is sure: he considered the basis for the answers to be a priori. Beyond that, it is likely that he did not consider this basis to be analytic or synthetic knowledge of the ordinary variety but to be *sui generis*.[50] In the case of the causal law, I shall simply treat his argument by elimination as an attempt to add a third, successful, explanation of the possibility of knowledge of this law to the two possibilities he believed Hume had shown to be unacceptable.

The basic outline of Kant's proposed explanation is familiar enough. Kant contended that the applicability of the category of cause to all appearances is a condition on the possibility of experience itself (where "experience" is understood as intersubjectively valid experience). Because the claim that the category of cause applies to experience is not justified by appeal to experience, but rather by analysis of the conditions on the possibility of experience, it allegedly does not fall prey to the problem of induction. Further, no claim is made about the applicability of the categories beyond the pale of experience, so their existence does not fall prey to the objections against the Cartesian intellect. Notoriously, Kant gave up claims to know "things in themselves" in order to secure claims to necessary knowledge within the domain of experience. He thereby adopted a position of "transcendental idealism," according to which human knowledge is limited to appearances, as opposed to things in themselves. Although this has seemed too high a price to some, it may not have seemed such a high price to Kant, since among his chief aims was to place theoretical knowledge of God and the soul beyond the grasp of ordinary metaphysics (failing, as Kant thought must happen, a proof of God and the soul within ordinary metaphysics—Bxxix-xxx).

It is interesting to consider in unreconstructed form Kant's discussion of the universal applicability of the category of cause. As Kant explained it, to say that the knower brings the category of cause to experience is to make a statement about the structure of the faculty of understanding, together with statements about the faculties of imagination and sensibility and their interaction with the understanding. The faculty of the understanding is imputed a category that, as "schematized" by the imagination (in Kant's terminology), is applicable to the manifolds of intuition provided by sensibility. The understanding actively "synthesizes" the elements of the manifold of intuition in accordance with the category, and thus in accordance with the causal law. The faculty of sensibility that provides such manifolds is passive, and so must be goaded into producing the sensory "matter" which is so synthesized; how this latter occurs is left unspecified.[51] The understanding then "takes up" and "goes through" the matter of intuition, bringing it to a unified representation.

To the delight of some but the embarrassment of others, Kant's "deduction" of the validity of the categories, as well as his subsequent elaboration of the "principles of knowledge" that attend each category, is laced with such psychological-sounding language. In developing his account of the possibility and limits of knowledge, Kant adopted the psychological vocabulary of his contemporaries, using such terms as "faculty of sensibility," "inner" and "outer" sense, and "sensory manifold," and he posited a set of activities, those of "synthesis,"[52] that allegedly account for the possibility of experience. Kant's account of the cognitive faculties and their operations is sometimes called his "transcendental psychology," a phrase that for some recognizes Kant as an ally in the development of the natural science of cognition,[53] but for others serves to indicate the opposite conclusion, that Kant's project is distinct from empirical psychology. It is the latter group that has found Kant's transcendental psychology to be an embarrassment. Because it is distinct from empirical psychology (as they, with Kant, believe it must be, in order to serve in the explanation of epistemic achievements), it is difficult to know what transcendental psychology is the psychology *of*. In his critical period Kant cannot (wish as he might) justify a commitment to an immaterial intellect as a "thing in itself," and so he cannot attribute the processes of transcendental psychology to an immaterial soul (even if his descriptions of transcendental "processes" in the mind sometimes recall his earlier, explicit immaterialism). And yet he does not conceive the processes of transcendental psychology as belonging to the mind or the body conceived as a natural object. They are not in the domain of empirical mental phenomena: they are what make these phenomena possible. The transcendental has its own queer status.

As Kant's response to Hume made clear, whatever Kant was doing in the *Critique* it wasn't empirical psychology (either in his sense of the term, or ours). But that leaves unanswered the question of what, precisely, the relationship between transcendental and empirical psychology was supposed to be. The beginning of an answer can be found by considering various ways in which transcendental psychology isn't related to empirical psychology.

At first sight, Kant's endorsement of the possibility of empirical psychology together with his insistence upon its inadequacy as an account of thought might appear to be a simple extension of Tetens' position. Tetens was willing to allow for physiological and psychological accounts of the lower faculties (the senses, imagination, and memory) but not of the higher faculties (the understanding and the will). Perhaps Kant then means to allow that the laws of psychology apply to sensibility and imagination but not to understanding,[54] which he is tacitly conceiving, like Tetens, as an immaterial agent with its own special powers.

But Kant did not adopt Tetens' position. Although he tended to ascribe the laws of empirical psychology, and particularly the laws of association, to sense and imagination as opposed to understanding (A115, B152), he did not restrict empirical psychology to the "lower" faculties, nor did he draw the distinction between empirical and transcendental description in terms of the lower and higher faculties. Kant made it quite plain that each faculty—sensibility, imagination, and understanding— has both an empirical and an a priori employment, which implies that each faculty may be investigated both in its empirical manifestation and in its transcendental mode.[55] For Kant, the laws of psychology apply to *all* the representations of inner sense; they must, therefore, apply to the representations that are brought to synthesis under the guidance of the understanding. Kant's transcendental story of the faculties does not begin where the naturalistic story leaves off—as if the conditions of knowledge pertained to a special substance that lies beyond the causal sequence described by psychology, and whose activity begins where the causal story ends. Rather, the transcendental account has its own separate beginning, from which it answers questions about sensibility, imagination, and understanding that are outside the scope of psychology: they are questions that empirically-based psychology cannot answer by its very nature.

A second candidate relation between empirical and transcendental psychology makes transcendental psychology an idealization of empirical psychology. Perhaps empirical psychology is concerned with various individual minds whose peculiarities can be known only through empirical sampling, whereas Kant is giving an account of Mind with a capital "M" that ignores the peculiarities of particular minds.[56] On this conception, Kant was aiming to reveal the common structure in all thought and the universal conditions for knowledge; he was not concerned with deviations from this structure or these conditions owing to local perturbations. Empirical psychology and transcendental psychology are coextensive in the phenomena of inner sense to which they apply, but transcendental psychology constitutes a necessary idealization beyond empirical psychology if the universality and generality required by transcendental philosophy is to be attained.

Taken as an appeal to the universality of the cognitive faculties and operations posited by Kant, this appeal to Mind with a capital "M" does not successfully set Kant's project apart from psychology regarded as a theory-driven empirical science. The theoretical psychologist might, in fact, eagerly embrace the project of uncovering the universal cognitive mechanism that is only imperfectly or "noisily" manifested in individual minds. If Kant's aim is distinctive only by its universality, the psychologist could bring to bear an analogy with the laws of physics as idealized universal laws whose operations are never cleanly manifested in our sublunary world. And indeed, Kant himself regards empirical psychology as entitled, at least in principle, to universal, exceptionless laws, as warranted by the applicability of the law of cause to inner sense. Nonetheless, not even the universal laws of empirical psychology could ground a transcendental investigation, for the laws of psychology are the laws of association, and as such are truth-neutral. Empirical psychology describes causal sequences of representations; it lacks the conceptual resources for answering questions of right, or questions of entitlement and justification.[57] Such questions ask not how representations in fact are connected in consciousness (what assertions in fact are made); rather, they ask with what right various representations are connected (with what right various judgments are rendered). Any universality claimed by empirical psychology would be conceptually beside the point from the transcendental point of view.

A third possible relation is that transcendental psychology is a separate and distinct enterprise that builds upon empirical psychology. According to this conception, empirical psychology provides an analysis of the mind into various faculties and their characteristic activities. Transcendental psychology, in its role as handmaiden to transcendental philosophy, then builds its account of the knower on the basis of the results of empirical psychology. It differs from empirical psychology in its aim. Whereas empirical psychology aims simply to describe the mind and its activities, transcendental psychology is put to service by the philosopher in order to explicate the grounds for knowledge as these are determined by the constitution of the knower. Transcendental psychology provides an analysis of the knower that philosophy uses to explain the possibility of objective judgment.

Although this possibility appropriately characterizes the Kantian project by emphasizing that talk of faculties and mental activities is in the service of epistemology, that is, of an explication of the conditions for knowledge, Kant would not have been happy with the idea that empirical psychology could provide any results of importance to transcendental investigation. For even though he appropriated wholesale the vocabulary of traditional psychology, he could not appeal to empirical psychology, or to any empirical investigation, to justify this appropriation. The rationale for introducing notions such as "form of sensibility," "pure and empirical intuition," "transcendental imagination," and "synthesis of apperception" must come from the role that such notions play in explicating the possibility of objective judgment. An appeal to empirical psychology to justify the positing of such faculties, representations, and operations could add nothing to the case because it would be conceptually beside the point in speaking to questions of right, and also because Kant's project was precisely to show how the assertion of universal generalizations under the guidance of experience is possible. Such assertions are the thing to be explained, and so they cannot figure in the explanation.

What, then, is transcendental psychology the psychology *of*? It is the psychology of the *knowing* mind, of the mind that makes objectively valid judgments. How is the knowing mind related to the mind studied in psychology? They are the same mind in that transcendental psychology is restricted in its concrete application to one and the same set of representations that constitute the object of empirical psychology. But transcendental psychology does not simply describe the sequences followed by these representations, rather, it prescribes what must hold of such sequences if objective judgment is possible. It differs from empirical psychology both in what it says about inner sense and in the basis on which it says it. It is, for instance, a finding of transcendental psychology that one and every representation is subject to the unity of apperception. Correspondingly, naturalistic psychology is committed to the doctrine that one and every representation falls under the laws of association, which brings about an empirical synthesis.[58] The claims of transcendental psychology regarding the unity of apperception are supported by transcendental argument, which is not itself based upon experience but which explains the possibility of experience. The discovery of the laws of association is an empirical discovery that must be justified through an appeal to inner experience.

Although the fact that transcendental and empirical psychology treat the same subject-matter from different perspectives may seem peculiar, it is characteristically Kantian, echoing the relation between nature and freedom as conceived in the first *Critique*. Kant regarded one and every action of human beings as subject to nature, or to the causal law; with respect to nature, he posited a complete determinism. At the same time, he regarded the human will as essentially free. He attempted to reconcile the two doctrines by restricting the scope of determinism to phenomena, or to the realm of appearances, and specifying freedom as a special type of causality that pertains to noumena, and thus lies beyond the boundary of experience, but which nonetheless has as its effect some portion of the chain of appearances (that portion which is said to result through the causality of freedom).[59] One and the same action on the part of a given subject is given two explanations: one in the language of freedom, the other in the vocabulary of natural determinism. Similarly, one and every representation appearing in inner sense is subject to the causal law, and, at the same time, is subject to the unity of apperception, an "act of spontaneity" of the faculty of representation. In the language of empirical psychology, representations are lawfully connected as cause and effect. In the language of transcendental psychology, representations are connected through the synthetic unity of apperception. One and every representation is subject both to association and to the transcendental synthesis produced by the understanding.[60] However, unlike the case of freedom, Kant does not, officially, ascribe the "spontaneity" of transcendental apperception to the noumenal self, but to the self as the vehicle of transcendental apperception—a self which he says he can "think" and can "be conscious of" but cannot know except as manifested in the empirical sequence of inner sense.[61] He was precluded from doing so, because a principal conclusion of the first *Critique* was that the noumenal self is beyond the reach of our knowledge, whereas he claimed to have knowledge of a special sort—"transcendental knowledge"—of the knowing self.

In sum, the doctrines of "transcendental psychology" must be established solely on transcendental grounds. Rather than being conditioned by empirical psychology, the relationship is, if anything, the other way around: empirical psychology is conditioned by transcendental psychology in such a way that the findings of the former cannot contradict the findings of the latter. Let us examine this relationship as found in Kant's discussion of space as the form of outer sense.

4 Spatial Perception and Geometry

Partly as a result of his continuing interest in the foundations of physics, Kant took an interest in space and spatial perception early in his career. From the 1740s into the 1760s he held a modified Leibnizian theory of space, rejecting Newtonian absolute space in favor of a relational theory. As he began the line of work that resulted in the first *Critique*, Kant's teaching on space and spatial perception shifted radically. By 1768 he had abandoned the relational theory of space and adopted the Newtonian theory, according to which space constitutes a framework or container distinct from the matter that occupies it. A year later he altered his Newtonian view: although he continued to treat space as something independent from the matter that occupies it, he rejected the doctrine that space has mind-independent existence. He ascribed to space the status of a set of a priori laws governing the representations constructed by the human faculty of sensibility.[62]

Within the *Critique* itself, Kant's doctrine of space was presented primarily in the Transcendental Aesthetic (although important elaborations occur in the Analytic of Principles and in connection with the Antinomies). Kant gave very little direction for interpreting the significance of the Aesthetic, or for fitting it into the structure of the *Critique* as a whole. Yet it played several important systematic roles. The Aesthetic drew upon and continued Kant's earlier work on the status of space by seeking to provide a third alternative to the spatial theories of Newton and Leibniz, thus preparing the way for Kant's solution to the first two Antinomies (that is, to the problem that the ontologies of Newton and Leibniz seem equally defensible on their own terms, but are mutually contradictory).[63] The Aesthetic also provided initial examples of or arguments for some important tenets of the Critical Philosophy, including the distinction between sensibility and understanding and the related distinction between intuitions and concepts. Perhaps most significantly, it provided Kant's initial arguments for transcendental idealism, and the first example of his transcendental method of argumentation, in the "deduction" of space as a form of sensibility.[64]

The specific teachings of the Aesthetic on space are reasonably clear. Kant argues (1) that our fundamental representation of space could not have been acquired through experience, (2) that this representation is not conceptual but "intuitive" (in Kant's technical sense), and (3) that these points provide the basis for explaining how geometry can constitute a body of synthetic a priori knowledge. Although it seems clear that Kant's overriding concern in the Aesthetic (taken by itself) is to explain the possibility of geometrical knowledge, it seems equally clear that the first two points do not depend upon the third one—and indeed Kant provided them with their own independent arguments. These points provide the framework for understanding Kant's arguments pertaining to geometry.

The first point is of special interest because Kant's claim that the ability for spatial representation could not be derived from experience became a focus of discussion for the philosophers, psychologists, and physiologists who responded to Kant's writings on space in subsequent decades. In the portion of the Aesthetic entitled the "Metaphysical Exposition of the Concept of Space," Kant gave three arguments against the notion that spatial representation could have an empirical origin.[65] First, he argued that our basic representation of space cannot be empirically derived from outer experience, "for in order that certain sensations be referred to something outside me (that is, to something in another region of space from that in which I find myself), and similarly in order that I may be able to represent them as outside and alongside one another, and accordingly as not only different but as in different places, the representation of space must be presupposed" (A23/B38). Kant would seem here to be blocking any radical empiricist attempt to suggest that the concept or the fundamental representation of space is acquired through experience. The perception of space includes a system of (actual or possible) relations among sensations. For these sensations to be represented as outside one another, the ability for spatial representation must already be in place. Thus our fundamental representation of space could not be acquired, for any experience from which it could be acquired (any experience of the "outside and alongside" relation) would already be an experience of space.[66]

Kant argues not only that an empirical derivation of our fundamental representation of space cannot be conceived, but, secondly, that "space is a necessary a priori representation." His argument is one sentence long: "We can never represent to ourselves the absence of space, though we can quite well think it as empty of objects," from which he concludes that space "must therefore be regarded as the condition of the possibility of appearances, and not as a determination dependent upon them" (A24/B38). At first, this looks like the adolescent thought-experiment of trying to picture the absence of space, and concluding from one's inability to do so that there is something "necessary" about space as a form of representation. However, it is likely that the real work in this premise is the contrast between the absence of space and empty space, and that the alleged fact that we can "think" space as empty of objects was intended to support Kant's attempt to establish his alternative to the Newtonian and Leibnnizian theories of space through an argument from elimination.[67] According to Kant, the Newtonian theory of space implied the "real existence" of space and time as independent substantial beings that are themselves neither substances nor properties of substances; he considered this position to be patently untenable. The Leibnizian theory that space is an idealized abstraction from the perception of relations among bodies was ruled out by the previous argument, with the consequent implication that space (as we represent it) cannot depend upon previous representations of objects. This leaves only the position that space is a "form of intuition" (in Kant's terms). The current argument supports the tenability of this remaining position by revealing that space can be "thought" independent of objects.

The second of the specific teachings about space, that our fundamental representation of space is intuitive, not conceptual, constitutes the basis for Kant's rejection of the Leibnizian doctrine that the sensory perception of space is really a confused form of intellectual perception. Kant maintained a general distinction between intuitions and concepts. Intuitions he considered to be singular representations that are "in immediate relation" to their objects; concepts serve as functions uniting intuitions, and are in this sense general representations that are only mediately related to their objects, that is, through intuition (A19/B33).[68] Intuitions (or, better, the content presented in intuitions) are concrete and singular, while concepts are rules or guides for connecting intuitions in accordance with their content. Intuitions are concrete and specific in that a given intuition does not represent anything other than its occurrent content; it is only by means of concepts that a given intuitive representation is put in relation to other representations.

In the Metaphysical Exposition Kant presents two arguments for the conclusion that our fundamental representation of space is intuitive, not conceptual. Each argument purports to reveal a way in which the representation of space differs from a conceptual representation. The first proceeds from the point that our fundamental representation of space is a unitary, as opposed to a discursive, representation. Concepts are discursive: they apply to a variety of individual instances. But, Kant contends, "we can represent to ourselves only one space." Space is not "a discursive . . . or general concept of relations of things in general," for its relations are already present in the representation of space; they do not arise as a result of a concept's being brought to bear on (and thereby "relating") a range of particulars. Far from the unity of space arising through a concept acting as a function among either objects or spaces considered as a manifold of constituent parts, it follows from the fact that "space is essentially one" that "the manifold in it, and therefore the general concept of spaces, depend solely on [the introduction of] limitations" into the one all-embracing space (A25/B39).

Kant's second argument follows up this last point. He contends that whereas concepts may include an infinity of particulars under them, as instances, the representation of space differs by including an infinite manifold within itself. Kant is contending that the "original" representation of space, or the basis for any concept of space that would include perhaps an infinity of spatial regions under itself as instances, is already a representation of a space into which an infinity of divisions could be introduced (as in the division of a line segment).[69] Particular spaces, considered as instances falling under a concept, must depend upon the division of a unitary space, for space "is thought" with all of its parts coexisting "ad infinitum." The concept of delimited spaces must, therefore, depend for its object upon divisions introduced into space as intuited; and so the intuition of space must precede any such concept.[70]

With these arguments in hand, we can turn to Kant's treatment of geometrical knowledge in the section of the Aesthetic entitled "The Transcendental Exposition of the Concept of Space." This section is crucial within the plan of the *Critique*, for it contains Kant's answer to the question posed in the Introduction, "How is pure mathematics possible?" (B20),[71] and so provides him with his first example of a transcendental argument. Kant explained the intent of his "Transcendental Exposition" as follows: "I understand by a transcendental exposition the explanation of a concept, as a principle from which the possibility of other a priori synthetic knowledge can be understood. For this purpose it is required (1) that such knowledge does really flow from the given concept, (2) that this knowledge is possible only on the assumption of a given mode of explaining the concept" (B40). The "transcendental exposition" of space will allow us to understand how synthetic a priori knowledge of space—in the form of Euclid's geometry—is possible. The explanation it introduces purports to be both necessary and sufficient for explaining the possibility of the type of knowledge in question.

It will be useful to quote Kant's exposition at length. The first paragraph runs as follows:

> Geometry is a science which determines the properties of space synthetically, and yet a priori. What, then, must be our representation of space, in order that such knowledge of it may be possible? It must in its origin be intuition; for from a mere concept no propositions can be obtained which go beyond the concept—as happens in geometry (Introduction, V). Further, this intuition must be a priori, that is, it must be found in us prior to any perception of an object, and must therefore be pure, not empirical, intuition. For geometrical propositions are one and all apodeictic, that is, are bound up with the consciousness of necessity; for instance, that space has only three dimensions. Such propositions cannot be empirical or, in other words, judgments of experience, nor can they be derived from any such judgments (Introduction, II). [B40-41]

The first thing to be noted is that Kant takes the object that is to be explained in this exposition—the science of geometry—to be given. He discusses its synthetic a priori status in the Introduction, and explains it more fully the Methodology; in each case he takes geometric knowledge, or the science of geometry, to be actual.[72] He is not out to ground geometric knowledge but rather to discover and to explain the grounding that he thinks it obviously must have.

In showing that his explanation reveals the necessary condition for the possibility of geometric knowledge, Kant considers three possible bases for geometry. It might be based on the analysis of concepts, i.e., analytic; it might be based on experience, i.e., synthetic a posteriori; or it might be synthetic a priori. In the above passage, Kant rules out the possibility that geometry is analytic by invoking his earlier contention that basic propositions in geometry "go beyond the concept." On Kant's conception of an analytic/synthetic distinction, the claim that propositions in geometry contain material in their predicates that is not "contained" in the concept of the subject, or, as is sometimes said, is not comprised within the meaning of the subject-term, is sufficient to establish the proposition as synthetic. Kant further explicates his claim that geometry cannot be analytic with the claim that from the concept of two straight lines alone one could never derive the proposition that two straight lines cannot enclose a space; material in the predicate, "cannot enclose a space," does not follow conceptually from the subject concept, "two straight lines" (A47-8/B65).[73]

The denial that the fundamental claims of geometry could be established analytically is bound up with Kant's conception of geometrical method. In his discussions of method, Kant contrasted the method of philosophy proper (i.e., of ordinary, nontranscendental philosophy) with the method of geometry. According to Kant, philosophy proper is systematic knowledge from concepts (one might think of Wolff's rational or deductive knowledge).[74] Ordinary philosophy could proceed by the principle of contradiction from propositions established analytically (i.e., through the analysis of concepts). Prior to Kant, a number of authors had regarded geometrical demonstration as based entirely upon this same procedure of applying the principle of contradiction to geometrical propositions. Kant admitted, of course, that geometry always proceeds in accordance with the principle of contradiction and all other principles of logic. In denying the analyticity of geometry, he denied that the axioms and postulates from which geometrical argument proceeds could themselves be established through the analysis of concepts together with application of the principle of contradiction. Thus, in effect, he denied that the proposition that two straight lines cannot enclose a space could be established by indirect proof, that is, by finding that negation of the proposition leads to a contradiction.

According to Kant, a number of crucial predications in geometry, including the parallel postulate and the equality between the three angles of a triangle and two right angles, could not be based upon a mere analysis of concepts, but only on the construction of a figure in intuition in accordance with the concepts. Indeed, he contended that geometrical concepts become determinate only through such a construction. Kant contended that the method of geometry differs fundamentally from that of philosophy because the appeal to intuition is essential for geometrical demonstration (A712-38/B740-66). As he explained, geometry establishes its propositions by direct appeal to intuition; it constructs its concepts—or the objects of its concepts—and then reads off the properties that enter into geometrical demonstrations. Here a special "truth-recognitional" ability is ascribed to the mind,[75] but in accordance with

Kant's skepticism about the real use of the intellect, this ability is stripped of the transcendent powers for apprehending the "thing in itself" claimed for it by previous metaphysicians; notoriously, Kant restricted the validity of geometry to the domain of appearances (thereby advocating transcendental idealism). This "truth-recognitional" ability is directed at the products of construction—according to Kant, we can "see" by inspection that two lines cannot enclose a space, or that the shortest distance between two points is a straight line.[76] The discernment of intuitively evident propositions in geometry requires a spatial extent, which may be supplied on paper or held in the imagination. It is because he considered geometrical demonstration ultimately to require an appeal to intuition that he rejected the possibility that geometry could be analytic; the requirement that a figure be "constructed"—an essential requirement in geometry, in Kant's view—entailed that geometrical argument must appeal to something beyond mere concepts, that is, to a figure in intuition.

The question arises as to whether the intuitions involved are a posteriori or a priori. Kant, of course, argued for the latter. In the passage quoted from the Transcendental Exposition he supported this claim by appealing to the epistemic status of geometrical propositions—their alleged absolute certainty—and then asking for the conditions under which a proposition based in intuition can be established with absolute certainty. He concluded that the intuitional basis for geometry must be a priori, which means, as he put it, that "it must be found in us prior to any perception of an object, and must therefore be pure, not empirical, intuition." Only in this way, he argued, can the necessity of geometric knowledge be explicated; for, as he tirelessly repeated, necessary propositions can never be established on the basis of experience (A24/B41, A40/B57).

Once we have been told that geometrical knowledge is based upon "pure intuition," the question naturally arises as to what we have been told. That the intuition of space "must be found in us prior to any perception of an object" has the ring of a psychological thesis. Being "in us" prior to perception sounds very much like being innate. And indeed, the rest of the Transcendental Exposition only reinforces this impression: "How, then, can there exist in the mind an outer intuition which precedes the objects themselves, and in which the concept of these objects can be determined a priori? Manifestly, not otherwise than in so far as the intuition has its seat in the subject only, as the formal character of the subject, in virtue of which, in being affected by objects, it obtains immediate representation, that is, intuition of them; and only in so far, therefore, as it is merely the form of outer sense in general" (B41). The language here speaks unmistakably of the characteristics of the knower.

How can a certain type of intuition exist in the subject prior to experience? By having its seat in the subject in such a way that when the subject is affected it obtains intuitions only of a certain character. The subject's sensory capacities are limited, or are positively determined, in a certain way, that is, so that the subject can have spatial intuitions only of the kind that are described by Euclid's geometry.

It is hardly surprising that a number of Kant's early interpreters took his enterprise to be based upon psychology. But among more recent interpreters, the standard philosophical response to this part of Kant's doctrine has been that in making these concrete ascriptions regarding the subject's sensory capacities, Kant was—or should have been—placing an epistemic rather than psychological limitation on the subject, and that he must or should have meant his talk of the priority of spatial intuition to be epistemic rather than temporal priority. Nearly everyone will—or should—grant the plausibility of shading Kant's exposition in this direction. Kant himself, in concluding this subsection of the Aesthetic, makes just this kind of claim for his argument. He claims that his explanation "is thus the only explanation that makes intelligible the possibility of geometry, as a body of a priori synthetic knowledge" (B41). Kant's formulation of his conclusion in terms of the possibility of geometrical knowledge reads as a return to the idiom of explicating the epistemic conditions for the justification of a certain type of knowledge. What Kant has done, in essence, is to make space "subjective," in a manner analogous to the way in which Descartes and Locke had made ideas of secondary qualities subjective. If we must choose between assigning spatial properties to things in themselves or explaining them by appeal to characteristics of the knowing subject, we should choose the latter alternative.

There is, however, an obvious difference between the story that Kant tells about space as a form of intuition and the standard story about the secondary qualities. Kant introduced his doctrine of the subjective character of space in order to explain the possibility of geometrical knowledge. This fact about the use to which he put the doctrine gives us a way of seeing that it pertained to transcendental, and not to empirical, psychology. An appeal to empirical psychology could not, in principle, play the role that Kant required of his doctrine of sensibility.[77] For the purpose of establishing Kant's conclusion about the possibility of geometric knowledge, the causal origin of the faculty of sensibility is irrelevant. Kant purported to establish not how representations of space come into existence but rather that the representations of space that come into existence must be of a certain character and no other. And he purported to establish this conclusion not through a theory of the causal processes that underlie spatial perception and its development but on the grounds that in this way, and in this way only, can the possibility of geometrical knowledge receive an explanation. Kant's doctrine of space may rightly be assigned to transcendental psychology because the force of the arguments that he used to support the doctrine derives entirely from their role in explicating the possibility of knowledge of space in general and of geometry in particular.

These conclusions regarding the possibility of knowledge of space are deeply bound up with Kant's transcendental idealism. Consider Kant's summary of the teachings of the Aesthetic:

> What we have meant to say is that all our intuition is nothing but the representation of appearance; that the things which we intuit are not in themselves what we intuit them as being, nor their relations so constituted in themselves as they appear to us, and that if the subject, or even only the subjective constitution of the senses in general, be removed, the whole constitution and all the relations of objects in space and time, nay space and time themselves, would vanish. As appearances, they cannot exist in themselves, but only in us. [A42/B59][78]

According to Kant's doctrine of transcendental idealism, space is empirically real and transcendentally ideal: we can have objective knowledge of spatial properties as these properties modify the objects of outer sense, but we cannot attribute spatial properties to things considered independently of the knowing mind (A35-6/B52-3). Our knowledge of space is valid only of appearances, not of things in themselves.

A deeper understanding of the transcendental status of space as a form of intuition may be gained by considering the further conclusions that Kant purports to derive from it. Thus far we have been told that only on the assumption that space is a form of intuition can we explain our knowledge of space. We have not been led to understand how the individual knower has access to space as the form of intuition, or how the knower has the kind of intuitions that underlie Kant's claims about space, for example, that space is intuited as an infinite given magnitude. These questions pertain to Kant's doctrine that we can have pure a priori knowledge of space, independent of sensory experience.

If we stand back and reflect upon the space of pure intuition, it seems clear that we must think of it as empty space, devoid of sensational "matter." This much falls out of the main program of the Aesthetic: to isolate sensibility from the understanding and then to "separate off from it [the faculty of sensibility] everything which belongs to sensation, so that nothing may remain save pure intuiton and the mere form of appearances, which is all that sensibility can supply a priori" (A22/B36). Kant asserted that we can "quite well" think of space as empty of objects.

However, Kant denied that we can gain immediate access to space as a form of sensibility in and of itself. A pure intuition, isolated from the understanding, would provide no knowledge. Kant held that pure intuition of space in itself could not constitute "experience" (pure or otherwise) of space. Space and time themselves, as the forms of intuition, are not objects of intuition: they are the conditions under which any object of intuition must be intuited. Kant stated this point quite explicitly: "The mere form of intuition, without substance, is in itself no object, but merely the formal condition of an object (as appearance)" (A291/B347). In order to have a priori knowledge in accordance with these forms, the subject must provide his or her own objects of intuition. The act of providing an object in the space of pure intuition is not something that sensibility, a passive faculty, is able to perform. The understanding, an active faculty, must cause an object to be imagined. Kant says that

> the mere form of outer sensible intuition, space, is not yet knowledge; it supplies only the manifold of a priori intuition for a possible knowledge. To know anything in space (for instance, a line), I must draw it, and thus synthetically bring into being a determinate combination of the given manifold. [B137-8][79]

If space is to be known a priori, the understanding must act to bring about a determinate object in the space of pure intuition. It may do so by drawing a line or other geometrical figure in imagination. This process of drawing, or of construction, is conditioned by the form of sensibility, which provides the laws in accordance with which any spatial figure must be rendered.[80]

The conception of space as a form of sensibility—that is, as supplying laws for constructing spatial representations—provides the foundation for extending Kant's doctrine of space from pure to empirical intuition. That Kant needed to be able to justify such an extension should be clear: he intended the *Critique* to provide an explication of the possibility of objective knowledge within the domain of possible experience. If his findings pertaining to geometry were limited to pure intuition in such a way that they did not condition empirical intuition, Kant would have divorced geometry from its application in empirical physics, a consequence he considered to be absurd.

Kant was quite explicit in extending his account of space as the form of pure intuition to space regarded as an object of investigation in physics: "Geometry is a science which determines the properties of *space* synthetically, and yet a priori" (B40, my emphasis). It was a part of Kant's conception of geometry that the object of geometry is space, including physical space as given through empirical intuition. He maintained that the applicability of mathematics to empirically given objects could be guaranteed only by extending his findings regarding pure intuition to include empirical intuition:

> Empirical intuition is possible only by means of the pure intuition of space and of time. What geometry asserts of pure intuition is therefore undeniably valid of empirical intuition. The idle objections, that objects of the senses may not conform to such rules of construction in space as that of the infinite divisibility of lines or angles, must be given up. For if these objections hold good, we deny the objective validity of space, and consequently of all mathematics, and no longer know why and how far mathematics can be applicable to appearances. The synthesis of spaces and times, being a synthesis of the essential forms of all intuition, is what makes possible the apprehension of appearance, and consequently every outer experience and all knowledge of the objects of such experience. Whatever pure mathematics establishes in regard to the synthesis of the form of apprehension is also necessarily valid of the objects apprehended. [A165-6/B206][81]

The doctrine of transcendental idealism, which limits Kant's findings regarding the possibility of knowledge of space to the case of possible human experience, makes possible at the same time establishment of certain unconditional, necessary, and universal conclusions regarding the possible objects of such experience: they will all conform to Euclid's geometry (A27/B43). Kant regards his position as explaining not only the possibility of pure geometrical knowledge but also the possibility of applying geometrical descriptions to all the phenomena of outer sense, that is, to the material world in general.

Kant's extension of his doctrine of space to empirically obtained representations illustrates his transcendental mode of argument. He was well aware that no appeal to previous experience, whether in the form of an appeal to the psychology of sense-perception or to the observations of the physicist, could establish that space as we experience it conforms to the space of geometry. He was, however, quite sure that we have geometrical knowledge of physical space. His explanation of the possibility of such knowledge appealed to certain "facts" about the constitution of the knowing subject. In accordance with transcendental argumentation, he can regard these facts as established once and for all, because they are necessary conditions on the possibility of knowledge that we do in fact possess. With these transcendental facts in his possession, he can then proceed to establish certain other doctrines with respect to the material world in general, now regarded as a possible object of experience. Among these are the applicability of geometrical description to all material objects. His ability to establish such doctrines depends upon the notion that the conditions for the possibility of geometry are at the same time the conditions for any intuition, a notion that requires acceptance of Kant's transcendental psychology, which identifies sensibility as the faculty of spatial representation in geometry with sensibility as the faculty of empirical intuition. The doctrine that geometry describes material objects is a necessary and universal conclusion based upon an analysis of the conditions for knowledge. It (allegedly) applies to every possible object of sensory experience.

I have characterized the basis of such conclusions as "facts" about the knowing subject. But these are no ordinary facts. Like brute facts, they admit of no further explanation.[82] However, they are not facts obtained through experience (empirical intuition) but through transcendental argument. They might be termed "transcendental facts." If transcendental findings about the faculty of sensibility are "facts" of any sort, we may well ask after their relation to empirically determined facts about that faculty. This raises the counterpart to the question that guided our investigation of the relation between empirical and transcendental psychology. We have seen that Kant considered the findings of empirical psychology to be irrelevant to transcendental psychology. But to what extent did he consider the findings of transcendental psychology to condition empirical psychology?

5 Transcendental and Empirical Psychology (Again)

The relation between transcendental psychology and empirical psychology may be understood by contrast with that between transcendental philosophy and empirical physics. Pure physics establishes that every sequence of events in nature occurs in accordance with the causal law, but it cannot determine the precise form the actual laws of physics take. Kant thought that the most general laws of physics could be derived from the empirical concept of motion, together with the a priori principles established in the Analytic of Principles.[83] The more specific laws of physics, such as the law of fall, require an empirical investigation. Transcendental philosophy, through the law of cause and other principles that hold for all of nature, thus describes a general set of a priori synthetic principles that condition any possible empirical finding. Empirical psychology may be regarded as conditioned in the same manner; for, as with empirical physics, it is conditioned by the analytic of principles. Just as pure or rational physics provides some general principles for the rest of physics, so too pure or rational psychology establishes the validity of the causal law as applied to empirical psychology.

This simple application of Kant's architectonic to the cases of empirical physics and empirical psychology masks the deeper relations that underlie the a priori conditions of both subject-matters. In the case of both physics and psychology, the body of a priori principles that condition all possible experience is derived as a consequence of the deduction of the categories. The necessity bound up with the unity of apperception explains how the causal law is not only possible but necessary a priori. This finding conditions physics indirectly, inasmuch as it establishes a body of principles that pertain to all possible objects of experience. It doubly conditions empirical psychology, which treats of the mind. Because the representations of outer sense must be connectable in accordance with physical causality, these representations must be associable: the law of association is thereby rendered necessary. But the sequence of representations available to inner sense, now taken as "natural" phenomena themselves, must in any case obey the causal law, again underwriting the law of association (independent of outer sense).[84]

The same type of reasoning may be extended to the findings of the Aesthetic. That space is a form of intuition cannot be established by or through empirical psychology, but once this has been established it conditions any finding that empirical psychology might put forward. Although this point is only implicit in the *Critique*, Kant made it quite explicit in a letter to J. W. Kosmann in September of 1789, in which he framed a distinction between a "psychological" and a transcendental deduction of space as the form of outer sense:

> We can attempt a psychological deduction from our representations inasmuch as we regard them as effects, [i.e.,] regard their causes in connection with other things; or too we can attempt a transcendental deduction inasmuch as if we have assumed grounds that are not of empirical origin we seek out purely the grounds of possibility for how they [can be] a priori but have objective reality. In respect to space it is not necessary to ask how our faculty of representation first came into use in experience, it is enough that when once we have developed it we can prove the necessity to think space, [i.e.,] to think space with these and no other determinations, by means of the rules of the use of this faculty, and can specify the necessity of the grounds of the same independently of experience, even though the matter is such that it does not allow itself to be developed out of a concept but is synthetic.[85]

The transcendental deduction of space as a form of sensibility, in its role of explaining the possibility of synthetic a priori knowledge, need not (and cannot) concern itself with the causal origin of the constraints on spatial representation. It starts from the necessity of Euclid's geometry for the mature mind and moves from there to the transcendental posit of a ground for this necessity in the rules for the use of the spatial faculty.

This amounts to the posit of space as the form of outer sensibility. Accordingly, no matter what the psychogenesis of the faculty of spatial representation, this faculty must be such that it determines (outer) intuitions in such a way that they provide instances of the space described by Euclid. It must also be such that its ability to do so is independent of previous experience (even if, as Kant says, such abilities are "awakened into action" [B1] through experience). Once this point has been established transcendentally, it holds good for any empirically supported account of the psychology of spatial representation.

There is, however, a difference between the psychological implications of Kant's doctrine in the Deduction and in the Aesthetic. The implications of the Deduction must, as Kant sees it, hold good for any finite understanding, whereas the implications of the Aesthetic could be peculiar to human beings. While the law of association might be a law of nature for any finite mind, the possession of space as the form of outer sense might be a fact peculiar to human psychology.

The status of Kant's conclusion that space is the form of outer intuition may be compared with his claim that the twelve categories constitute the "form" of the understanding. According to Kant's doctrine, the twelve categories must be postulated in order to account for the transcendental unity of apperception in any finite intellect.[86] They constitute the universal conditions for the possibility of a unified consciousness—that is, they underlie the very possibility of understanding or of thought itself. Not so with space as a form of intuition. From the perspective of accounting for the possibility of absolute certainty in geometry, Kant maintained that the space described by Euclid's geometry must be postulated as the necessary and universal form of human outer sense. But from the perspective of ourselves as finite intelligences, it is merely contingent that we experience things in space. For, notoriously, Kant opened the possibility that, for all we know, other finite intelligences might have different forms of sensibility.[87] Thus, even though the necessity that all our spatial intuitions should have a certain character is grounded in epistemological considerations, the fact that we have spatial intuition at all, rather than some other form of intuition, has the status in Kant's system of a brute fact. One is tempted to say that it has the status of a brute fact of human psychology, but it is not one that would be studied empirically or justified by appeal to specific empirical results. It is, in the terminology of the previous section, a "transcendental fact."

It may seem that, by assigning space a merely contingent status as the form of sensibility, Kant has undercut its role in explaining our synthetic a priori knowledge of geometry. This is not the case. The "contingency" in the claim that space is a contingent form of sensibility pertains to a comparison with other possible finite intelligences; since other finite intelligences might have other forms of sensibility, there seems to be no ground for supposing that some particular finite intelligence, *qua* finite intelligence, must have a particular form of sensibility. Yet given what we know about the body of synthetic a priori principles that are available to human subjects in the form of Euclid's geometry, it can be asserted that all human knowers, *qua* human, necessarily experience outer things as spatial. Kant, in fact, sought to show just how a mere limitation on human sensibility could yield universally valid knowledge:

> If we add to the concept of the subject of a judgment the limitation under which the judgment is made, the judgment is then uncondi- tionally valid. The proposition, that all things are side by side in space, is valid under the limitation that these things are viewed as objects of our sensible intuition. If, now, I add the condition to the concept, and say that all things, as outer appearances, are side by side in space, the rule is valid universally and without limitation. [A27/ B43]

Given that space is an a priori form of human sensory intuition, then if the domain of discourse is restricted to possible human sensory intuition, a priori synthetic knowledge regarding that experience is attainable.

Transcendental psychology stands in a prescriptive relation to empiri- cal psychology: the conditions prescribed by transcendental psychology must be met with in empirical psychology. But the scope of these prescriptions remains unclear. For example, the conditions on space as the form of sensibility must surely have implications for psychological discussions of the faculty of spatial representation, but what are their implications for the specific senses, such as vision and touch? Do they rule out Berkeley's notion that we learn to see in three dimensions, thus aligning Kant with nativist explanations of visual depth perception? I shall argue that the implications of Kant's transcendental psychology for the psychology of the senses are more limited in scope than has been realized.

6 Kant's Psychologies of Space and Vision

Histories of perceptual theory treat Kant as a nativist with respect to the psychology of vision; they impute to him the position that the ability to perceive three-dimensional space by means of sight is inborn. To a number of writers, including Helmholtz, the case for placing Kant among the nativists has seemed clear, given Kant's doctrine that all human experience is conditioned by space as the form of (outer) intuition. Surely the belief that we experience objects in space because of the constitution of our form of sensibility is the same as the belief that our ability for visual space-perception is innate.[88]

Classifying Kant as a nativist because of his doctrine of outer sense confuses two issues: (1) the question of whether our faculty of sensibility is such that we must come to experience objects in space (whether this space be connected with a particular sense such as touch or vision, or experienced merely in the imagination), and (2) the question of whether we have the innate ability to localize objects in three-dimensional space by means of vision. The first question asks whether the ability to represent spatial relations in general could be acquired through experience, and it is clear that Kant answered it negatively, contending that the faculty of sensibility must be such that it determines outer intuitions in accordance with Euclid's geometry. The second question pertains to nativism with respect to visual space-perception, and therefore to the controversy about innateness (see chapter 2).

In order to reconstruct Kant's implicit position with respect to visual space-perception, I need to flesh out his empirical psychology. The sources for Kant's views on empirical psychology come from student notes to his lectures on the subject, from his lectures on anthropology, published under Kant's supervision in 1798, and from scattered remarks on empirical psychology in the *Critique*. In general, Kant maintained that each faculty listed in transcendental psychology—sensibility, imagination, and understanding—has its empirical counterpart. The transcendental notion of synthesis—the activity of combining representations—also has its empirical counterpart in the laws of association. The representations that are combined in a synthesis may be either pure or empirical. In the latter case, they are termed "sensations," which constitute the empirical "matter" to be organized in accordance with the "forms" of intuition. Sensations, or the matter of experience, are one and all intensive, but not extensive: they vary in both qualitative character and degree, but they are not spatial—as Kant put it, space and time "are not to be met with" in sensations (A166/B208). Like space and time, sensations depend upon the constitution of the knowing subject; unlike space and time, they are not the subject of a body of a priori knowledge. Kant held that we are able to anticipate, a priori, only one feature of sensations: that they must possess some intensive magnitude (degree). But we are unable to anticipate the qualitative character of sensations prior to experience—a person who had never experienced red could never frame a representation of that color (A167-76/B209-18). In any event, according to Kant's teaching the empirical matter of intuition is constituted by nonspatial sensations varying only in quality and intensity. They are brought to empirical synthesis through the reproductive imagination in accordance with the laws of association.

The importance of the faculty of imagination in both transcendental and empirical psychology cannot be overly stressed. Both in his account of the possibility of empirical knowledge, and especially in his few scattered remarks on perception in the *Critique*, Kant ascribes a central role to imagination. Of special interest here is Kant's doctrine that the perceptions of the senses cannot be explained by an appeal to the passive faculty of sensation alone. Kant believed that perceptual images of objects are not given from or produced by the senses alone but are constructed by the imagination. He contended that psychologists had not sufficiently appreciated this fact:

> Psychologists have hitherto failed to realize that imagination is a necessary ingredient of perception itself. This is due partly to the fact that that faculty has been limited to reproduction, partly to the belief that the senses not only supply impressions but also combine them so as to generate images of objects. [A121, note]

According to Kant, in perception the faculty of imagination functions "to bring the manifold of intuition into the form of an image" (A120). The imagination constructs perceptual images from the nonspatial sensations that serve as the manifold for the synthesis that underlies empirical intuition.

Kant's constructive account of perception might seem to provide evidence for those who have classified him as a nativist. For in its constructive activity, the imagination must surely be guided by the laws of sensibility, so that in the case of outer sense imagination will be constrained to construct spatial images that are in accordance with Euclid's geometry. But although this surely must be the case, it does not entail nativism with respect to vision, and for two reasons.

First, that space is the form of sensibility provides by itself only very weak a priori constraints on the perceptual images that could be formed by imagination (leaving aside, for the moment, any constraints that derive from the understanding by way of schematized concepts). The images that are constructed must accord with the rules of Euclid's geometry, but this fact does not imply any specific rules for mapping sensations into perceptual images. The shapes met with in the perceptual images must only conform to Euclid's geometry; but this implies nothing about the relation between the shapes themselves and the objects that affect the sense organs and cause the empirically given visual sensations, whether these objects be conceived as things in themselves (in which case Kant would have us eschew questions about the relation) or as empirical objects affecting the empirical eye. Compare the (empirical) situation to the production of images on a television screen. Given the way the screen and the electron gun is set up, its activity must of necessity produce a spatial pattern of light and dark that must conform to certain topological properties (e.g., except at the edges, each small region will be between two others, and on a given row, to reach a point that is not next to a given point will entail passing through yet other points). Any pattern that results on the screen must be describable in accordance with this topology. But that says nothing about the rules of mapping the points from the scene in front of the camera to the screen. Convention has it that two-dimensional relations are preserved, so that a perspective image of the the scene is produced on the screen; but none of the rules of screen-topology would be violated if the camera were so constructed to achieve another mapping, one that appeared random from the point of view of the perceiver. So too with the mapping of the matter of intuition into the form of space. If we restrict ourselves to the constraints provided simply by the fact that space is the form of sensibility, any mapping yielding a product that conforms to Euclid's geometry will do; it need not yield information about the spatial characteristics of objects in the environment.

It is apparent, however, that if we consider doctrines beyond the Aesthetic, this picture will not do. For Kant claimed to establish through the Deduction and the Analogies that experience must present a world of objects ordered in space and time. And this surely requires that the senses, including vision, present us with a world that is not merely in conformity with Euclid's geometry but contains re-identifiable objects that can be fit into a coherent spatiotemporal framework. These stronger constraints surely require that the relation between objects and the eye (to continue the description of the senses as empirically-known organs) must be such as to produce experience of an ordered world. An arbitrary mapping will not do; the production of visual images must stand in a regular relation to objects. Moreover, the deliverances of vision presumably must be in harmony with those of the other senses. Thus, Kant's extended teaching does require that the visual faculty must, of necessity, and independently of experience, be such as to contribute to the experience of a world of objects ordered in space and time. But his doctrine of space by itself does not.

The second reason that Kant's account of perception does not entail nativism builds on the distinction between psychogenesis and transcendental constraints, found in the letter to Kosmann (previously quoted). Kant there distinguished between the psychogenesis of the ability to have spatial representations and the transcendental fact that the spatial representations produced must conform to the rules of Euclid's geometry. Here we may distinguish between the transcendental fact that the sense of vision (or any spatial sense) must contribute to the experience of a spatial world, and the psychological facts about the psychogenesis of our visual abilities. The visual faculty need not have the innate ability to perceive a three-dimensional visual world when presented a two-dimensional pattern of stimulation. It is enough that the visual faculty develop in such a way that it acquires that ability. The causal basis of this development is irrelevant; transcendental philosophy only requires, and therefore can only establish, that the sequence of development must yield an ability for spatial perception, subject to the constraints described in the previous paragraph.

Kant was at least implicitly aware of the independence of his doctrine of the form of outer sense from any particular theory of visual space-perception, for he in fact adopted the standard eighteenth-century position that the construction of three-dimensional percepts in vision is an acquired ability. His position is revealed both in his lectures and in his published works. The evidence from the lecture notes is quite strong. Kant argues that vision originally represents the world only in two dimensions, so that a sphere would appear as a circle, and that touch educates vision, so that the percipient comes to see the shape of things in depth, rather than their mere projection: "Only through the mediation of touch are we able to construct concepts of the shape and size of bodies. The eyes present us objects only as in a plane."[89] This point is confirmed by the *Anthropology*, in which Kant says that our perception of solid (three-dimensional) shape would be impossible without the sense of touch; he makes vision depend upon touch for its ability to perceive objects in depth, thereby implying the standard Berkeleyan account.[90]

Evidence from the *Critique* itself is less direct but no less unequivocal. Kant gives a brief account of perceptual illusion (used metaphorically to illustrate "transcendental illusion") that is similar to the standard position as summarized by Gehler. According to this position, illusion is not due to the senses but to the inference of the understanding (in the case of vision, in inferring the third dimension from a two-dimensional image). Adherents of the view that we "learn to see" explained away the fact that we do not notice the psychological operations underlying vision by maintaining that they become habitual and therefore unnoticed. Kant adopted the same position:

> A distinction is commonly made between what is immediately known and what is merely inferred. . . . Since we have constantly to make use of inference, and so end by becoming completely accus-

tomed to it, we no longer take notice of this distinction, and frequently, as in the so-called deceptions of the senses, treat as being
immediately perceived what has really only been inferred.[91]

The result of a mistaken inference becomes confused with immediate
perception itself. Kant refers to the example of the moon illusion, in
which the moon appears larger at the horizon than it does in the heavens
because it has been judged to be farther away (it is judged larger to
account for the fact that it subtends the same visual angle in each case).[92]
Presumably, these inference patterns are acquired habits; if they were
"given" with the transcendentally established constraints on the form of
intuition, custom and habit would have no role to play.

This brief excursion into Kant's conception of perceptual psychology
makes explicit the contrast between Kant's point that space is the form of
outer sense and the nativist belief that infants are innately endowed with
spatial vision. On the one hand, Kant believed that our cognitive faculties
are so constituted that all sensory intuitions are conditioned by the rule
that they must be disposed in space as described by Euclid's geometry.
On the other hand, he accepted the typical eighteenth-century doctrine
that we must learn to see a world of three-dimensional objects. Contrary
to first appearances, Kant's belief that space is the form of outer sense is
compatible with the doctrine that spatial vision is an acquired ability.

The compatibility of Kant's doctrine of space with the notion that we
learn to see depends upon the fact that in the Aesthetic Kant was
speaking of a generalized capacity for having a certain kind of representation—one that exhibits spatiality—whereas writers on vision had been
concerned with the specific perceptual ability of localizing objects in
space on the basis of optical stimulation. Kant's doctrine of the apriority
of space merely maintains that our sensibility must be such that our
perceptions must occur within the framework of three-dimensional
space. This transcendental fact does not guarantee that the sense of vision
is innately disposed to generate spatially elaborated perceptions of a
three-dimensional world at the onset of optical stimulation (or after a
brief period of maturation). A number of different relations could be
thought to hold between the disposition for spatial intuition that Kant
envisioned and the psychological process of generating representations
of spatially organized bodies from sensory impressions. One could hold,
as Kant seems to have held, that in vision a representation of only two
dimensions is originally given—that is, that within the three-dimensional framework of original intuition, our visual representations at first
vary only in two dimensions—and that we learn to infer the third
dimension. Or one might adopt a fully nativist position, that the visual
system is innately disposed to generate representations of three-dimen-

sional bodies. In either case, the capacity for having spatial representations, indeed, the necessity that all representations are conditioned by space, is distinct from the mechanisms that map individual color sensations into this space during visual perception.

One might conjecture that Kant's general position on the relation between visual space and space as a form of sensibility was as follows. Prior to birth the individual is determined so that its faculty of sensibility must develop in such a way that it yields spatial intuitions of the appropriate sort. The form of sensibility in itself, however, does not constitute a spatial representation. One means of awakening the form of intuition to space is through the act of pure intuition, in which the understanding acts upon sensibility through the imagination to provide a pure intuition of space. Empirical matter may be introduced into the space of intuition through vision, or rather, the faculty of sensibility and the imagination may be awakened to construct spatial intuitions as a result of visual sensation. As a matter of transcendental fact, the faculty of sensibility must develop in such a way that these sensations come to be ordered within a three-dimensional framework. Yet this a priori condition on the form that any representation of outer sense may take does not guarantee that our original visual representations will be representations of the three-dimensional world of adult experience. It is up to empirical psychology to ascertain how this or that arrangement of color sensations gets mapped into intuitional space. Nothing can be known of the mechanisms for mapping color sensations into intuitional space simply from the fact that this intuitional space is itself three-dimensional. Kant's doctrine of outer sense was mute with respect to the genetic psychology of visual space-perception.

7 Conclusions

Kant insisted upon a distinction between psychology as the investigation of inner sense and transcendental philosophy as the investigation of the faculty of thought. Psychology belongs to natural science; it is no longer the science of the soul, but neither is it the science of thought. Kant introduced this distinction between naturalistic psychology and transcendental philosophy without appealing to a deity or to the mind's participation in the divine. He separated the natural from the normative not by appeal to the supernatural but from a stance immanent to human experience. According to Kant, the naturalistic stance could never account for the highest achievements of theoretical cognition, the discovery of universally valid laws in natural science. These achievements must be accounted for from within the transcendental stance, which is inherently normative, or concerned with questions of right.

For Kant, two irreconcilable vocabularies were applicable to the same set of phenomena, the phenomena of inner sense. As a natural phenomenon, the inner play of representations is subject to the causal law. But considered from the standpoint of the spontaneity of synthetic judgment, the synthesis of representations must be explained in the normative language of transcendental philosophy. Indeed, Kant gave primacy to this second language, maintaining that any empirical description of the mind must accord with transcendental philosophy. Of course, the force of the prescription is only as strong as the arguments that underwrite transcendental philosophy. In succeeding chapters, I shall evaluate this force in connection with numerous challenges to Kant's doctrine by nineteenth century German physiologists, psychologists, and philosophers.

Chapter 4

Spatial Realism and Idealism: Kant Read, Revised, and Rebuffed

Kant's critical philosophy came to dominate the German intellectual scene with astonishing completeness. Within a decade of the publication of the first *Critique*, there arose a formidable scholarly apparatus aimed at presenting, interpreting, and either refuting or extending Kant's work. Dictionaries of the critical philosophy were produced; journals devoted to its propagation or demise were founded.[1] While German philosophers writing at the end of the eighteenth century might disagree with Kant, they could hardly afford to ignore him.[2] Kant's work inspired responses not only in the central areas of the critical philosophy, from metaphysics and theory of knowledge to ethics and aesthetics, but also in psychology, physiology, and biology. And his influence continued throughout the nineteenth century. Even as various lines of thought claiming to develop and supplant his work played themselves out, they were put to rest with calls for a return to the true Kant.

Two streams of thought are likely to appear to us today as Kant's legacy to German philosophy in the century after his death: German Idealism as developed by Fichte, Schelling, and Hegel; and the brand of "critical positivism" that found its champions in scientifically trained individuals who did not hold chairs of philosophy, such as Helmholtz and Mach.[3] German Idealism constituted the predominant response to Kant prior to 1850. Fichte gave the movement its original direction by seeking to remove the remaining elements of "dogmatism" from Kant's philosophy—the references to the "thing in itself"—that he considered to weaken what he took to be Kant's fundamental insight, the doctrine of transcendental idealism. Fichte, Schelling, and Hegel all developed the side of Kant's thought that emphasized the role of the knowing subject in constituting knowledge (hence, their "Idealism"). They were themselves criticized for construing the ego as a "thing in itself," thereby engaging in their own form of dogmatic metaphysics.

By the time of the mature Hegel, Idealism no longer seemed a direct continuation of Kant's work, but its Kantian roots were recognized by friend and foe alike, the latter criticizing it for deviating, to its own detriment, from Kant's true intentions. This critical response to Idealism fueled the second wave of (allegedly) Kantian philosophy, under the banner of going "back to Kant." Helmholtz and others claimed that they were the true Kantian faithfuls; they emphasized the antimetaphysical elements in Kant's thought as a palliative against the metaphysical excesses of Hegel and his followers.[4]

Although these two responses to Kant stand out from our present vantage point, they do not fully describe the German philosophical scene as its inhabitants understood it throughout the nineteenth century, and as we should understand it if we wish both to know how Kant has been read and to appreciate the background to Helmholtz. Perhaps as a result of scholarly argument waged at the turn of the present century, when psychology and philosophy were in direct conflict for academic resources (each asserting authority over at least one domain claimed by the other: the investigation of cognition),[5] the various psychological, or, as they came to be known, "psychologistic" readings of Kant have faded from historical-philosophical consciousness; if they are recalled at all, it is only as a warning against one way of getting Kant wrong.[6] Yet virtually every reading of the *Critique* in the first hundred years after its publication was "psychological" in one way or another.[7]

Fichte and Hegel both characterized Kant's method as "psychological," meaning that they attributed to him, de facto, a method of introspection.[8] As we now know, care must be taken in interpreting the meaning of this term. Indeed, the work of Fichte and Hegel itself might be characterized as psychological, inasmuch as they investigated the "phenomena" of the ego or of consciousness.[9] In any event, most readings of and responses to Kant in the early nineteenth century may be straightforwardly classified as "psychological," if not always in a natural-scientific sense of that term. Such readings and responses ranged from Fries's attempt to provide an "empirico-psychological" or "anthropological" confirmation of Kant's doctrines, to Herbart's metaphysical alternative to what he regarded as Kant's mistaken attempt to base philosophy on psychology, to the attempted refutations or extensions of Kant through physiology and psychology by Steinbuch and Tourtual. These authors all agreed that Kant's doctrines regarding human knowledge, particularly those pertaining to space and time as the forms of sensory intuition, were essentially psychological, even if unwittingly so; they differed in their reasons for accepting this conclusion, and in their beliefs about what needed to be changed or rejected in Kant's work.

Notwithstanding the attitude of dismissal that has been taken toward psychologistic readings of Kant in more recent times, such readings are not without interest. They are not as immediately susceptible to the standard objections against them as one might expect. Moreover, the issues raised by such readings are quite wide-ranging, including the foundation of transcendental knowledge itself and arguments about whether to adopt transcendental idealism or, on the contrary, to assert the reality of spatial properties. In order to understand more fully the philosophical issues at stake in giving a psychological reading to Kant, I shall examine the relevant portions of Fries and Herbart, before turning to the natural-scientific approaches of Steinbuch, Tourtual, and a third physiologist, Johannes Müller. The chapter ends with Lotze's combined physiological and metaphysical approach, developed just prior to Helmholtz's own response to Kant.

1 Foundations of Transcendental Knowledge in Fries and Herbart

In their respective evaluations of Kant's grounding of transcendental knowledge, Fries and Herbart both concluded that Kant's doctrine of the categories and the forms of intuition was at root psychological. Both contended that, despite his protests to the contrary, Kant had attempted to ground his doctrine of transcendental knowledge in empirical psychology. But Fries thought Kant's unavowed appeal to psychology was on the right track, and he attempted to continue in the same vein, whereas Herbart denied that psychology could ever ground the fundamental principles of philosophy.

1.1 Fries: Transcendental Idealism and Psychology

Doctrinally, Fries remained closest to Kant of all the authors who have been mentioned. His major philosophical work was entitled *Neue oder anthropologische Kritik der Vernunft*[10] in order to mark this doctrinal proximity. Like Fichte, Fries regarded transcendental idealism as Kant's fundamental teaching;[11] further, he accepted the doctrine of the categories and forms of intuition with only minor revision (and some flattening). Fries took himself primarily to be correcting Kant's failure to realize that his investigation of the knower was really an empirical investigation through inner sense (and theoretical reflection upon the givens thereof). Instead of recognizing the empirical basis of the claim that the law of cause is a law of thought, or of the postulation of separate faculties of sensibility, understanding, and reason, Kant had attempted to prove a priori that the mind is possessed of certain faculties and laws of thought. Indeed, Fries argued that Kant's blindness to the psychological foundations of his own position resulted from his obsession with proof, an obsession that Fries found comparable with the dogmatic failings of rationalist metaphysics (1:20-26, 28-30). This failing affected the very heart of Kant's transcendental doctrine, the deduction of the categories.

Fries contended that in his attempt to provide a proof of the objective validity of the categories, Kant had been forced to give a circular argument: Kant's deduction simply presupposed the objective validity of sensory experience, and then introduced the categories as the one possible explanation of this validity. This charge rested on the claim that Kant had based his assertion that sensory knowledge is objective on the presupposition that sensations have an external cause (2:97-8).[12] Fries believed that he could avoid this difficulty by eschewing the demand that a deduction have the status of a proof; he would stay within the bounds of transcendental idealism, which meant that he would base his own deduction on an empirically based description of the faculty of cognition. The attribution of the law of cause to our faculty of thought would be based on inner sense and reflection, not, as had allegedly been the case with Kant, on assumptions about the objective external cause of inner representations.

At first sight, most of Fries's criticisms and revisions of Kant may seem easy to dismiss. His charge that Kant merely assumed the objective validity of outer experience is perhaps nearest the mark, though the force of this criticism depends upon how one describes Kant's own aim (specifically, upon whether Kant's aim was to ground empirical knowledge of nature, or instead to prevent extending the principles underlying such knowledge to the supersensible). But Fries's central methodological claim to base transcendental philosophy on empirical psychology seems most easily dismissed, on at least two grounds. First, it is dubious that aspects of Kant's analysis of thought could be based on inner sense. Kant explicitly denied that the transcendental unity of apperception, a notion which Fries rightly described as central in Kant's analysis of thought, is available directly to inner sense. Second, Fries was simply confused in attempting to provide an empirical justification from inner sense for philosophical principles such as the law of cause, since he agreed with Kant in ascribing necessity and universality to such principles.

In fact, Fries was aware of both these problems and attempted to develop a methodology that could address them. He did not seek a "proof"—empirical or otherwise—for principles such as the law of cause but developed an argument to show that, because they were universal laws of thought, they should therefore be accepted as philosophical principles. Fries distinguished between the law of cause regarded as a metaphysical principle pertaining to nature, and the law of cause regarded as a principle of the understanding; he labeled knowledge of metaphysical principles "philosophical," to distinguish it from "transcendental" knowledge of the principles of thought (1:29). He held that transcendental investigation does not lead to the discovery of new philosophical knowledge; it simply brings to consciousness what has been at work in actual instances of such knowledge. Transcendental investigation discovers the principles of the mind that make philosophical knowledge possible. However, Fries avoided the claim that transcendental knowledge proves the validity of philosophical knowledge; he developed his own notion of "deduction" according to which a deduction "exhibits" but does not "prove" philosophical principles such as the law of cause; it does so by deriving them from laws of thought that have themselves been established through empirical investigation of the mind, based on inner sense. In the end, the aim of transcendental knowledge is to set us free from the empiricist skepticism of Hume and the dogmatic claims of Wolff, and to leave us with the recognition that we possess universally valid philosophical principles.

To achieve this aim, two things would need to be accomplished: establishing what he called the "theory of reason," which would include the laws of thought, and "exhibiting" that the universal validity of metaphysical principles such as the law of cause follows from this theory. Fries's project was to develop a universally valid theory of reason (an "anthropological," as opposed to philosophical, theory) through induction, which he did count as a form of proof.

Fries was sensitive to the charge that he had simply pushed one step back the Humean problem of achieving universal validity through the observation of particular instances, from the law of cause to the theory of reason. His explicit response was to compare his enterprise to theory-making in the natural sciences, noting that the universal laws of science are proposed and tested on the basis of particular observations and are then accepted as universal premises for demonstrative arguments in explaining further particular phenomena.[13] In the (more recent) jargon of the schools, he characterized his theory of reason as having a hypotheticodeductive status (a move that begs the question of whether inductively based premises may count as "proven"). But further, and more importantly, the actual "inductive" arguments for the particular parts of his theory were transcendental arguments in the Kantian style. These typically started with some "fact" of cognition—such as the existence of universally valid claims in mathematics and physics, or the "idea of a world," which is the idea that our experience is of one world— and proceeded to the characteristics of our thought that must be presupposed if these "facts" are to be explained.

Fries's style of argument may be appreciated by examining briefly his "inductive" proof of the unity of transcendental apperception and his subsequent "deduction" of the law of cause. Fries maintained that the presence of synthetic unity in our knowledge proves the unity and necessity of transcendental apperception. The latter unity must be more than mere association, more even than the mere subjective unity of all of one's representations in a single consciousness; according to Fries, the unity of apperception consists in the fact that all of one's representations constitute parts of a whole that is determined through one and the same cognitive activity. The unity lies not in the mere connectedness of the representations, but in the fact that they make one knowledge. Such unity, he maintained, can be guaranteed only through the activity of a single cognitive process that is applied to every representation and acts always the same. Fries contended that a unified cognitive process for bringing form to the matter of intuition, a process which he labeled "original formal apperception," must be posited in order to meet the requirements imposed by the transcendental unity of apperception. The principle of original formal apperception is for him the highest principle of empirical psychology (or philosophical anthropology). It is required in order to explain the "facts" canvassed above; in providing this explanation, it forms the basis for the subsequent deduction of philosophical knowledge (2:48-55, 60, 65).

Fries's claim that his deduction makes no pretense at "proof" depends upon his characterization of it as an "explanation." He posited certain laws of thought as presuppositions to cognitive achievements that he considered to be actual, and he then used these laws to explain the possibility of universal and necessary laws of philosophy. The universality and necessity of the axioms of geometry, or of the causal principle, were explained by arguing that the categories and forms of intuition, in accordance with original formal apperception, are the basic laws of the cognitive process of bringing form to the matter of intuition. They apply universally and necessarily to all representations, and thereby ground the apodictic knowledge available in intuition or through the application of concepts to intuition.[14] Fries here takes an unambiguous stand on how to interpret the Kantian notion of the possibility of experience as grounding synthetic a priori knowledge: the conditions for knowledge are specific processes for "filling" space and time with matter (2:102-6).

Fries's approach to "deducing" the categories and forms of intuition may not seem more obviously "psychological" than Kant's (if we give the same weight to Kant's transcendental psychology Kant himself gave it). Kant's arguments crucially depended upon a characterization of the mental processes of the knowing subject. However, analysis of the source from which his arguments drew their force revealed that his investigation of these processes could not have been based on empirical psychology, for they started from "transcendental givens," such as the actuality of geometry, or of intersubjectively valid experience. Fries's own inductive "exposition" is ultimately based on a similar set of givens, and his account of mental processes may reasonably be regarded as fleshing out Kant's own sketchy account of the transcendental synthesis of the imagination.

But it now may seem somewhat puzzling that Fries was so insistent on characterizing his approach as "psychological" by contrast with Kant's. Indeed, this term could easily lead to a misperception of Fries's arguments. His approach to philosophy was not "psychological" in the sense of pursuing a "naturalistic" approach to mind—for example, by attempting to reduce reason to sense and imagination by reducing their activity to mechanistic laws of association. It would, therefore, be a mistake to assume that Fries's approach can be dismissed as a simple case of psychologism, understood as an attempt to derive normative or prescriptive force from naturalistic description. Indeed, it would be misleading to say that Fries attempted to apply natural-scientific findings to the adjudication of logical or even epistemological matters. He did, in fact, call his theory of reason a "physical" theory, but that meant only that he considered it to rest on inductive grounds, in the same manner as physics (2:72, 1:25-6).

There are two senses in which Fries's *Neue Kritik* was "psychological" by contrast with Kant's own *Critique*. First, Fries claimed that he was proceeding by an empirical methodology, in contrast to Kant's allegedly a priori method: where Kant had attempted to give an a priori transcendental "proof" in the form of a deduction, Fries would give a transcendental "exposition" based upon an inductively proven theory of reason. The fact that this expository deduction allegedly rested upon empirical findings gleaned from inner sense did not commit Fries to an introspective approach to discovering the categories or forms of intuition. His appeal to such notions as "original formal apperception" was "empirical" only in that it was justified through its applicability in explaining a wide range of data, as are the universal generalizations of other empirical sciences. The "data" to be explained in the case of the theory of reason were all thoroughly normative: they were the data of our epistemic practices. But the concepts Fries used to explain them were less clearly normative, and this constitutes the second sense in which Fries's approach may be characterized as "psychological" by contrast with Kant's. Fries explained the possibility of mathematical knowledge, or of the unity of all knowledge, by appealing to certain laws of thought that he conceived as concrete causal processes for "filling" the forms of intuition. Rejecting Kant's appeal to notions of objective validity, he sought a subjective explanation of the unity of thought in the causal processes responsible for generating particular thoughts. He put forward the universal status of these laws as grounds for the universal (though subjective) validity of human knowledge.

Fries believed that his attempt to explain human knowledge by an account of the structure of the human mind had a chance at success because of the severe restrictions he put on the knowledge that he would explain. He contrasted two conceptions of truth: transcendental truth, which defines truth as "the correspondence of a cognition with its object," and empirical truth, or truth for the understanding, which he explicated in terms of consistency within a given consciousness. In accordance with transcendental idealism, he adopted the empirical and, as he put it, "subjective" conception of truth. He believed that his anthropological approach allowed him to establish universally valid a priori truths because it limited its aspirations to a subjective conception of truth (1:343, 346-52). Knowledge of the universal laws of the causal genesis of experience (the laws of original formal apperception) provide the basis for universally valid judgments that hold good for all experience.

Fries was in agreement with the driving force behind Kant's celebrated "Copernican revolution." A motivating force in Kant's acceptance of transcendental idealism was his belief that no adequate justification could ever be given for positing a "harmony" between the concepts and forms of intuition of the knowing subject and the properties of things in themselves. With respect to space, this meant that there could be no adequate justification for ascribing spatial properties to things in themselves. According to Kant's Copernican revolution, any correspondence between mind and object could be explained only on the supposition that "objects must conform to our cognition."[15] A "harmony" could be posited between the forms of intuition and the properties of objects, but only at the price of limiting the extent of human cognition to possible objects of experience, or to objects considered as objects for a subject, that is, as objects of representation. Fries accepted Kant's teaching on space and geometry without quibble, arguing that an appeal to pure intuition is sufficient to ground the universal validity of geometrical demonstration precisely because, in accordance with transcendental idealism, no claim is made that the spatial properties described by geometry apply to things as they are in themselves.[16]

In the end, Fries's transcendental idealism became indistinguishable from the skepticism he hoped to avoid. It is true that he purported to explain the possibility of intersubjective agreement on geometry and the causal law by contending that each person has the same original formal apperception. But this could explain only the possibility of agreement on pure geometry or on the abstract principles of nature. Fries was forced to limit the application of these laws to the domain of each subject's consciousness; his notion of empirical truth pertains "only to the inner history of my cognitive acts [*meines Erkennens*]" (2:346). Although he might believe that his work could convince each reader that the same held for him or her, and thereby that each was entitled to universal and necessary truths within the domain of the "history" of his or her own mind, he explicitly forsook the attempt to establish the possibility of intersubjective knowledge of objects.

Fries may be seen as attempting to resolve a tension present in Kant's own thought. Kant had asserted the possibility of intersubjectively valid knowledge of objects for a "consciousness in general," by which he meant a consciousness limited to no particular subject, and thus potentially assumable by all subjects. He envisioned the possibility of objectively valid knowledge of a shared world. This claim has seemed difficult to establish from Kant's perspective, by virtue of Kant's own acceptance of transcendental idealism. Kant may have thought he was entitled to claims of a common objective world because of his commitment to the existence of things in themselves. His conception that the thing in itself is available at the epistemic boundary of knowledge and his consistent tendency to speak of the thing in itself as the cause of our subjective sensory manifolds open up the possibility of a common world shared by subjects who causally interact with it. Whether or not we treat this position as Kant's considered view, Fries rejected not only this strong version of objective grounding but even the weaker view that empirically given outer objects can be known to stand in a causal relation to the organs of sense. As a result, Fries re-interpreted "consciousness in general" to mean consistency within a single consciousness, contrary to Kant's usage.

Fries's wholly subjective conception of truth, and his corresponding treatment of space as merely subjective, were rejected by the remaining authors to be examined in this chapter, each of whom sought to establish that the subjective space of individual perceivers corresponds to the objective space of the external world, thereby establishing a "harmony" between our forms of intuition and the determinable properties of material objects.

1.2 Herbart: Metaphysics and Realism

Herbart was at Jena simultaneously with Fries and shared the latter's displeasure with Fichte's brand of Kantianism.[17] Like Fries, Herbart was critical of Kant on methodological grounds. But whereas Fries portrayed himself as developing Kant's latent psychological tendencies, Herbart considered Kant's (alleged) attempt to found philosophy on psychology to be among his greatest failings. Herbart believed psychology had a role to play in philosophy, but not that of foundation; indeed, he contended that psychology itself must be based upon metaphysics. Further, whereas Fries's methodological reformulation of critical philosophy led him to

change little of the positive doctrine, Herbart proposed a radical alternative to the content of Kant's transcendental metaphysics, adopting a
realist stance toward "things in themselves."[18]

Herbart's work is best approached through his claim to a new methodology. Herbart characterized the aim of philosophy as the reform—
literally, the "reworking" (bearbeiten)—of concepts.[19] Philosophy begins
with the "givens" of experience, which include a spatiotemporally
ordered world of objects undergoing alterations. Ordinary consciousness—perhaps guided by past philosophical systems—applies various
concepts (such as "cause" or "space") to these objects and their alterations. Philosophy goes to work on these concepts, finding the latent
contradictions in them and reworking them into a consistent picture of
the world sufficient to account for the givens of experience.[20] In so doing,
philosophy discovers and is guided by the limits to what can be thought
as possible, or can be seen to be impossible. Discovery that some
conceptions (e.g., that substance is not permanent) are impossible lets us
apply the principle of contradiction to establish the necessity of the
opposite.[21] This process is not a mere analysis of concepts; it is the
reworking of concepts into an articulated system that can account for the
givens of experience. Through the repeated application of the principle
of contradiction, some concepts will be replaced, altered, or seen to have
a meaning other than that which was originally supposed—e.g., the
causal relation is found not to be a concept, but a "necessity of thought,"
whereas "alteration" is the proper concept to apply to sequences of
change within experience.[22]

In taking concepts for his primary object of philosophical investigation, Herbart was careful to set them apart from objects of psychology. In
effect, Herbart invented the notion of "psychologism" without coining
the term. He contended that the notion of "thought" may be regarded
from two points of view: as the activity of the mind, and as what is
thought thereby. Concepts (Begriffe) apply to the latter, which is the object
apprehended through the concept (das Begriffene). Philosophy in particular, and logic as its subdiscipline, concern themselves with the
Begriffene. According to Herbart, logic must abstract from the manner in
which we feel, produce, or reproduce thoughts; it must ignore psychology, and divorce itself from what he termed "the natural history of the
understanding" (that is, it must reject not only, let us say, Hume and
Lossius, but also, by Herbart's lights, Wolff, Kant, and Fries).[23] Logic itself
pertains to the "formal" relations among concepts as these are manifested in judgments and syllogistic inferences.[24] Philosophy reworks the
concepts of ordinary consciousness into a coherent picture of the world.
In effect, it reworks our conception of the objects of these concepts,
producing a conception of the world itself. Herbart saw no reason not to

treat that conception as a description of the world as it is in itself.[25] Herbart's attitude seems to have been that if we employ our minds to the best of our ability, we should take the product to be knowledge of the world, not knowledge of the merely transcendentally ideal.

Comparison of Herbart's project with Kant's rejection of empirical psychology as a foundation for philosophy suggests that Herbart was not so different from Kant as he may have hoped. Herbart's project may be compared with the "objective" side of Kant's investigation, that is, his examination of the "objects of pure understanding," in order to "render intelligible the objective validity of its a priori concepts."[26] Both authors sought to discover a set of principles not justified solely on the basis of experience but nonetheless applicable to all experience. They sought to determine principles that apply to objects, or, at one step removed, to determine the epistemic status of such principles; neither took the knowing subject as his primary object of investigation. Nonetheless, they differed fundamentally on the place of "subjective" considerations in philosophy. Even though Kant did not make the transcendental subject his primary subject-matter, he nonetheless found it necessary to refer to various characteristics of the transcendental subject in his deduction, relying particularly upon the doctrine of (transcendental) faculties and their characteristic operations. Herbart attacked Kant at just this point. He considered the doctrine of the faculties to be a psychological assumption surreptitiously imported into philosophy. Herbart set out to base his philosophical findings on only the "necessities of thought" as applied to the interplay among concepts, the givens of experience, and the perception of what is possible and impossible. He turned to psychology only after his metaphysics had been laid down.[27]

So much for Herbart's method in the abstract. When applied, it led to some startling results. Although Herbart began with the spatiotemporally ordered world of ordinary consciousness, his reworking of concepts took him far afield. His analysis of the concept of "the real" underlying existence and change led him to believe that the universe is populated with atomistic beings, each originally possessing but a single quality. He named these individual beings, or essences, "Reals."[28] He postulated the need for such "Reals" through a series of arguments reminiscent (as he was the first to point out) of Eleatic arguments for a permanent substratum underlying change.[29] Following Leibniz, he argued that the "Reals" underlying the aggregate of things found in experience must themselves be simple, akin to Leibniz's monads (as, again, he was the first to observe). But he disagreed with Leibniz in that he posited real causal interactions among Reals in order to explain their internal changes of state. His monads, it would seem, had doors at least, and perhaps windows, too.[30]

Herbart's conception of what is real led him to reconceive both the place of geometry in philosophy and the status of spatial representation as representative of reality. He chastised Kant for taking geometry as a given for the purposes of metaphysics, and he argued that metaphysics must first evaluate the status of space and its geometrical description before geometry can be taken as a description of the real.[31] Herbart was not questioning the validity of geometry as a description of the continuous, nor was he questioning the self-sufficiency of geometry as science. What he placed in question was the relationship between geometry and the real. Only an answer to whether the "real" is actually continuous could indicate whether geometry actually describes the real or is instead limited to its own special subject-matter.

As it turns out, Herbart maintained that space is not an inherent property of the Reals; in themselves, they have neither spatial nor temporal determinations.[32] Moreover, he contended that the "given" concept of space is not of a continuum, but of a set of discrete points; the continuum is a derivative intellectual construction. Continuous space and its description, geometry, are mere "helping concepts" used to describe the actual relations among the Reals; these actual relations are causal, not spatial.[33] But these helping concepts are not mere fictions; space captures actual relations among Reals mirroring the order of "next-to-ness" defined by actual or potential causal interactions among Reals.[34] Herbart considered it necessary to use continuous mathematical functions to describe these causal relations, leading him to conclude that the continuous must necessarily apply to these relations among Reals. For this reason, he described what might be termed the "causal space" of the Reals as an *intelligible space*, to which the mathematics of the continuum applies. He contrasted this intelligible space, as postulated in metaphysics, with the *sensible space* of phenomenal experience, as described in psychology.

In Herbart's view, the "given" concept of space with which philosophical speculation must begin posits discrete parts: it is the concept of the "outside one another," and its parts are points.[35] Herbart was also convinced that continuous space cannot be constructed from points and that the concept of space appropriate to both geometry and physics requires continuity because its points "flow into one another." His commitment to the latter arose from reflection on the foundations of the calculus and on the application of continuous functions to causal processes. According to Herbart, the continuity of both intelligible space and sensible space arises through a consideration of continuous functions.[36] Or rather, the metaphysical notion of continuity arises through the contemplation of continuous functions, and the psychological representation of space arises through the instantiation of continuous functions in the interaction among sensations.[37] Thus, even though Reals themselves are atomic, so that a sequence of them would not form a continuous line but would be composed of simple "points," nonetheless the continuity of geometrical space describes an actual feature of Reals: their ability to be more or less "next to" one another in continuously varying degree of causal influence.[38]

Herbart's treatment of space brings his difference from Kant into relief. He accepted the distinction between intelligible and psychological (or subjective) space as found explicitly in Kant and implicitly in Leibniz, but he sided with Leibniz against Kant on the question of the existence of intelligible space. Whereas Kant had held that "intelligible space"—a space existing independently of human perceivers and populated by things in themselves—is for us unknowable, Leibniz (according to Herbart) thought that monads exhibit real spatial relations among themselves.[39] Herbart interpreted the notion of "intelligible space" not as a spatial container, nor as a set of real spatial relations, but as a description of actual or potential causal relations among Reals. This intelligible space serves as the ground of the space we intuit, i.e., psychological space.

Herbart disagreed with both Leibniz and Kant on the apriority of space as a form of representation. Leibniz, he argued, was committed to the apriority of space by virtue of his doctrine of preestablished harmony: since monads have no causal interaction with other monads, their perceptual representation of the spatial relations obtaining among fellow monads can occur only because the a priori ground of these sensory representations is found in the monad itself.[40] Kant, who denied the knowability of any such "harmony," was committed to the apriority of space as a form of representation because he needed a foundation for the necessity he found in Euclidean geometry.[41] Herbart based his denial that space is an a priori form on his doctrine of the simplicity of Reals. He maintained that those Reals we call souls (those having perceptual capabilities) are, like all other Reals, essentially simple, without any manifoldness. At least originally, a soul-Real can intuit in a single moment only a single quality (a tone, a color, etc.) of determinate intensity. But since a truly simple representation can include no representation of relations within it, it cannot include a representation of spatial relations.[42] Therefore space cannot be an a priori form of intuition.

On Herbart's view, space is an acquired form of representation. Whereas Kant and Fries had rejected any attempt to describe the relationship between mental representations and things in themselves, Herbart the realist pressed ahead with just such a description. He maintained that the ability for spatial representation is acquired as a result of causal interactions among soul-Reals and other Reals in intelligible space; psychological space is built up through interaction with the real (causal) relations found among things in themselves. He supported his position with metaphysical arguments from the simplicity of Reals, relegating the subsidiary task of describing the actual process of acquisition to psychology.

Herbart's psychology is noted for his firm rejection of faculties; indeed, he compared the notion of a faculty in previous psychology with that of phlogiston in eighteenth-century chemistry, and forecast a similar demise.[43] Instead of assuming certain faculties innate in the mind, he maintained that all psychological activities must be built up from the properties of the soul-Reals. The system Herbart developed relied heavily upon principles of association (or principles of reproduction, as he called them) and thus had affinities with association psychology. But there were two fundamental differences between Herbart and associationism. First, consonant with his rejection of mental faculties or powers, Herbart derived the tendency of mental representations to become associated with one another—so that the occurrence of one representation leads to the reproduction of a second representation—from the representations themselves and their instantiation as states of a Real, rather than from a power of association supposed to inhere in the mind itself. Second, his appeal to association did not constitute an attempt to explain the mind in the spirit of methodological naturalism (as discussed in chapters 2 and 3); Herbart had no intention of reducing thought to sensibility and imagination, or indeed to any other psychological process. Concepts are logical ideals that are not to be identified with naturally occurring representations. He conceded that, through a process of association in which the variable and the particular cancels itself out, a complex of representations could come to play the role of a general concept; but no complex of representations may be equated with a general concept, except as an imperfect approach to an ideal.[44]

Herbart's conception of mental processes was cast in the language of mechanics. The atoms or elements of his mental mechanics are representations (*Vorstellungen*), which are states of Reals. His metaphysics led him to conclude that representations are in conflict with one another: because soul-Reals are simple, they can be fully conscious of only one representation at a time. Yet the physiological state of the nervous system may affect a soul-Real in many ways simultaneously, and thus arouse many representations in a soul-Real at once. These representations all strive to come to consciousness. In so doing they inhibit one another. Herbart found it convenient to treat each representation as a force, and to express mathematically the relations among simple representations striving to realize themselves in consciousness. He characterized this striving as a raising and lowering above the threshold of consciousness, and he described the interactions among representations as a mental mechanics.[45]

Herbart proposed that the laws governing these interactions are those of complication and fusion. In reality, he contended, only one representation can be present in consciousness at any given instant. But a variety of representations may, in that instant, be "striving" to enter consciousness; they constantly inhibit one another, one driving another below the threshold of consciousness or else being driven below the threshold itself. Among the various "laws of reproduction" governing the interaction of simple representations, the most important for the development of psychological space is that governing the formation of series. The general law involved here is fusion (*Verschmelzung*): if two representations of similar qualitative character are presented in temporal contiguity, they become connected or fused together without either one's completely losing its identity. Further, such connections can be chained. Thus representations *a* and *b* might become fused; but then *b* might become fused with *c*, forming a series *abc*, such that *a* is connected with *c* only by virtue of its connection with *b*; the fusion of *c* and *d* would add a fourth element, and so on.[46]

Herbart maintained that the key aspect of space, and therefore the key feature of the representation of space, is extension along a dimension. The psychological basis of the representation of an extended dimension is the formation of a series with a reversible order: a series that has been connected so that the introduction of *a* leads to the reproduction of *b*, *c*, and *d* (in that order), while the introduction of *d* leads to the reproduction of *c*, *b*, and *a*. Through building up various side chains of representations, a two-dimensional and, eventually, a three-dimensional manifold can be developed, such that *a* is connected not only with *b*, *c*, and *d*, but also with *b'*, *c'*, and *d'*, as well as *b''*, *c''* and *d''*, each of which also is connected with a reversible series. The interconnections are consistent with the spatial relations that would obtain among each representation if it were an atom contained in a three-dimensional manifold.[47]

Herbart described a psychologically real process by which spatial representations are constructed. If a soul-Real cannot originally represent spatial manifolds, it follows that the eye cannot originally provide representations of a spatial manifold in either two or three dimensions. Herbart stated this outright: "*The original apprehension of the eye cannot be spatial. For the perceptions of all colored locations converge in the unity of the soul and in this every trace is lost of right and left, above and below,* etc., which found a place upon the retina of the eye."[48] Arguing from the original nonspatial simplicity all representations, he rejected the claims of those who believed that the optical system provides spatially extended mental representations just because the retina and the nerves leading to the brain are spatially extended.[49] The soul is unitary, and it is only by fusing simple representations into series that a spatial manifold can be represented. The actual operation of this process of fusion results from the movement of the eye (or the finger, or some other organ of sense).

> But in seeing the eye moves; it changes the center of its visual field. Along with this movement there is constantly connected [1] a fusion of the representations obtained, [2] an incitation of those which are strengthened by perceptions of what lies outside the middle of the field of vision, and [3] an innumerable multitude of reproductions interlacing one another.[50]

The eye, in sweeping a stationary visual field and retracing its steps now and then, forms reversible series of associations that serve as the foundation for the apprehending of a two-dimensional manifold.

Causal commerce between soul-Reals and other, independently existing Reals (among which intelligible spatial relations obtain) serves to establish the correspondence between intelligible space and psychological space. It is because the eye (or finger) sweeps over an actual spatial manifold, which admits of being swept back and forth, or stroked, that various representations can become fused in reversible series. Herbart summarized his teaching on the relation between intelligible space and psychological space as follows:

> In connection with the whole of metaphysics, it can, by the way, specifically be maintained that we perceive external objects as spatially ordered because they actually are spatially ordered. For this law of reproduction [underlying spatial representation] depends upon multi-gradated fusions; the fusions depend upon perception; now whence comes the perception of these gradations? From general metaphysics it is known that nothing can be pre-arranged in this regard, that on the contrary perceptions are governed through disturbance of the soul by substances [Wesen] different from it; and that in these disturbances there can be no other regularity than such as might be grounded external to and independent of the soul; finally, it is known (also from metaphysics) that these beings must be granted an intelligible space in which they move, and that in accordance with these motions there occur perturbations among themselves and also those disturbances that the soul undergoes. Through these means the spatiality that belongs to the substances (though not as a real predicate) is determined, as well as the apparent spatiality which the soul is obliged to assign to its sensory representations.[51]

The law of reproduction, by which the series that serves as the foundation of spatial perception is formed, operates only if the requisite regularity is found in the representations actually experienced; in other words, series of fused representations result only if elemental sensations present themselves in proper order. But metaphysics teaches us that the causal ground of such actual chains of representations cannot lie in the soul, but must be external to it, in other beings. Metaphysics also teaches that these other beings themselves in fact do possess an intelligible spatial order among themselves (grounded in their causal interactions). The regularity found in actually experienced chains of representations must be accounted for by the properties of the beings that act upon the soul and cause it to have those representations. Although spatial predicates do not properly pertain to the "Reals," Herbart ascribed them an intelligible spatial order in virtue of their causal connectedness; it is (only?) this spatial order that provides opportunity for the experiencing soul, through the movement of an organ back and forth, to obtain reversing series of representations, which can then be fused in order to determine a spatial manifold.

This elaborate mechanics of spatial perception may seem to belie two of my claims about Herbart. I claimed that Herbart did not attempt to naturalize the mental, yet what else could a "mechanics of thought" be, besides a naturalistic psychology? And second, Herbart's claim to explain the "correspondence" between psychological space and intelligible space may after all be on the same shaky ground as the attempt to establish geometry empirically.

Neither of these objections hits the target. The mechanics of thought was indeed a species of naturalism: it was an attempt to describe the natural laws which govern how the mind—or, better, the soul—works. But it does not serve a foundational role with respect to the analysis of thought. In Herbart's view philosophy and logic are prior to psychology; the philosophical concept of space is developed through the reworking of common or ordinary notions in connection with the givens of experience. Psychology is left to fill in certain questions that metaphysics left unanswered, questions regarding the natural occurrence of spatial representations. But psychology is irrelevant to the evaluation of logical reasoning or philosophical propositions. Furthermore, the mechanics of spatial representation is not intended to *justify* the claim that sensory space represents intelligible space. Although it provides an explanation of how the relation between the two could be established, the justification (and interpretation) of the claim that there is an intelligible space distinct from and represented by sensory space comes from metaphysics and is independent of psychology.

This is not to say that there are no apparent problems with Herbart's account. In particular, the attempt to derive spatiality from nonspatial representation is problematic. Herbart denied all innate structure to the mind, including faculties and laws of association. He attempted to derive the laws of association from the properties of the single soul-Real itself: the tendency of its states (representations) to preserve themselves in consciousness against the encroachment of other representations that are themselves seeking to come to consciousness. It is through the law of reproduction that the series of representations underlying spatiality arises. But doesn't this amount to building a space-producing process into the notion of reproduction, thus making a sham of the purely empirical derivation of spatiality? Herbart would have replied, I think, that the law of reproduction is not a special law of spatiality, and that only when a series becomes connected in a special way—becomes a reversible series—does spatial representation occur. And this, he would have argued, happens only when reversing chains of representations are experienced, an event which, if it is to occur other than fortuitously, depends upon contact with external beings that possess an actual spatial order among themselves.

This reply does not render the empirical derivation of spatiality plausible. For even if we grant Herbart's argument that interaction with a spatial world is a necessary condition for explaining the etiology of reversing chains of representation, why should reversing chains in themselves be sufficient to produce a representation with the *phenomenal character* of space? Such chains could easily *represent* spatial relations, just as a series of tones—or any series whose members can be brought into one-to-one correspondence with a set of spatial locations—might convey information about spatial relations by, say, pitch and sequence. But although the pitch and sequence of the tones would bear information about spatial location, and in that sense "represent" spatial configurations, the experience of these tones would not in itself reproduce the experience of space. Why should the members of Herbart's reversing series be any different? Why should they be perceived as not only temporally contiguous but also as being simultaneously next to one another (as seems to be required of a properly *spatial* representation)?

Herbart might well have accepted the implications contained in this objection and still stood his ground. According to his account, the representations do not simultaneously co-occur as a state of the soul-Real. Relying upon the notion of infinitesimals, Herbart explained that the soul-Real takes on the various states in a reversible series in sequence, and does so in a time smaller than any assignable time. The result is a succession of representations that "is not a represented succession."[52] The appearance of a simultaneously co-occurring spatial order is a kind of illusion. Perhaps for this reason, Herbart maintained that the representation of next-to-ness is better called a concept (*Begriff*) than an intuition

(*Anschauung*).[53] Spatial apprehension (*Auffassung,* as Herbart tended to call it, rarely even using the word *Anschauung*) is the result of a constructive mental process. In this manner, Herbart denied the traditional Kantian distinction between intuition and concept. He assimilated spatial representation to the same constructive processes found in the psychological counterpart to concept formation.

The treatment of the phenomenal extendedness of space as an illusory appearance is counterintuitive (and doubly so). In the end, the acceptability of this and other results must turn upon the grounding of Herbart's metaphysics. Herbart attempted to justify his metaphysical method and its results through a long series of arguments, but here two remarks must suffice. First, the appeal to perceptions of possibility and impossibility in developing a metaphysics of "Reals" may seem like a regressive step toward the real use of the intellect, the position against which Kant had argued so effectively. But, second, it may not be such a step, for Herbart did not propose to base metaphysics on a direct intuitive grasp of the essences of things; rather, he proposed a "reworking" of concepts, starting from the "given" concepts of common sense. This "reworking" was guided by the ability to find "contradictions" or "impossibilities" in ordinary notions, an ability that remained unexplained. Even so, Herbart's project constitutes a step away from the real use of the intellect. Moreover, contrary to Hume, or even Fries, Herbart did not simultaneously step toward psychologism. His work is an early expression of the idea that philosophy is "conceptual analysis" in which the ontology of concepts is unimportant and the mode of investigation is neither a priori (in either the traditional sense of the real use of the intellect, or in the Kantian sense) nor empirical (in the way that naturalistic psychology purports to be empirical). It may itself be seen as a "reworking" of the commonplace conception, expounded by Kant, that philosophy is the analysis of concepts.

The contrast between Fries and Herbart effectively crystallized the distinction between those who would base philosophy on psychology and those who would give it its own independent basis. Herbart was by far the more influential, providing a viable alternative to the dominant Hegelian philosophy in the decades prior to 1850. But those who made use of his work did not always adopt his methodological dictum that philosophy and metaphysics are logically prior to psychology. Indeed, his followers and imitators typically endeavored to develop his psychology, conceived as a branch of natural science, into a general science of human thought that would be propaedeutic to philosophy. This was true not only of his avowed followers, such as Theodor Waitz, but also of philosopher-psychologists such as F. E. Beneke, who openly contested the extent of Herbart's influence on their philosophical thought while at the same time applying and extending his psychological system.[54] Beneke, in fact, aligned himself with Fries philosophically in contending that observation through inner sense is the primary method of philosophy. But he went further than Fries in comparing the results of this observation to the findings of natural science. He maintained that philosophy should be founded upon natural scientific psychology; indeed, he went so far as to contend that psychology properly belongs at the center of philosophy (like the sun at the center of the solar system) and that the other branches of philosophy are, therefore, instances of applied psychology.[55] The Herbartian side of Beneke's work expressed itself in the character of this psychology, which was built up from associative laws of "fusion."[56]

Thus, while some philosophers attempted to apply the sure techniques of natural science to the foundations of philosophy, others posited a strict separation between psychology on the one hand and the theory of knowledge on the other. Physiologists and psychologists joined both camps.

2 The Physiology and Psychology of Spatial Realism

As the cases of Fries and Herbart alone would reveal, Kant was a dominating presence in philosophy during the decades following his death, even if his work provoked a variety of responses. He was hardly less visible within physiology and psychology—two disciplines[57] that were grappling with their own conceptual foundations during that period— and the responses to his work were hardly less varied. Some physiologists and psychologists sought either to refute or to extend his work directly; others endorsed philosophical positions opposed to Kant's, specifically Herbart's.

Early nineteenth-century German physiology and psychology stood in a twofold relation to Kant. First, sensory physiologists and psychologists read Kant's doctrine that space is a form of intuition as a psychological thesis, or at least as a thesis that could have psychological (and sensory physiological) implications. They then attempted to extend, or perhaps to rebut, Kant's teaching as they understood it, regarding Kant as the purveyor of an influential doctrine pertaining to the senses.

But, second, sensory physiologists and psychologists owed another, and in many cases a deeper, debt to Kant. This debt pertained to the growth of physiological thinking itself, and in particular to issues involving teleology and reduction. In the eighteenth century it had been commonplace to see the natural order as a divine creation, to understand the organization of living bodies as the product of design, and perhaps to attribute immanent rationality to a vital power that directs bodily processes.[58] This conception came into tension with the growing acceptance of Newtonian cosmology, which, when strictly adhered to, seemed to demand that all natural phenomena be reduced to mechanics. But excluding considerations of design and purpose from the natural order proved to be problematic for physiologists, who found it impossible to explain the organization of living things as the result of blind mechanism.[59]

The problem of reconciling purpose in and lawfulness of nature had been addressed by Kant in his *Critique of Judgment*.[60] He contended that although we do not have epistemic access to God's purposes or intentions (or even his existence), we nonetheless cannot do without the notion of purpose or design in seeking to understand living things. He proposed to treat finality as a problematic but indispensable aspect of our understanding of nature, divorced from any direct consideration of the purposes of a deity.[61] Teleological considerations would be brought to bear on natural phenomena to the extent that they seemed to be demanded by and justified through the investigation of actual natural forms and processes. The investigator would treat things in nature *as if* they were the product of design, and would attribute to organisms a "formative force" that directs their development. This force is over and above purely mechanical laws, although all the processes that it directs within the organism are nonetheless subject to the laws of mechanics.[62] Teleology and mechanism are distinct but compatible aspects of nature.

Kant was not alone in attempting to reconcile teleology with mechanism. German physiologists contemporary with Kant, and especially J. F. Blumenbach and his circle, were deeply concerned with the problem.[63] Some contended that the organization of living things could derive only from divine design, while others conjectured that it might ultimately be explained through the special "elective affinities" of organic matter; but (nearly) all parties agreed that it was best to avoid speculation about origins in the actual study of functionally organized systems.[64] Organization could be observed, and it could be observed to be passed from generation to generation; investigation would be limited to these manifest phenomena and their organismic basis. The organization itself was to be understood in terms of functioning systems and subsystems, each playing a part in the economy of the organism as a whole. Functional organization was accepted as a primitive or irreducible feature of living things; it was an irreducible component of nature, and of a naturalistic viewpoint. The job of the physiologist was to analyze this functional organization.[65]

Physiologists and psychologists of the senses commonly spoke of the "function" of the sense organs, and analyzed the manner in which this function was carried out.[66] But what, precisely, is the function of the sense organs? This question was usually answered unambiguously but abstractly: the function of the sense organs is to provide representations of the external world in order to guide the organism toward benefits and away from harms. Vision was typically praised for its power in performing this role, for it provides contact at a distance (through the mediation of light).

The general function of the senses is and may have been clear enough, but the specific manner of performing this function by a given sensory system required and received deeper analysis. The anatomical and physiological structures manifestly subserving a function, such as the optical apparatus of the eye or the neural pathways underlying single vision, were investigated, but so were the mental mechanisms underlying the achievements of the sensory systems. Although the question of the status of mental functional organization was explicitly addressed only on occasion, it implicitly conditioned the physiology and psychology of vision throughout the nineteenth century. The functional status of spatial representation is of particular interest. For if the function of vision is to put us in contact with distant objects, then spatial representation and spatial localization must be regarded as functional achievements subservient to this end. It then becomes reasonable to ask what explanation should be given to the mental side of this function.

In the first quarter century after Kant's death, two works attempted to explain an aspect of the functionality of the senses that Kant would have deemed inexplicable: the relationship between space as a form of intuition and the spatial properties of objects conceived as things in themselves. In his *Beytrag zur Physiologie der Sinne* (1811), Steinbuch proposed to explain the acquisition of spatial representation and spatial sensing wholly in terms of the effects of experience. His account of spatial representation and spatial perception was radically empirical, or empiristic. By contrast, Tourtual, in his *Sinne des Menschen* (1827), developed an account of spatial representation and spatial perception that attributed these abilities to innate physiological and psychological structures. While Tourtual regarded his own radically nativist account as an extension of the physiology of the senses along Kantian lines, Steinbuch took himself to be refuting Kant.

Their aims differed from Johannes Müller, a physiologist who resisted the temptation to regard the physiology of the senses as having a direct bearing on questions of transcendental idealism and realism. Although Müller has been read as a kind of physiological Kantian, especially with respect to his famed doctrine of specific nerve energies, in reality he had a more sophisticated attitude toward the relation between physiology and psychology than the one usually attributed to him. Like Steinbuch and Tourtual, he opposed transcendental idealism and was a realist. Unlike them, he produced a philosophical argument in support of his realism.

2.1 Steinbuch's New Theory of Vision

Steinbuch's *Beytrag zur Physiologie der Sinne* was the most comprehensive attempt to develop a physiological and psychological theory of the senses in Germany during the first two decades of the nineteenth century. Working within a framework of association psychology, he focused especially upon the active use of the sense organs in connection with their musculature, and he provided a more thorough analysis of the contribution of muscular activity to spatial perception than had yet been achieved. Although later writers would quibble with details of his theory of muscle sense and some would shy away from the wide explanatory scope he attributed to it, the basic elements of that theory were nonetheless accepted by authors who disagreed with his general account of spatial vision. His theory provided the basic framework for the psychology of vision for the decades following its publication.[67]

Steinbuch studied medicine at Erlangen from 1797 to 1800, also taking courses in anthropology and metaphysics from J. H. Abicht, a self-avowed follower of Kant. With F. G. Born, Abicht edited the *Neues philosophisches Magazin* (1789-91), which was devoted in part to advancing Kantian philosophy through empirical psychology. Abicht himself developed an argument for transcendental idealism based upon a comparison of the "structure" and "organization" of the human sense organs with those of other animals.[68] Steinbuch agreed with his teacher that psychology was relevant to the fortunes of Kant's philosophy, but he did not think that, when properly applied, it would advance Kant's cause. He contended that a proper psychology of the senses would show that space is not an original but an acquired form of representation, and that the process of acquisition depends upon the activity of the motor nerves controlling the musculature of the sense organs.

Steinbuch was perhaps the most radically "empirist" sensory psychologist of his century. Indeed, it may at first be difficult to see just how radical his theory of spatial perception was. He attributed all spatial abilities to learning. In the Kantian terminology Steinbuch himself adopted, he contended that space as a form of intuition is acquired. Thus, not only did he hold that we "learn to see," he also maintained that we must learn how to represent things spatially at all; according to him, all sensations are originally nonspatial, and the mechanisms or process by which they are formed into spatial representations are wholly the product of experience. He denied that a representation of the two-dimensional retinal image is primitive in vision. Indeed, he believed that no primitively spatial representations are received by any sense, including touch, and that the ability to represent two dimensions in any manner whatsoever must be acquired.

Although Steinbuch proposed a radically empirist theory of the genesis of the abilities for spatial representation and spatial localization, in other ways he did not alter the central features of the standard analysis of vision developed in the eighteenth century. What made his theory different was that he required an explanation for various representations and processes that other theorists had accepted as given and had treated as theoretical primitives; he attempted to derive the "givens" of previous theory from even simpler elements. Thus, although he accepted the assertion, common among authors from Berkeley to Kant, that the visual perception of depth and distance is acquired by associating visual sensations with tactual perceptions, he refused to accept the spatial ability of touch as simply given. Further, even though his account of vision gave prominence to the representation of the two-dimensional visual field, he required that the existence of this representation be explained. Finally, although his account of the use of "cues" in the perception of the three-dimensional visual world followed previous theory with little change, he rejected the standard "judgmental" account of such processes as recounted by Gehler and adopted a pure associationist theory. Moreover, while recognizing that many previous theorists had considered the laws of association to be primitive, not admitting of deeper explanation, he speculated about just such an explanation.

Steinbuch's relentless search for the simplest possible theoretical primitives was an extension of the dominant mode of natural-scientific explanation in the eighteenth century: the explanation of natural objects and events by analyzing them into a few simple elements governed by a few simple laws. Hartley, Hume, and Lossius had already applied this type of explanation to mental phenomena. Steinbuch carried it to an extreme. According to him the mind's bare capacities included only the ability to have simple, nonspatial sensations, and to associate them according to the usual laws. This aspect of his theory, his radical empirism, cannot be explained simply as a result of his "analytical" explanatory bent, for others who accepted the analysis of sensory processes into single sensations and processes for combining them did not adopt radical empirism.[69] The origin of Steinbuch's radical empirism is to be found in his brand of physiology; more particularly, in his subscription to organic materialism. Steinbuch subscribed to the physiological program of Johann Reil, who sought explanations for all physiological phenomena through the organization of matter. In applying this program to psychology, Steinbuch contended that the faculties of the mind must be explained through the development of the body. He did not deny the existence of the mind or soul, but he did deny that it was "preformed" (just as he denied bodily preformation).[70] Originally, the mind has only the bare power of receptivity—that is, the power to receive impressions from the brain. All its other capacities and abilities are to be explained through the interaction of this original power with the developing body, interaction that stimulates the mind and provides it with *in utero* training. The development of space as a form of sensibility occurs through this training, contrary to Kant's teaching that the form of intuition is given a priori.[71]

The novel portions of Steinbuch's theory of vision can best be understood against a slightly deeper analysis of previous theory. In German sensory physiology and psychology, the distinction between the spatial form of perception and the nonspatial matter of sensation was widely accepted. Anyone who adopted the position that vision begins with a two-dimensional representation was (at least implicitly) committed to an innate mechanism for organizing the "matter" of sensation into this spatial form. Steinbuch himself believed that this standard commitment to a primitive two-dimensional representation continued to survive because it was mistakenly equated with a widely accepted assumption about the relation between nerve fibers and sensations: the assumption that single sensations arise from the stimulation of single nerve fibers. Steinbuch accepted this assumption but he rejected the further assumption that the spatial organization of sensations in perception is determined by the spatial organization (in the brain) of the fibers causing the sensation. The latter assumption was at work in Gehler's *Dictionary*, to which Steinbuch referred.[72] According to this type of account, the original two-dimensional apprehension of an image in point-for-point correspondence with the retinal image results from a point-for-point neural projection from the retina into the brain; the spatial relations in retinal stimulation are preserved as they are transmitted into the brain, there giving rise to a spatially articulated set of sensations. Similarly, the spatial organization of tactile nerve endings in the brain were presumed to preserve information about the relations of their endings in the skin, and thereby to produce spatially organized sensations.

Steinbuch objected that this account confounded different levels of description and analysis by equating a physically extended brain image (at the termination of the optic or tactile nerves) with a mental representation of qualities arranged in space. He protested that "the difference between the physical-spatial action of the brain and spatial representations in the mind is just as large as the difference between brain activity and ideas in general" (147). A spatially organized pattern of brain activity could not acceptably explain spatial representation in the mind.

Steinbuch believed the key to explaining spatial organization would be found through analysis of the motion of the sense organs and their attendant musculature during perception. He argued that for each distinct position of a sense organ, such as the eye, or the limbs in their role as tactile organs, there must be available to consciousness a distinct idea representing the state of activation of the musculature. These ideas serve a dual function: they are the cause of muscular innervation, and as such are considered to be simple acts of will; but since a qualitatively distinct simple idea must exist for each state of muscular contraction that falls under voluntary control, these same ideas can also serve as representations of the state of the musculature.[73] From these qualitatively distinct motor ideas, which are themselves originally nonspatial but are correlated with spatially distinct muscular positions, Steinbuch would construct a psychological explanation of spatial representation and localization.[74]

His story of the development of spatial representation begins with the fetus in the womb. Through random firings of the motor nerves at the appropriate stage of development, a set of phenomenally distinct motor ideas, or (he termed them) "*Bewegideen*," arise in consciousness. (Steinbuch accepted the common assumption that any mental representation is "conscious," even if unnoticed.) These *Bewegideen* are not merely phenomenally distinct; their qualitative differences vary systematically. First, *Bewegideen* belonging to the same muscle are more similar to each other than they are to the *Bewegideen* of any other muscle; thus, for each voluntary muscle, there arises a phenomenally similar group of (individually distinctive) *Bewegideen*. Second, within the group of *Bewegideen* belonging to a single voluntary muscle, the degree of similarity or affinity among the various ideas is ordered according to the concurring degree of contraction of that muscle. For each degree of contraction of a muscle there is a corresponding *Bewegidee*, and these are related in a qualitatively continuous series; thus, if the fetus happened to move its arm steadily in one direction by contracting the deltoid muscle—an action that Steinbuch assumed would require a continuously increasing degree of contraction—the corresponding series of *Bewegideen* would exhibit continuity in their successive phenomenal characters (33-35).

The systematic variation in phenomenal character of these *Bewegideen* provides the basis for the development of spatial representation in the course of *in utero* fetal motion. To explain how these originally nonspatial *Bewegideen* yield spatial representations, Steinbuch appealed to the usual laws of association. The intuition of the continuous series of *Bewegideen* that would arise from the contraction of (say) the deltoid muscle is conditioned by the associative law of similarity, which prescribes that ideas occurring simultaneously in consciousness will become associated in virtue of their qualitative similarity (35).[75] The ideas produced by steady contraction of a single muscle vary continuously in quality; therefore, they "must become present to consciousness as a continuous series."

> Such a series of closely related *Bewegideen* floating in consciousness can be for the mind nothing other than the original concept of *extension in length*. That is, each point of this series appears in intuition, through its own characteristic quality, by itself, that is, as present outside of and next to the other points of this series, and indeed as present outside the element of neighboring similarity according to the successive ordering, whereby these points must be intuited as lying next to one another along a direction or dimension. [36][76]

The associative law of similarity accounts for extension in one dimension. The law of simultaneity, another associative law, explains the development of spatial representation in two dimensions. According to this law, if two or more simple ideas come to consciousness simultaneously, they become fused together into a single idea. The products of such fusions are phenomenally simple, and they may enter into further activity as a unified element.[77] But the elements entering into the fusion retain their causal functions, so that the mind need produce only one compounded but phenomenally simple *Bewegidee* in order to cause the muscle contractions corresponding to its component ideas. The combination of simple ideas from neighboring muscles results in a qualitatively intermediate product which, when ordered in accordance with the law of similarity, falls between the linear frames of reference provided by the unmixed ideas arising from innervation of the individual muscles.

Here is how the fetus psychologically constructs an inner two-dimensional space in the case of vision. Imagine a hemisphere lying in front of the eye. It is formed by a sweeping motion of the optic axis (an imaginary line to which an arbitrary length has been assigned) as the eye randomly assumes all possible directional fixations. The first stage of constructing "inner" visual space amounts to the development of an inner two-dimensional surface corresponding to this external hemisphere. The elements in the process of construction are the *Bewegideen* of the four rectus muscles of each eye: they are combined according to the associative laws of similarity and simultaneity. To each point on the outer hemisphere there corresponds a specific directional position of the eye (although this correspondence is originally unbeknownst to the percipient); to this position there corresponds a degree of contraction of the four recti; and to this degree of contraction there corresponds either a simple or a compound *Bewegidee*, depending upon how many muscles are simultaneously at work. Thus for any given point on this external hemisphere, there is a phenomenally distinct *Bewegidee* that, if activated, would direct the optic axis toward that point.

A two-dimensional inner hemisphere arises through the associative connections among the *Bewegideen* produced by random *in utero* ocular motions. Four series of phenomenally continuous simple *Bewegideen* are produced by each of the recti, as each goes through its entire range of contraction; the associated *Bewegideen* of the four recti form four radial lines diverging from a central intersection and corresponding to increasing degrees of contraction of each muscle. Compound *Bewegideen* (which are phenomenally simple) have a qualitative character falling between those of the simple *Bewegideen* from which they have been formed, and varying continuously depending upon the strength of contraction of each muscle (so that the quality of the dominant muscle dominates the compound). The areas between the four radial lines are filled in with these compound *Bewegideen* that result from the contraction of two or more muscles. A two-dimensional array is thereby formed (158-62).[78]

The visual space that develops *in utero* through this process can be no more than a surface, for the eye's four recti require only such *Bewegideen* as vary qualitatively in two dimensions. The situation is analogous to that of the three shoulder muscles whose function is to rotate the arm about the shoulder and whose *Bewegideen* provide the foundation for tactual two-dimensional extension. The arm, however, has muscles whose function is to bend the elbow; their *Bewegideen* produce a three-dimensional tactual space *in utero*. Steinbuch maintained that there are no analogous muscles for the eye, so that visual space at birth is only two-dimensional (64).[79]

The newborn child possesses an inner visual space (an intuited extended surface) as well as a set of nerves in the eye that produce sensations of light and color when stimulated. Nonetheless, when it opens its eyes it does not experience light and color arranged in a surface. The child possesses both the form and the matter of vision, but it as yet lacks a mechanism by which the material sensations of light and color come to be intuited in the inner visual space. What remains to be established is a one-to-one correspondence between points of stimulation on the retina and points of the inner visual space; that is, a mechanism is needed to "place" sensations of light and color in the inner visual space in a manner corresponding to the physical pattern of light on the retina that produces the sensations. Having rejected Gehler's physiological account of this mapping, he attributed it to an acquired psychological mechanism.[80]

In order to understand how the point-for-point correspondence between retinal elements and regions of intuitional space arises, consider a case in which a single source of light, say, a candle flame, is present in the visual field. With the eye muscles contracted so that the eye is focused directly ahead, the *Bewegidee* lying at the center of the space of inner intuition stands out phenomenally within visual space. If we assume that the eye fixates the candle, an isolated point of illumination falls upon the fovea, stimulating a central foveal element. According to Steinbuch's account, this stimulation gives rise to two things: a "specific sensory activity" peculiar to that retinal element, and a sensation of light. The specific activity, which remains peculiar to that location, is associated with the *Bewegidee* responsible for fixing the eye on the candle, in accordance with the associative law of simultaneity. The sensation of light produced by the specific sensory activity of the illuminated retinal point thus appears as being "in" the simultaneously appearing *Bewegidee*, and it is surrounded by an otherwise empty visual space. The outer situation of a candle flame in the center of the visual field is copied by an inner luminous point at the center of visual space.[81]

Now suppose that a *Bewegidee* contiguous to the previously willed *Bewegidee* is activated in such a way that the eye is moved just slightly (say, by the slight contraction of the upper rectus). This second *Bewegidee* now stands out in consciousness. In it is localized the light sensation resulting from the stimulation (by the previously supposed stationary light source) of a retinal element contiguous to (in fact, just above) the previously stimulated retinal element. The light sensation becomes associated with this second *Bewegidee*, so that a correspondence is formed between a second retinal element and a second point of inner visual space. This second connection having been established, if the two retinal elements are now stimulated simultaneously, the resulting sensations are represented in inner visual space as outside and next to one another, just as their causes lie in the external world (i.e., on the retina). With each of the *Bewegideen* that defines a direction of the optic axis, this same procedure can be imagined. Thus are associative connections established between each retinal location and a corresponding point of visual space (174-6).[82]

The correspondences established through this process preserve spatial contiguities. The end result of the associative process is that sensations produced by the stimulation of foveal retinal elements are intuited at the center of visual space, while sensations produced by stimulation of successively peripheral points are intuited successively toward the periphery of visual space.[83] In virtue of this point-for-point correspondence between the retinal image and its mental representation, the relative size, situation, shape, and motion or rest of the pattern of light found in the retinal image will come to consciousness "as if the mind saw immediately the very image present on the retina" (181). Thus while the mind has no direct contact with the outer world of things in themselves, including physical retinal images, through the development of associative connections it comes to have mental representations of such images.

This much of Steinbuch's elaborate theory of *Bewegideen* and spatial representation merely brings us to the starting point of psychologies of vision preceding his. The rest of his account of spatial perception follows familiar lines (up to a point). The sense of touch develops its own inner space and its own mapping of the skin and the three-dimensional space surrounding the body. It then educates vision about the third dimension through association of the six familiar "cues" for depth or distance (see chapter 2) with tactually determined values for depth or distance. Once this process is complete, the observer comes to "know" (*erkennen*) distance immediately, seemingly unaware of the associative connections underlying the experience of spatiality and the localization of objects in ambient three-dimensional space (207-20). Steinbuch agreed with previous theorists that, once psychological development is complete, the perception of distance and size is psychologically *mediate* but phenomenally *immediate*: although both distance and size are perceived through the mediation of associative processes as outlined, the experience of space is direct.

In one respect, however, Steinbuch did deviate from the commonly accepted description of the psychology of vision in his time: he denied that the processes that yield the perception of size and distance should be characterized as inferential or judgmental. In his view, the correct theoretical description of our visual achievements is that we know distance through "intuition," not through inference (218-29). Steinbuch's account of illusion illustrates his explicit distinction between associative and judgmental accounts: he ascribed illusion to defective intuition, rather than to false or mistaken judgment. He considered the small appearance of distant objects to be such an illusion. In the case of near objects, he reasoned, we get the distance right and so the size appears correctly (size constancy), but with distant objects the empirically-given signs for distance are weak (or we are poor at using them—Steinbuch thought hunters and farmers would do better than city dwellers), yielding an undervaluation of distance and therefore of size. The deficiency, he intoned, belongs to the process of intuition and ought not to be characterized as a mistake in judgment.[84] In distinguishing the associative processes of spatial perception from intellectual acts such as judgment, he departed from textbook accounts; apparently unbeknownst to himself, in so doing he joined company with his contemporary, Herbart, and participated in a type of explanation that had been recognized by Berkeley, Hume, and Kant himself.

Although Steinbuch considered the perception of depth and distance to be phenomenally immediate, he also characterized the space in which distant objects are experienced as an "inner visual space" (*innerer Sehraum*), in order to emphasize his point that this space is a psychological construction (165-6, 264-5). He has said nothing yet about the perceiver's coming to treat these "inner" mental representations as representations of an independently existing material world. Steinbuch did not maintain that the observer localizes the "inner space" of perceptual experience inside his or her head, for the body and head are themselves part of the visual and tactual scene represented in "inner space." Throughout its period of development, the infant has no notion its "inner space" is "inner." In Steinbuch's account, the mature individual as well as the infant simply takes the intuited world to be the world in which she lives.[85] Her body is represented in her intuited space, and other objects are represented as external to it. The representation of intuited objects as external to the intuited body is the basis of "externalization," of the attribution of sensations to objects separate from one's own body. The actual outer physical world, containing the individual's physical body and the real physical objects with which it interacts, is, of course, never directly intuited. The spatial relations of this world are, however, faithfully represented or "copied" in the intuited spatial relations of the "inner" world.[86] The individual simply takes this inner world for the outer world, or rather for the only world there is.

The positing of this "copy" relation between inner and outer space contains the grounds for Steinbuch's realism regarding spatial properties; the position that space is an acquired form of intuition, Steinbuch thought, let him establish such realism in contrast to Kant. Kant had been unable to explain the relation between space as a form of intuition and the spatiality of "things in themselves," Steinbuch believed, because he regarded the spatial form of intuition as innate. Kant was therefore faced with the problem of explaining how a *preestablished* harmony between space as an innate form of intuition and the space of things in themselves could arise. Steinbuch denied that space is innate and so was no longer faced with this problem. However, since he believed that our forms of intuition capture the real spatial attributes of things in themselves, he still needed to explain their fit. His strategy was to reverse the Kantian conception of the relationship. Kant had supposed that the forms of intuition condition the only world we can ever know, the world of appearance. Steinbuch countered that the forms of intuition can mirror the spatial attributes of things in themselves because the forms are acquired through interaction with the external world. The construction of internal space takes place through interaction with external bodies.

Steinbuch maintained that the essential role played by the external world in his theory of spatial perception constituted an argument in favor of his realist position.

> Although we do not . . . become aware of the objects of the world in itself through our senses, but only through the pictures [*Bilder*] of these objects produced by the senses with guidance from the objects, nonetheless we are obliged, in accordance not only with our feelings, but also with the scientific view here proposed, to acknowledge an outer cause corresponding to the innerly sensed stimulation of the organ of perception. [205-6]

We must acknowledge an independently existing external world, in Steinbuch's view, not only because the sensory perceptions we have are not controlled by will (a standard post-Kantian argument) but because it is required by our best theory of the senses:

> Consequently, my theory presupposes the existence of external objects of the world, or of an external world in itself, as an essential condition of our *Weltanschauung*. Our faculty of perception cannot possibly bring into being this *Weltanschauung* by itself, through its own powers, since there belongs to it no more than a mental faculty which, by hypothesis, can be developed only through a great deal of contact of the sense organs with the world (in itself) and, conversely, also of the world with these sense organs. [206-7]

Steinbuch did not simply presuppose realism: he attempted to argue for it on the grounds that the best psychology of the senses available requires a realist commitment. Physiology requires empiricism—or at least organic materialism, which precludes preformed mental faculties or functions, requires empiricism—and empiricism requires that the organism causally interact with an independent world. Whether his arguments achieved their end and, in particular, whether they successfully provided a response to Kant are another matter.

Some problems internal to Steinbuch's empirist derivation of space as a form of intuition arise independently of its status as a response to Kant. In explaining the development of "inner space," he applies different laws to apparently equivalent phenomena without further explanation. Simultaneously appearing *Bewegideen* of a single muscle are organized into a line according to the law of similarity, whereas the simultaneously appearing ideas of two muscles are first fused according to the law of simultaneity and then associated with yet other ideas on the basis of their character after fusion. In the first case, simultaneously appearing ideas are associated *next to* one another, while in the second, they become *amalgamated*. Steinbuch supplies no additional details. We are left to conclude that he simply assumed what he needed in order to get the result he wanted.

The problems internal to Steinbuch's account of spatial representation become deeper with his claim that he has shown how intuitional space can be constructed out of nonspatial elements. If he intended to show that intuitional space can be generated from nonspatial elements by a faculty of representation that contains no foundation of spatiality, then clearly he failed. In asserting that the associative law of similarity orders elemental representations *next to* one another, his account begins already provided with a law for generating spatial representations. Without this associative principle so formulated, there would be no explanation why sensory elements representing various motions ever develop beyond a one-to-one correspondence between external spatial locations and discriminably different nonspatial *Bewegideen*. The law of association that is responsible for the intuiting of these *Bewegideen* as next to one another does the real work in Steinbuch's derivation of space as a form of representation. But this law itself receives no further explanation.[87] But then, inasmuch as the mind is innately disposed to order mental representations according to the law of similarity, it is innately disposed to produce spatial representations, contrary to Steinbuch's claims for his analysis.

Considered as a response to Kant the most serious problem with Steinbuch's theory lies in the very conception of an empirically-based rebuttal of Kant's work. In attempting to use his empiristic theory of vision to refute Kant's doctrine of a priori spatial knowledge, Steinbuch came close to fitting the charge Kant raised against Oswald, Beattie, and Reid for their attack on Hume: "by always taking for granted what he was doubting, and, on the contrary, proving, with violence and often with great unseemliness, what it had never entered his mind to doubt, they so mistook his hint as to how to improve matters that everything remained as it was, as if nothing had happened."[88]

The problem with Steinbuch's response to Kant is that it assumes that Kant intended his doctrine of space to ground a particular theory of vision. But Kant's remarks on space as a form of intuition provided minimal constraints on visual theory, and Kant himself most probably accepted a weakly empirist account of visual spatial perception; indeed, he could, without inconsistency, have accepted a radically empirist account of vision. In attempting to prove that we must "learn to see," Steinbuch was "proving what it had never entered" Kant's mind to doubt. Steinbuch's boldest claim, however, was that he could explain the alleged correspondence between spatial representation and the spatial properties of things in themselves. Here, Steinbuch seems to have taken for granted the very thing Kant had doubted. For Kant explicitly considered and rejected the possibility of an empirical basis for the claim that the form of intuition mirrors the relations among things in themselves. He rejected this possibility not as a psychological hypothesis but as a conception of the evidential or justificatory grounds for asserting that geometry applies to external objects conceived as things in themselves. In answering Kant's argument about justification with a psychological theory, Steinbuch missed the point of Kant's argument. Short of a solution to the problem of induction, Steinbuch's empiristic theory of space as a form of intuition must fail to provide adequate justification for the apodictic nature attributed to geometry in both its "pure" and "applied" forms.

Steinbuch did not simply equate the "inner space" of his theories of touch and vision with the pure space of intuition that grounds geometry. He regarded pure space as something that must be attained through coordination among the various inner spaces developed in connection with vision, touch, and so on. The various spaces must be built into a single, isotropic space, which abstracts from everything peculiar to any given sensory modality. His account of the development of this space begins from the point at which a generalized space has been abstracted from the *Bewegideen* of a group of muscles (in this example, torso muscles):

> The abstract concept of space belonging to the torso cannot be what we adult humans call an abstract concept of space, by which we understand a pure extension without any attendant special qualia. The latter first arises through repeated acts of abstraction on the part of the mind, in which this abstract torso-space is repeatedly compared with other similar concepts of space from other organs of motion, such as the eyes, and the common properties of the two, distinguished from the particular, are intuited or represented for themselves. Through this second act of abstraction the remaining qualitative character pertaining to the torso in general, and so those of the eyes, are both distinguished from pure extension, and the latter is intuited or represented as a pure abstract sensory space. [50-1]

This is Steinbuch's entire account of pure extension. Although it contained the seeds of a response to Kant's doctrine of pure intuition, they did not germinate.

Steinbuch characterized the development of an intuition of pure extension as an "act of abstraction." From the time of Aristotle onward, the notion of abstraction founded accounts of geometrical knowledge that were opposed to those based on sense-independent rational intuition. So the notion of abstraction at least pertains to the same problem area as Kant's theory of spatial intuition. Nonetheless, in order for an abstractionist approach to prove acceptable, it would need to provide a convincing argument that "abstraction" is immune to Kant's general challenge to empiricism. However, Steinbuch provided no such argument nor even acknowledged its need. He did not even connect his derivation of the concept of pure extension with the problem of understanding the basis for geometrical knowledge—the problem that had been the focus of Kant's theory of space. Indeed, the quoted account was virtually a throwaway remark. It is lost beside Steinbuch's primary response to Kant, which he avowedly conceived as an attempt to refute transcendental idealism through a theory of the psychological and physiological development of spatial representation (and not through a theory of abstraction). Differently put, Steinbuch himself regarded his main correction of Kant to arise from a refutation Kant's nativism. Yet nativism is beside the point in assessing Kant's transcendental claims. An empirically-based response to Kant is not thereby ruled out (see Helmholtz), but any psychological response that makes nativism the main point of contention must come under serious scrutiny, for it most likely has missed the point of Kant's work.

2.2 Tourtual's Transcendental and Empirical Physiology

Caspar Theobald Tourtual was a visual physiologist and psychologist of some note during the second and third decades of the nineteenth century. He studied medicine at Göttingen from 1820 to 1822, where he had Blumenbach for comparative anatomy and physiology. The following year he studied philosophy, mathematics, physics, chemistry, anatomy and philology at Berlin, where the physiologist C. A. Rudolphi was among his teachers. He was on the surgical faculty at Münster in 1827 when he published his major work, *Die Sinne des Menschen in den wechselseitigen Beziehungen ihres psychischen und organischen Lebens: Ein Beitrag zur physiologischen Aesthetik,* a comprehensive theory of the senses.[89] As his use of the term "physiologischen Aesthetik" might suggest, Tourtual saw this work as a physiological contribution to a Kantian theory of the senses. The title also suggests that Tourtual considered the physiology of the senses to have two subdivisions: physiological or organic, which analyzes bodily structures and functions, and psychological or "*seelisch,*" which analyzes mental structures and functions.

Tourtual's theory of vision was deeply indebted to Steinbuch's work on spatial representation, including his conception that spatial representations are formed through an active process of construction. He agreed with Steinbuch in analyzing the psychology of spatial perception into nonspatial elements and laws for their combination. Nonetheless, he rejected the theory of *Bewegideen* and posited an innate basis for localization, independent of the the muscle system. He thus opposed the defining feature of Steinbuch's work, the commitment to radical empirism, and for several reasons (223-8).

Empirically, he appealed to arguments from clinical observation and comparative physiology to show that spatial perception could occur in the absence of associated motor-ideas. Conceptually, he found Steinbuch's derivation of the form of spatial representation from nonspatial elements unconvincing. Other aspects of his position allowed and even required a favorable attitude toward nativism. His methodological position permitted nativism with respect to mental characteristics: he argued that mental processes should be placed on equal footing with bodily processes for the purposes of physiological investigation. From this perspective, it was equally legitimate to speak of the innate functional organization of the mind as to speak of the innate functional organization of the body. Furthermore, he was inclined toward nativism by the belief that his Kantian attitude toward space as a form of sensibility entailed a commitment to that position.[90]

It is revealing to compare Tourtual and Steinbuch in their relations to Kant. Tourtual did not conceive his task to be that of providing an empirical confirmation of Kant's doctrines in answer to the previously attempted empirical refutation. Rather, he assumed as given certain features of Kant's doctrine of sensibility—precisely those features that served within Kant's system to explain the possibility of geometrical knowledge—and then proceeded to fill in the theoretical details that were missing if Kant's doctrine was considered as a proper theory of the senses. However, despite his alignment with Kant on geometry, he agreed with Steinbuch in rejecting Kant's "subjective" position on the reality of spatial properties. But whereas Steinbuch had attempted to *argue* for the reality of spatial properties by appealing to sensory psychology, Tourtual *assumed* a commonsense realism and tried to show how it could be accommodated with the theory of the senses. Unlike Steinbuch, he did not attempt to answer Kant empirically and thereby miss Kant's point, but rather he countered Kant's transcendental idealism with a pragmatic attitude toward the reality of the physical world (lx, 144).[91]

In the preface to his *Sinne des Menschen*, Tourtual divided the theory of sensory representation into two parts: physiological and transcendental. The first, "empirical," part included all the activity of the "bodily and mental organs" involved in sense perception, beginning with the outer stimulus and ending with the completed intuition; here he included everything that physiology and the theory of the mind (*Seelenkunde*) can attain through observation and experiment. The second, transcendental, part contained the "metaphysical consideration of the content of our sensory representations, insofar as this content is preformed in [the faculty of] sensibility, through the generation and development of life, preceding all external influence."[92] Tourtual made it clear that, although he was pursuing the physiological side of the division, transcendental considerations could not be avoided altogether: he rejected the conception that science could be based purely on outer experience, independent of philosophical reflection (vi-vii).[93] He did not, however, regard transcendental considerations to be free from empirical constraint; transcendental physiology would rest upon the "empirical foundation of our consciousness, in that it can be derived only from the facts of inner experience through careful comparison and consistent inference" (v-vi). In its analysis of the innate structure of our representational faculties, transcendental physiology is constrained by but not reduced to factual considerations.

Two aspects of the avowedly philosophical portion of Tourtual's work are of especial importance. First, he attempted to place the theory of the mind on equal footing with the theory of bodily processes for the purposes of sensory physiology. His strategy for doing so did not follow the organic materialism of Reil and Steinbuch; although he was interested in unifying psychology with physiology, he did not believe that mental processes could be identified with or wholly explicated through bodily processes.[94] Rather, he treated mental function and bodily structure as separate but equal, an attitude evident in his position regarding the use of the comparative method in physiology. He contended that it is not the comparative *shapes* of bodily organs that holds the key to an understanding of the sensory representations that they produce but the *functional activity* associated with the organs (xiv).[95] His point goes further than the still-familiar distinction between anatomical structure and functional activity; for the functional similarities that he sought within and across species included both bodily and mental processes (vii-x, 3-4).

The second relevant piece of Tourtual's "transcendental" ruminations pertains to his general orientation toward the metaphysics and epistemology of sensory perception. Tourtual himself identified three basic positions in the history of philosophical consideration of the senses—the objective, subjective, and "middle" standpoints—which he identified with three metaphysical theories of truth (xxxiii-xl). The objective standpoint he identified with empiricism, which he characterized as the view that objects directly cause sensory representations and are thereby presented immediately to the mind without any significant contribution on the part of the knowing subject. The subjective standpoint he identified with rationalism, which he characterized as the view that the mind constructs the world; he placed Kant's transcendental idealism under this rubric. The middle way, which Tourtual claimed for his own, gives a role to both subject and object in the production of sensory representations.

In agreement with the "subjective" position, Tourtual distinguished between sensory objects and actual objects, or between the world of subjective representations and the world of things in themselves (4-11). He rejected the naively realist view of the *objective* position, according to which objects are known immediately. His acceptance of mediating representations did not arise from epistemological arguments from sensory relativity and the like but from an appreciation of the constructive activity that must accompany the production of a sensory representation. Naive realism treats the senses as passive conduits that make no contribution to the perceptual act. Such a position is counter to the physiologist's appreciation that the functional characteristics of the senses, including both bodily structure and mental function, cannot be assumed but must be explained. Nonetheless, Tourtual found unacceptable the defining feature of the subjective view—the belief that the subject simply constructs a world that is essentially illusory when compared with things in themselves. He confessed to an unshakeable conviction that the outer world is real, a conviction he characterized as that of "everyman" (lx). His "middle way" acknowledged that the subject's sensory apparatus conditions the sensory representations that it produces, but it also affirmed that the causal influence of external objects upon the sensory apparatus yields accurate representations of those objects.

Tourtual's middle way stressed the interactive nature of sense perception. The relevant notion of "interaction" goes beyond the idea that sometimes the subject is relatively passive in receiving impressions (as when one simply opens one's eyes), and sometimes relatively active (as in touching and manipulating an object). The object is always active inasmuch as it has a causal influence; correspondingly, the subject is always active inasmuch as the mind must act to produce representations as a result of sensory stimulation.[96] The mind produces these representations according to its own innate laws of sensibility. Since every sensory representation is to some extent an active creation of the mind (even if a lawfully constructed creation), the perception of objects is always mediated by these laws of sensibility.

In Tourtual's theory the innate laws of sensibility serve not only to explain sense perception, they also account for the functionally appropriate structure of the sense organs. According to Tourtual, the soul precedes and conditions the construction of the body during embryological development (one is reminded of Kant and Blumenbach's *Bildungstrieb*). This process of construction directs the development of the sensory organs in accordance with their functional role in receiving stimulation, just as other structures in living things develop so as to be appropriate to other aspects of the animal's environment. The assimilation of function to environment goes for both mind and body: the senses are accommodated to their sensible objects, and so is the mental process of intuition, "in its immaterial way" (liv; also, 3-9). That the mental processes of intuition are functionally accommodated to the environment constitutes the substance of Tourtual's claim that the senses can be regarded as producing representations that correspond to external reality. However, Tourtual was merely attempting to show that realism with respect to spatial properties is tenable: he was not out to establish realism empirically. He admitted that, conceivably, the spatial relations in nature do not correspond to our subjective space; the functionality of the senses does not guarantee their epistemological validity as organs for perceiving actual objects, or things in themselves.[97]

Tourtual's empirical and conceptual arguments against empirism and for nativism arise along with his actual physiological analysis of sensory processes. According to Tourtual, there is one space of intuition. It is innate and is common to vision and touch. There is thus an innate basis for the comparison of the spatial intuitions received by these two senses,[98] but this innate basis ensures only that all intuitions of vision and touch will be spatial. It does not determine which of the various possible geometrical shapes will be constructed when one or another external stimulus impinges upon the organ of vision or touch. The features of the stimulus are responsible for determining which particular geometrical form is intuited.[99] A minimum degree of "realism" requires that the same stimulus give rise to the same representation on various occasions. The representation the mind creates from the material of sensation can do that only if the mental application of subjective geometry to the material of sensation is subject to hard and fast laws. The task of a physiological and psychological account of spatial perception is to discover those laws according to which representations are constructed in intuitional space on the basis of the (actual) spatial configurations presented to the sense organs.

The "hard and fast laws" of sensibility are the basis both for Tourtual's understanding of geometry and of his theory of spatial perception per se. He agreed with the general Kantian position that geometry has its basis in the spatial form of sensibility; as he put it, "geometry is the analysis of the laws of subjective space."[100] The laws of space are the laws by which sensibility organizes the matter of (outer) intuition; they necessarily are applied to each and every outer representation. Tourtual resisted assigning this organizing activity to imagination and intellect, as Kant had suggested and Fries had explicitly maintained. He was convinced that it would be incorrect to treat the geometry of spatial perception as arising from an hypothesis or an act of reasoning on the part of the perceiver. The geometrical laws that guide sensibility are not to be identified with intellectual principles or treated as axioms of perceptual inference. Like Steinbuch, Tourtual maintained a sharp distinction between sensibility and intuition on the one hand and intellect and reasoning on the other. By his lights, the investigation of spatial perception, including space as its subjective form, was an investigation of the faculty of "lower intuition," which provides the materials for understanding without itself being conditioned by the activity of the higher faculties (xi-xii, 306).

Tourtual's conception that the laws of sensibility are not intellectual principles conditioned his arguments for the innateness of these laws. His arguments differed significantly from the transcendental arguments of Kant and Fries, which had been aimed primarily at explaining the apodicity of geometry. Although Tourtual understood that Kant's own focus had been the explanation of the possibility of geometrical knowledge, he did not use Kant's arguments to support his own position. His general conviction that space is an innate form of sensibility stemmed from his belief that no adequate explanation of how we acquire spatial representation through experience could be formulated. He regarded Steinbuch's attempt to derive spatial representation from nonspatial *Bewegideen* as the best effort in this direction; however, he contended that unless the associative law of similarity already constitutes a principle for organizing nonspatial elements into spatial representations, spatial relations would never arise. A series of qualitatively distinct *Bewegideen* would remain just that—a series of nonspatial representations—unless an innate principle for ordering them was posited (224-7).[101] Similarly, he contended that knowledge of optical principles, which he believed Fries attributed to the visual system, could itself be acquired only through visual experience. If vision occurred through hypotheses regarding visual geometry, directional localization would depend upon a knowledge of the refractive laws of the eye; but this position implies that we originally learn to see through the study of physics. The relevant physical principles could be known only on the basis of experience, and such experience would already presuppose the ability to have spatial perception. Therefore, he concluded, there must exist innate principles of sensibility that are responsible for spatial perception and that serve as the basis for the understanding's empirical discovery of the physical laws (some of them expressed geometrically) governing nature (229-32).[102]

Tourtual's project of providing a physiology and psychology consistent with his understanding of Kant's theory of the forms of intuition required that he specify in detail how the innate principles of sensibility underlying geometry could be responsible for all spatial perception. Given that he, along with Steinbuch and Kant, regarded the "matter" of intuition as nonspatial, his first task was to establish that the "form" for organizing this matter is innate. Then he had to explain the process by which sensations are localized in the space of intuition during sense-perception itself. Tourtual maintained that during any individual act of perception the mind constructs a representation of the external object in internal space. By his account, spatial perception occurs through an ordering of simple parts into spatial representations. His theory of spatial perception therefore required both an account of "the elementary representations (*Elementarvorstellungen*) in themselves" and "the process by which the mind, from a multitude of these punctiform representations, constructs the representations of lines, planimetric and stereometric forms" (175). Tourtual's account of planimetric vision can serve to illustrate the nature of these elementary representations and of the processes that operate over them.

The problem of planimetric or surface-relational intuition consists in explaining why the mind localizes the various sensations that result from a single retinal image in such a way that their intuited spatial relations represent the spatial relations of the stimulus. Gehler and Steinbuch had already addressed this very problem of why each sensation is intuited in relation to the phenomenal axis of vision so that its phenomenal position corresponds to the position of the (physical) retinal impression causing the sensation. Tourtual agreed with Steinbuch that the explanation of a planimetric representation corresponding to the retinal image could not lie simply in the physical arrangement of the fibers in the retina that project into the brain. His solution was similar to Steinbuch's inasmuch as each retinal element was assigned a discriminably different "sign" that manifests itself in the quality of every sensation aroused by stimulation of that element. It could therefore serve as a "specific sign" or "specific characteristic" for the location of the element: the specific sign (*specifische Merkmal*) "is that which gives the sense the impulse to determine the [spatial] relation of the sensation according to the appointed laws" (185-6).[103]

Tourtual now needed to explain how the various nonspatial elemental representations, each possessed of its own "specific sign," are put together to form a spatial representation. As Tourtual saw it, any solution that relied on simply placing representations next to one another was unworkable; for, assuming that the nonspatial elements really are nonspatial, an infinity of them would be insufficient to form a line segment (the problem of composing the continuum). He proposed that the basic activity in forming spatial representations is a kind of attentional "drawing" (*Zeichnung*). Just as a physically moving point can move the length of a line segment in finite time and yet nonetheless pass through an infinite number of points, so too the mind can produce a representation of a line segment by continuously "moving" (in its activity of referring sensations to regions of intuitional space) from the sign of one spatial point to that of another (186, 188).[104] In the case of planimetric vision, the attentional activity of soul is guided in its drawing by the phenomenal character of the various specific signs. The specific signs vary in quality in direct correspondence with the spatial relations of the fibers in the retina that produce them, so that physical relations of "next-to-ness" become phenomenal relations of "most similar to." Accordingly, the qualitative character of the special sign of each fiber of the retina can provide the basis for the mind to arrange the sensations associated with each sign in intuitional space in an order corresponding to the position of the fibers in the retina.[105]

Tourtual had now reached the beginning point of most previous accounts of vision (with the exception of Steinbuch). He could proceed to explain the mechanism by which we assign a distance to each point in the two-dimensional field so that the world of appearance attains its characteristic stereometric form. Tourtual believed that the ontogenetically primary means of distance perception is accommodation (which he thought occurs through equal contraction of the four recti to change the shape of the eye, and especially the curvature of the cornea: 213-14). He distinguished two ways in which accommodation acts in the perception of the spatial proportions of objects. First, the eye moves so as to pass over various parts of an object; the perceived direction of each successively fixated point is determined by the muscle sense as the eye moves, and distance is determined through the muscle activity underlying accommodation. Second, the "attentional activity" of the mind is directed here and there within a stationary field of vision, and the accommodation of the eye varies with the attended distant point; in this manner, the depth relations of a static scene may be "drawn."[106] Since the accommodative act obtains for the whole eye at once, this second means of stereometric perception is most useful only when the object in the field of vision alters in depth uniformly in all directions. In the case of a hemisphere in the center of the visual field, for example, the accommodative act might be supposed to change systematically so as to bring into focus successive latitudinal circles on the surface of the sphere; in this way a representation of the sphere could be constructed in intuitional space. Finally, in cases when accommodation is inadequate, such as for objects beyond a certain range and for the peripheral field of vision, the familiar cues for distance are used, including convergence, apparent size (when actual size is known), and brightness. Unlike accommodative muscle activity, which produces a sign for distance that leads to localization immediately (without learning), these secondary criteria become effective through a process of associative learning (216-23).

Tourtual's account of spatial vision stood exactly opposite Steinbuch's in its radical nativism—he regarded as innate both intuitional space and the laws by which points are localized within it. Nonetheless, his account showed remarkable similarities to Steinbuch's, inasmuch as both authors believed that the fundamental elements of visual intuition are nonspatial and that these elements are referred to spatial locations through principles of sensibility, divorced from the judgmental activity of the understanding. Steinbuch and Tourtual similarly conceived the nonspatial "signs" that allowed this localization: both assigned a specific sign to each nerve fiber (respectively, a "specific sensation-activity" and a "specific characteristic") to allow for localization of sensations specific to that fiber. They differently conceived the processes that led to the localization of sensations connected with these signs. Steinbuch attempted to show how his *Bewegideen* could serve as an acquired spatial matrix for localizing the signs with their attendant sensations, whereas Tourtual posited innate principles of localization.

In the not too distant past, readers trained in psychology would likely have regarded Tourtual's position as inherently inferior to Steinbuch's, on the grounds that nativism is in itself a patently bankrupt position. According to this reasoning, nativism cheats by simply assuming what it needs to produce the explanation it wants. In simply assuming an innate ability, nativism refuses to undertake the honest work required to show how an ability could be acquired through experience—work of the sort that must have gone into the theory of *Bewegideen*.[107]

Nevertheless, Tourtual's methodological justification of innateness should not to be dismissed out of hand. His argument that mental function should be placed on a par with bodily function as a legitimate theoretical posit could find support both from scientific practice at the time he wrote and retrospectively from our post-Darwinian standpoint. Prior to Darwin, the functional organization of the body could not be denied and yet no apparent explanation (outside design) was available. Nonetheless, functionality was admitted into biological analysis. If for bodily organs, why not for mental processes? After Darwin an explanatory schema was available for explaining how the functional structure of bodily organs has arisen. Again, if for the body, why not for mental processes, too?

2.3 Müller's Physiological Subjectivism and Scientific Realism
Johannes Müller was the dominant figure in German physiology during the second quarter of the nineteenth century. He had wide-ranging interests, contributing to developmental embryology, to the study of subjective sensory phenomena, and to discussions regarding the fundamental concepts and basic methods of physiology in general. Nonetheless, compared with the previous two authors, his work on spatial perception was simply an elaboration of the received position, although carried out with greater critical acumen. At the same time, Müller's philosophical discussions were the most extensive and sophisticated among the three.[108]

Methodologically, Müller cut a careful path among the various positions that were then vying for predominance.[109] He rejected French positivism as the mindless collecting of facts, and he was no more receptive to what he termed "falsche Naturphilosophie," which he criticized for a total abstinence from facts (PG, 3-15). He argued that physiology must be based on observation, but that observation must be guided by ideas; the true physiologist will go from idea to experience and back to idea. However, he carefully separated himself from the "physiology of the understanding" as practiced by Reil and his followers. According to Müller, Reil's choice of the understanding as his sole guide to vital phenomena guaranteed that his concept of the living would include only the causal processes acknowledged by the understanding: those described within physics and chemistry. But, Müller contended, such processes are inadequate to the phenomena of life, as even Reil had been forced to admit when he introduced an "imponderable matter" or an "unknown force" into to his explanations, to take up where the known phenomena of chemistry left off (PG, 15-17; HP 1:26-7). Müller himself preferred to posit a special vital substance to account for the peculiar properties of living things; he was a proponent of classical vitalism. Although he accepted the general applicability of physics and chemistry to the causal processes in living things, he doubted that physical and chemical concepts were sufficient for grasping the *ground* of vital phenomena. For the latter, the investigator must bring a special concept of the vital to his experience. The ordinary concepts of the understanding would allow him to find no more than an inert ("dead") set of connections among the facts. But the goal of the physiologist should be the ability to *think* the living, by which Müller seems to have meant the ability to find concepts that bring the whole range of vital phenomena into ordered connection. He found this connection in the concept of a special vital force, akin to Blumenbach's *Bildungstrieb*, irreducible to physics and chemistry, and manifested in the commonalities and variations across the range of living things (PG, 17-24; HP 1:18-25).

Müller's methodological perspective is reflected in his division of the previous history of sensory physiology into three broad periods (PG, vi-xviii). The first, or "dogmatic" period lasted from the time of the ancient Greeks to the "revival of science." It was characterized partly by the framing of "bare hypotheses" that yielded no explanation of the phenomena at all. The early period was also characterized by positions that possessed philosophical content. Among these Müller discovered a murky recognition of one of his own theoretical tenets: that the sensation of light must have its basis in the subject itself and cannot be explained wholly in terms of the external stimulus. The second, or "physical," period sought to rectify the poor state of empirical knowledge regarding light and the eye, and it achieved notable success. Followers of this physical approach, however, mistakenly believed that an account of the retinal image was equivalent to a complete explanation of vision. According to Müller, followers of the physical approach sought to explain the sensation of light or sound wholly through a "specific sensibility" for external light and sound, without acknowledging the contribution of the senses themselves (xii-xiii). Müller boasted that physiologists in Germany had entered a third period, the "physiological or theoretical," by contrast with investigators in other nations (and especially in France) who were afraid to go beyond mere facts (xiv-xv). Among the most important discoveries of this period was the contribution the subject makes to perception. The researches of Goethe, Troxler, Steinbuch, Purkinje and others had shown that light, dark, and color are essential energies immanent to the visual sense (xv-xvi). On this work plus his own observation of subjective visual phenomena, Müller posited his famed "specific nerve energies," which differ from mere "specific sensibility" by attempting to account for the qualitative content of sensations that the subject contributes. In positing such "energies" Müller offered the kind of connection of thought with experience, of observation with theory, that he considered to be the highest goal of physiological research.

In 1826 Müller published his *Vergleichenden Physiologie des Gesichtssinnes*, which contains his first discussion of the phenomena of spatial vision, with brief discussions of the origin of spatiality and of the concepts of apparent size and apparent place.[110] It did not, however, contain a general theory of spatial vision (for example, it contained only a parenthetical discussion of distance perception).

Müller's discussion of the origin of spatiality arose in connection with the problem of the differentiation of subject and object in sensory intuition. He contended that, although space is a necessary form of intuition and our original sensations are spatial, we do not originally refer these sensations to an external cause; they begin as sensations of the extension of our own body. "At the outset of sensibility," Müller maintained, "the individual senses only himself spatially extended, only himself filling out space" (PG, 40).[111] Through the interplay of touch and vision, the individual comes to distinguish sensations produced by his own body from those produced by other bodies. He comes to notice that a certain object (his own body) is a constant feature of his visual field. He recognizes this object as his body when he discovers that voluntary motions systematically produce a change in the part of his visual field corresponding to it; he can now separate those parts of the visual and tactual worlds that belong to himself from those that do not. This distinction arises not from sensations alone but through judgments regarding the cause of those sensations.

According to Müller, the spatial sensations of vision derive literally from the sensation of the extension of the retina. He stated this position quite explicitly: "In each visual field the retina sees only itself in its spatial extension and in its state of affection" (PG, 55).[112] The literalness with which Müller maintained that the spatial sensations of vision are a sensing of the retina is exemplified in his treatment of apparent size. The "apparent size" of an object is equated with the actual size of the object's projection on the retina; thus, the largest apparent size that an object can have is limited by the physical size of the eye. Equating apparent size with the size of the retina was, of course, restricted to the case of pure sensation, divorced from any judgment regarding the external cause of the retinal state.

In 1826 Müller had not yet gone beyond a treatment of the sensations of vision; he discussed only the individual's sensation of the state of his own retina, eschewing discussion of the perception of an object's actual size, distance, and shape. In the *Handbuch der Physiologie des Menschen* (1834-1837) he provided a fuller account of spatial vision. He qualified his position that visual sensations are sensations of the extended surface of the retina by admitting that the seat of visual sensation might just as well lie in the central part of the visual apparatus located in the brain (HP 2:349-52). Whatever the seat of these sensations, however, their content is equivalent to the retinal image: visual sensations originally appear as depicted in a surface. According to Müller, our experience of depth is an idea (*Vorstellung*) rather than a sensation; depth perception is a product of the understanding. The perception of size and shape depend upon combining the surface-like visual sensations with a judgment of depth. He described this process in the usual manner: the judgment of true size results from a combination of apparent size (equivalent to retinal image size) with an idea of distance (2:352-363). His general account of spatial perception, then, did not differ essentially from Gehler's and other standard accounts; he merely took positions on matters with regard to which reference works stood mute, as in his explanation of apparent size.

Müller's *Handbuch* also contained an entire division on the phenomena of the mind, which included his views on general epistemology and metaphysics (as distinguished from the methodology of physiology). From the first appearance of his work, Müller was regarded by some of his near contemporaries as a follower of Kant, on the ground that in making space as a form of intuition depend upon the structure of the sense organs he thereby assigned it a "subjective" status.[113] However, although Müller regarded spatial sensations as subjective in the sense that they depend upon the subject, he at the same time ascribed them an objective basis in the real extension of the retina or other sensitive surface. According to him, we experience space because the sense-organs really are spatial; subjective space is determined by objective space. This understanding of the "necessity" of spatially-ordered experience, a necessity based upon the spatiality of the "sense-organ in itself" (so to speak), could not be further removed from Kant's arguments for space as a necessary form of sensibility.

Although others read Müller as a Kantian, he rejected the Kantian position that sensory processes provide a basis for the necessities of geometry. Further, he explicitly sought to distance his own epistemological position from both rationalist nativism and Kant's apriorism. He regarded the question of the "harmony" or "correspondence" between thought and objects as among the central problems of philosophy. Originally, he maintained, this epistemic "correspondence" had been explained in one of two ways: on the basis of sensory experience or through a "preestablished harmony" between the mind's innate endowment and the world. But these explanations had been undermined by Hume and Kant (HP 2:517-18). The empiricist approach fell when Hume successfully argued that the associative combinations available through experience could yield merely subjective necessity, thereby precluding empirically based claims to objective knowledge. Kant had countered by arguing that the actuality of mathematics shows that there must be pure a priori concepts and pure a priori forms of intuition (which Müller distinguished from merely inborn representations); but he had also argued that their applicability was limited to possible experience, thereby foreclosing knowledge of the thing in itself.

Müller rejected the positions of both Hume and Kant and suggested that we do get necessity through experience. Explicitly aligning himself with Herbart, he contended that the "power of abstraction" enables us to derive general concepts from experience, including universal concepts such as change, being, infinite, finite, shape, size, quality, space, time, motion, force, matter, object, subject, self, causation, existence, and nonexistence (HP 2:519-20). In other words, the basic concepts of science and philosophy are abstracted in this way. The degree of fixity accorded to such concepts varies in accordance with their scope. Thus, the concept of cause is among the most binding of such concepts, for it is "adequate" (in Müller's terms) to all sequences of change, both mental and physical. Other concepts are less "fixed," although their degree of fixity may alter as they are discovered to have a more nearly universal scope (2:520-22).

Although rejecting Kant and Hume, Müller did not fall into the simplistic position of dogmatically asserting that certainty is to be had despite the problem of induction. For he also rejected the assertions of philosophers such as Bruno, Spinoza, Schelling, and Hegel (his list), according to whom an absolute knowledge of things in themselves can be attained. In the spirit of Herbart, he contended that through speculation we can approach knowledge of things in themselves but we should not expect to meet apodictic standards in any but the simplest of cases. Thus, in the case of pure mathematics, where explicit definitions can be attained, we can achieve absolute knowledge; he insisted that our knowledge of size, shape, and their relations meets these standards. In other cases, full definition, or knowledge of essences, is beyond our grasp. For example, he contended that we can derive the paths of celestial bodies from the law of gravity but that the essence of gravity itself remains hidden, and that Ampere had shown that electromagnetic phenomena could be treated axiomatically while the essence of electricity remained unknown. And, he continued, psychology itself regarded as a natural science might come to achieve a deductive framework of principles, even though the essence of the mind remain hidden (HP 2:521-2). Müller thus adopted a mitigated form of scientific realism, one that treated the achievements of science and mathematics as real knowledge about the world "in itself" but also recognized the limited ability of science to meet the standards of essentialist knowledge as demanded by the old metaphysics, against which Kant had developed his subjectivist position.

Müller's balanced discussion of Hume and Kant had little effect on subsequent interpretations either of Hume and Kant or of Müller's own work. Specifically, it did not foreclose the oft-repeated but erroneous assertion that his work was Kantian in spirit because of its attribution to the sensory system of innate "specific nerve energies" and innate spatial feelings. The latter two aspects of Müller's teaching became widely accepted during the 1830s and 1840s. Owing in some measure to Müller's influence, the traditional view, challenged by both Steinbuch and Tourtual, that spatial perception begins with a "given" spatial representation to which the third dimension is added through learning remained the standard textbook doctrine.[114] The work of Steinbuch and Tourtual did not go unnoticed, but only the specialized publications of visual theorists attended to it.[115] With the publication of Helmholtz's *Physiological Optics,* the radical empiricism embraced earlier by Steinbuch attained preeminence in presentations of visual theory (albeit briefly).

Prior to Helmholtz, however, two developments helped weaken Müller's case for spatial sensations. The first was E. H. Weber's work on the sense of touch.[116] By applying points of a compass to the skin, he found that the distance the points must be separated to produce discriminably different sensations varies greatly depending upon which part of the body is touched (e.g., a much wider spread was needed on the back than on a fingertip). He hypothesized that sensations from the skin are mediated by a mosaic of "sensory circles," each corresponding to the terminal area, or end-zone, of a single nerve fiber. Only one sensation can be produced at a time by stimulation within any one sensory circle, so that if both points of the compass fell within a single circle, a single sensation would be produced; two sensations could be produced only if each compass point landed in a separate sensory circle. Weber applied the same analysis to visual acuity, explaining the difference in acuity between central and peripheral vision in terms of the size of the sensory circles.[117] He therefore conjectured that in both touch and vision each sensory fiber produces but a single sensation at a time, and this sensation is itself simple (without spatial differentiation). The representation of spatial attributes, as occurs in the localization of distinct points of stimulation on the skin, can come about only through the development of a representation (*Vorstellung*) comprising the individual sensations. This process of developing spatial representations occurs through mental activity that goes beyond the mere reception of sensations. He attributed the bare ability to produce spatial representations to an "inborn mental disposition or mental force" that leads us to organize originally nonspatial sensations into a three-dimensional spatial representation.[118] He believed that our ability to localize sensations within this space in a manner corresponding to the organization of the stimulus (an object impinging on the skin or reflecting light into the eye) should be attributed to experience.[119]

Although the principle that single fibers produce single sensations had been expressed many times before (it was, for instance, implicit in Gehler's account of the sensations produced by retinal stimulations), Weber's work had the effect of making this view an established doctrine. The textbooks of the 1850s and 1860s taught that single fibers produce single sensations, and that spatial representations depend upon the combination of numerous elemental sensations.[120] Accounts of the mechanism by which sensations become related differed. Some gave a special role to the physical order in the nerve fibers themselves (though without asserting that this order is directly sensed), while others pleaded ignorance of the physical relations among nerve fibers and postulated various innate or acquired psychological processes to produce spatial representations.[121]

The second development that helped diminish the acceptance of spatial sensations was Lotze's theory of local signs.[122] Lotze introduced the term *Lokalzeichen* to name a theoretical entity whose role in explaining spatial vision was akin to Steinbuch's *Bewegideen* and Tourtual's "specific signs." A local sign is a qualitative feature of the sensation arising from the stimulation of a particular place on the retina or skin. Its quality is constant and this quality is manifested only when that location is stimulated. Local signs accompany the sensations of color that arise from the stimulation of any given retinal point and, by providing a qualitative "sign" tied to that one retinal location, allow for the ordering of various sensations spatially in such a way that the order found in the light hitting the retina is represented in the resulting complex of sensations. The concept of local sign was not new with Lotze. However, he placed it within a comprehensive account of the metaphysics, epistemology, physiology, and psychology of spatial apprehension.

3 Lotze: Philosophy and Psychology in Balance

Hermann Lotze was well known as a physiologist when Helmholtz began his career, and he rose to prominence within German academic philosophy even as Helmholtz came into his own as a physiologist and physicist. Lotze was also the last of the great German metaphysicians in the traditional sense of that term. Like other major modern metaphysicians, he contributed to both science and philosophy. He studied philosophy and physiology simultaneously at Leipzig, earning his philosophy degree in 1838 with a dissertation on Descartes and Leibniz and completing a medical degree in the same year with a work on the future of philosophical principles in biology. After practicing medicine for a year he started teaching medicine at Leipzig in 1839. In the following year he gained the right to teach philosophy, and in 1844 he became professor of philosophy in Göttingen, remaining there until called to Berlin in 1881 where he promptly caught pneumonia and died. His major works included the *Medizinische Psychologie oder Physiologie der Seele* of 1852, the wide-ranging *Mikrokosmus*, which appeared in three volumes from 1856-64, and his *System der Philosophie*, which appeared as a *Logik* (1874) and a *Metaphysik* (1879).

Lotze's scientific and philosophical work was persistently informed by the paired convictions that philosophy must take into account the findings of natural science, and that science does not have final authority in matters philosophical. Thus, Lotze gained early prominence for arguing that, on scientific grounds, "vital force" should be excluded from physiology,[123] but that purpose and finality cannot be excluded from metaphysics. His metaphysics was a metaphysics of value, in which moral and aesthetic considerations become ultimate conditions on the conception of the real.[124] In psychology he extended a scientific approach to the mind as far as it would take him, but this approach was not incompatible with his postulation of an immaterial soul: his "physiology of the soul" soon left the domain of physical processes to consider processes he attributed to an immaterial substance. He approached such processes in the spirit of a scientist making and justifying a theoretical posit.[125]

Lotze's strategy in writing was to sort through the variety of positions on a given issue, seeking to understand the motivating impulses behind the various contenders and finally giving his support to the position that fared best under sympathetic but thorough criticism. His writings on the physiology, psychology, epistemology, and metaphysics of mind and space carefully consider the positions, or near-relatives of the positions, expressed in the works of Fries, Herbart, Steinbuch, Tourtual, and Müller. Lotze did not mention all these authors by name, but that is not surprising, for he gave few citations in his physiological writings and still fewer in his philosophical works. Nonetheless, his writings explore the conceptual landscape of questions in which physiology, psychology, epistemology, and metaphysics intersect with meticulous care, and he in fact canvassed all the major positions of nineteenth-century German physiology and metaphysics, without mentioning names.

Lotze's works in the period from 1846 to 1856, and especially the *Medizinische Psychologie*, provided a thorough and insightful discussion of the theoretical issues involved in spatial perception. Working explicitly within a dualistic metaphysics, he argued that since the soul is an immaterial being, it cannot be affected by the spatial configuration of physiological events in the brain but can react only to the intensity of neural activity. This led him to conclude that the elements of mental life are nonspatial representations varying only in qualitative character and intensity (MP, 203).[126] He then argued persuasively that, if this view of the elements of mental life is accepted, spatial localization can be explained only on the assumption that each sensational element is accompanied by a "local sign" (*Lokalzeichen*) whose qualitative character is dependent upon the specific location of the sensory nerve whose stimulation produced the sensation.[127] This argument proceeded simply from the assumption that sensational elements are nonspatial. It was thus binding upon any author who accepted this starting point, independently of whether that author embraced dualism. The necessity of postulating local signs of one sort or another was later accepted by individuals who varied widely in their views on mind and body, including Wundt, who adopted a position of psychophysical parallelism, Helmholtz, who chastised both materialism and dualism as "metaphysical hypotheses," and Hering, who was explicitly materialistic.[128]

Lotze, however, did not merely argue for the necessity of postulating local signs; he discussed various hypotheses pertaining to their causal origin. According to one account, which he did not favor (MP, 333), the sensation of a given nervous fiber receives its local sign simply by virtue of the fact that it has been produced by that fiber; each nerve fiber "marks" all of the sensations that it produces (cf. Tourtual). Lotze's own preferred account agreed with Steinbuch's in advocating a close connection between the muscle system associated with a sense organ and the local signs that accompany the sensations produced through stimulation of the sensory nerves of that organ. But he explicitly rejected an associational account of this connection (such as Steinbuch had proposed), and explained the origin of local signs through an innate anatomical mechanism (MP, 337-41, 355-89).[129] In the case of vision, the explanation went as follows. The visual system is innately disposed to fixate a single luminous point that appears in the visual field; if a point of light falls in the periphery of the retina, the eye will rotate so as to fixate it. He conjectured that this rotation is effected by a direct anatomical connection between each nerve fiber leaving the retina and a motor nerve fiber, such that when the motor nerve fiber is stimulated the eye will perform the appropriate rotation to achieve fixation. On his view of motor innervation, the production of a motor impulse simultaneously produces a sensation in the soul; this sensation, which will be unique to each motor fiber, "colors" the sensations produced by the attendant sensory fiber, producing a "local sign." Since each fiber of the optic nerve is, by hypothesis, connected to its own motor fiber, each has its own local sign, which is produced by virtue of the conjunction of sensory and motor fibers, even in the absence of actual motor impulses (the resting eye perceives spatial extension).[130]

Having argued for the necessity of postulating local signs, Lotze had yet to explain the actual formation of spatial representations. For Lotze, as for Steinbuch and Tourtual, local signs are no more than qualitative markers of spatial differentiation; if one were to perceive the local signs themselves, they would be perceived as distinct from one another without being placed in spatial relation to one another, like so many simultaneously discriminable (but unlocalized) tones (MP, 331-6). Thus, some further process or mechanism was required to form spatial representations in response to qualitatively distinct signs. Lotze (again in agreement with the majority of his predecessors) distinguished two aspects of this further process: the ability to form spatial representations at all, and the ability to localize sensations on the basis of stimulation.

Lotze contended that any sensory system that develops spatial representations must inevitably (and trivially) be ascribed an innate ability to have such representations.[131] He therefore focused his attention on the processes by which local signs mediate the localization of sensations, which, in the case of vision, meant localization within the field of view.

Lotze here explicitly addressed a difficulty that had existed in the theory of spatial vision for over a century but had usually been ignored. This difficulty pertained to the unconscious character of the mental states upon which operations are performed to produce spatial perceptions. From Berkeley's time onward, various writers had spoken of the suggestion of one mental state by another or of judgments rendered about the immediate object of perception or of *Bewegideen* whose qualitative character enters into spatial localization, but they had then hastened to add that these mental states and associational or judgmental operations usually go "unnoticed," or occur "without our being conscious of them."[132] These qualifications could parry the request for introspective evidence regarding the mental states and operations referred to. Having "explained" why these states and operations usually are not found in consciousness, the writers tended to brush over the question of whether they are *essentially* different from those that are open to introspection. Previous authors had usually left unspecified whether our *usually* not being conscious of the psychological processes underlying spatial perception in fact meant that we could *never* become aware of them, and thus that introspection could never provide evidence for or against the operations supposed to underlie perception. Lotze owned up to this problem directly. He accepted the commitment to unconscious mental states in the theories he favored, while he granted that the properties and alterations of such states could never be introspectively fathomed (MP, 179, 336-7). He then proposed properties for these unconscious states by working back from the variations in conscious states these properties were to explain, subject to considerations of theoretical elegance. He thus posited a variety of states that were unconscious and yet mental, outside of awareness and yet differentiated by their qualitative character.

Lotze considered the processes by which spatial localization occurs to be mental and unconscious. Perhaps for this reason he has usually been assigned to the "empirists" of the mid-nineteenth century, that is, to the group of authors who held that we "learn to see."[133] It certainly is the case that Lotze denied that humans are possessed of an innately given two-dimensional spatial sensation of the sort proposed by Johannes Müller (a paradigm nativist), for he denied that we have any primitively spatial sensations at all. Yet as the case of Tourtual has made clear, it was conceptually open to nineteenth-century sensory physiologists and psychologists to posit nonspatial sensations along with innate mental processes for localizing them. And indeed Lotze adopted one version of such a position: he posited an innate mental mechanism for localizing visual sensations within a two-dimensional representation on the basis of the qualitative character of their local signs, in such a way that a representation is produced that conforms to the retinal image.

> The only spatial intuition that derives from the structure [*Einrichtung*] of our organization and the physical-psychical mechanism, without assistance from the [actual] course of representations, is the surface-form ordering of points in the visual field; vision knows spatial depth only mediately according to the instruction of experience, while touch produces all of the dimensions of space equally solely through an [empirically-driven] concatenation of its individual sensations. Long practice has made this adjudication of depth so familiar to the percipient, that he thinks he sees it immediately. [MP, 418][134]

Lotze was a nativist with respect to two-dimensional visual form, but an empirist with respect to visual depth and all tactual localization.[135] His reasons for attributing two-dimensional localization to an innate mechanism were twofold: first, reports that the newly-sighted blind are immediately able to localize objects in the visual field, a reason he cited frequently but to which he claimed to attach little weight (MP, 360, 369, 419-20), and second, the lack of theoretical elegance of the associational account (MP, 355-7, 361), which must posit the formation of an individual associational connection for each retinal element (as Steinbuch did).

When Lotze turned to assessing the validity of space as a form of representation (the question of the ideality of space) and to explicating the grounds of geometrical knowledge, he maintained that the physiological and psychological considerations he had analyzed with great care were without relevance. His first extensive discussion of the epistemic status of spatial representation and of the grounds of geometry occurred in the *Mikrokosmos*. He adopted a position agreeing with Kant; in particular, he affirmed that mere associative connections do not amount to judgment, and he endorsed the ideality of space. He claimed that association cannot amount to judgment because it does not settle the question of right: "the mind is not content to have connections of ideas imposed on it by the mechanism of perception and memory; as an abiding critical energy, thinking seeks to test all of these by the grounds of right that condition the connection of the connected and prove the coherence of the co-existent."[136] Further, he contended that the question of whether a given representation, or form of representation, is acquired or inborn does not speak to its validity. Whether we perceive spatially as the result of an inborn mechanism has no bearing on the question of whether things in themselves are spatial, for whichever is the case our sensations would be the same; that is, even if things in themselves, including the nervous system, really were spatial, the sensations entering the mind would be the same, for the spatial organization of the nerves is not preserved in the sensations they produce, which are nonspatial.[137]

Although Lotze endorsed the ideality of space, his reasons differed from those he attributed to Kant. He objected both to Kant's arguments for transcendental idealism and to Kant's understanding of the foundations of geometrical knowledge. Lotze understood Kant as having grounded his commitment to the ideality of space on the argument that the spatial form of intuition must be innate if the necessity and universal validity of geometry are to be explained. Lotze thus interpreted Kant as equating innateness with apriority and necessity.[138] But Lotze denied that innate spatial representations could play any role in establishing the validity and certainty of geometrical knowledge, and for two reasons. First, he denied that spatial representations apply to things in themselves, on the grounds that things in themselves are essentially spiritual and hence nonspatial.[139] Second, he maintained that judgments pertaining to the postulates of geometry are grounded in the evidence of intuition itself. He argued that it is simply *given* to us in intuition (by inspection) that between any two points only one straight line can be drawn; he referred the "first principles" of geometry to the same basis in intuition.[140]

Lotze provided a more extensive discussion of his views on the grounds of geometrical knowledge in his *Logik*. There he agreed with Kant that geometrical knowledge is synthetic a priori, and that intuition plays an essential role in revealing the self-evident truths, such as the parallel postulate, from which other geometrical propositions might be deduced.[141] The knowledge attained by such intuitions he regarded as "a priori," even though it depended upon the experience of an occurrent intuition; for, although he denied that geometry could have a purely logical basis in the principle of identity or of contradiction, and he regarded the appeal to intuition as essential, he conceived the role of intuition as that of revealing the self-evident, not that of providing a basis for induction through repeated experience.

In his discussion of the basis of geometrical knowledge Lotze sought to tease apart the innate and the a priori. In contending that space must be an innate form of intuition (since the form of space could not be acquired through repeated experience of nonspatial elements) Lotze was willing to grant the terminological concession that this innate spatial form of sensibility might be termed an "a priori" form of intuition. However, he considered this sense of the term "a priori" to be irrelevant to the question of a priori knowledge. "It is not because the idea of space is innate in us, that we are in a position to frame universal propositions in geometry, which once thought are valid always," he argued. ɪndeed, if it were possible to explain psychologically the acquisition of space as a form of sensibility, that would not preclude our perception in intuition of the self-evident truths of geometry. For the "immediate apprehension of the universal truth" contained in spatial relations "would not be more inexplicable (though it would be equally inexplicable)," than if the spatial form were innate. The basis of geometrical knowledge according to Lotze was the self-evidence of the fundamental propositions of geometry, as revealed in the immediate apprehension of spatial relations in intuition; the causal origin of the intuited relations was irrelevant. Lotze therefore announced that, in his discussion of knowledge, he would "apply the term 'a priori' in a restricted sense only, viz. to indicate cases of knowledge that are not derived by a process of induction or summation from particular instances which exhibit them, but are thought to begin with as truths of universal validity, and are thus prior to the particular instances in the sense of being rules by which they are determined."[142] Such rules are not temporally prior to and causally determinative of their instances, as Fries would have it; they are "logically prior" to their instances: they are universal propositions that we recognize as universal in and through apprehending their instances.

Lotze thus completely divorced questions of the psychological genesis of spatial representation from questions of the status of geometrical knowledge of spatial relations. In so doing he was in accordance with Kant's own conception of the relation between innateness and the a priori (as discussed in chapter 3). Helmholtz would challenge the claim that the status of geometry is independent of the psychological genesis of the perception of space, maintaining that his empiristic theory of spatial perception spoke directly against the Kantian understanding of the basis of geometrical knowledge.

Chapter 5

Helmholtz: The Epistemology and Psychology of Spatial Perception

Helmholtz was a scientist first and a philosopher second. An anatomist and physiologist by training, he was a physicist by vocation. Schooled in medicine to finance his education, he was an army surgeon (a position he held under the terms of his financial aid agreement) when his celebrated "memoir" on the conservation of energy was published in 1847.[1] His first academic positions were in anatomy and physiology; he became professor of physics only in 1871. During the 1850s and 1860s he contributed to several areas of physiology, but especially to the physiology and psychology of sight and hearing, on which he published two major books: *On the Sensations of Tone* (1863),[2] and *Physiological Optics* (appearing in parts between 1856 and 1866).[3] He did not abandon work in sensory physiology after 1871, but he devoted himself primarily to thermodynamics and electrodynamics. From 1887 until his death in 1894, he administered the Physical-Technical Institute in Charlottenberg.[4]

Although always a scientist by profession, Helmholtz wrote on philosophical topics throughout his career, from the philosophical introduction to his paper on the conservation of energy to the last major paper published before his death, "The Origin and Correct Interpretation of Our Sense Impressions" (SW, 501-12). He took his philosophical writing very seriously.[5] He shared with Herbart and Lotze a firm conviction that natural science is important to philosophy, but he differed from them in attempting to banish metaphysics from philosophy once and for all. Helmholtz held that epistemology alone is the proper subject-matter of philosophy. He believed he was thereby calling for a return to the legitimate teachings of Kant, whose chief systematic goal was to undermine traditional metaphysics by placing limits on human reason. He also rejected the idealistic metaphysics that Fichte, Schelling, Hegel, and others had propagated in Kant's name. Helmholtz was a prominent member of the "back-to-Kant" movement of the second half of the nineteenth century, a movement that formed the background to early twentieth-century positivism.[6]

Helmholtz's epistemological interests, although expressed in the memoir of 1847, were most fully developed in connection with his work on the senses. His most enduring philosophical interest focused on the epistemological status of sensations as "signs" of the external world, a topic he took to be Kantian in inspiration. In treating sensations as mere signs of their external causes, Helmholtz considered himself to agree with Kant's epistemological modesty regarding the relation between appearances and the thing in itself. He did not, however, agree with Kant's epistemology on all counts. Helmholtz's work on problems in applied mathematics led him to reflect on the foundations of geometry, a topic on which he soon departed from Kant. In particular, his researches into n-dimensional manifolds—undertaken during the 1850s in search of an appropriate mathematical representation for mixtures of colors—led him to consider non-Euclidean geometries and to question the necessity of Euclid's axioms as a description of physical space. Subsequently, he combined his interest in the epistemology of sensational signs with his work on the foundations of geometry (both of which he continued to situate in relation to Kant) to produce his most finished piece of philosophical writing, "The Facts in Perception," which he delivered to the faculty of the University of Berlin in 1878 in his capacity as rector.[7] The paper approached its epistemological subject-matter from the standpoint of natural science; Helmholtz applied his own extensive work in the psychology of perception to the problems of epistemology as he understood them.

In physiological work during the 1840s Helmholtz had endeavored to prove that vital phenomena are subject to the laws of physics and chemistry and thereby to secure their position within natural science. One might see his work in epistemology as a similar attempt to extend the results of natural-scientific investigation to an understanding of the knowing mind, for indeed Helmholtz adopted a naturalistic attitude toward the mind. He was committed to the existence and explanatory legitimacy of laws of mental activity, whose acceptance he believed could be justified empirically. However, his efforts in psychology and epistemology differ importantly from biophysical work of the 1840s: although his attitude toward vital phenomena was reductionist, his attitude toward the mind was not. With respect to mental processes, he was a *methodological* rather than a *metaphysical* naturalist. He treated the laws of psychology as autonomous from physics or physiology; he had little patience with what he termed the "dogmatism" of materialists, or with the vague statements of physiological reductionists (PO 3:432). His naturalism was embodied in the methodology and modes of explanation he brought to the study of mental phenomena. The laws he adopted in psychology mirrored his ideal for lawful explanation in the physical sciences: they could be stated as simple generalizations governing the interaction of elemental states of the mind in accordance with the variation of such states along a few dimensions.

Helmholtz's commitment to psychological laws was most manifest in his work on the physiology and psychology of spatial perception and especially in the third or "psychological" portion of his *Physiological Optics*. In this part of the *Optics*, which he found the most challenging (he complained of frequent headaches while composing it),[8] he attempted to provide explanations of a variety of the phenomena of spatial perception by bringing them under universal psychological laws; he also sought to extend his naturalistic account of the mind to the domain of "higher" cognition. Fundamental to his picture of the relationship between psychology and epistemology was the idea that the process of experimental investigation in science and the resulting scientific inferences are simply an extension of the most ordinary psychological processes, such as those underlying perception of the size, shape, and distance of objects in the field of view. As he enjoined on more than one occasion, scientific inference is the psychology of perception writ large.[9]

The comparison of perceptual with scientific inference reveals two fundamental tenets of Helmholtz's theory of perception: he characterized the psychological processes underlying perception as unconscious inferences, and he emphasized the role of active experience in the formation and testing of such inferences. He first announced both tenets in his Kant Memorial Lecture of 1855,[10] and they were included (though not without revision) in the last paper he published, in 1894. Helmholtz emphasized the central role experience plays in both scientific inference and perceptual psychology. Just as experimental scientists develop and test laws on the basis of experience, the perceptual system acquires the "laws" of spatial perception on the same basis. Unconscious inference is involved in both learning and applying these laws, which are acquired through reasoning by analogy (cf. Whewell and Mill). Once acquired, in perception they serve as major premises guiding inferences to the size, shape, and distance of objects on the basis of retinal stimulation, and in science they serve to explain and predict observed phenomena.

Helmholtz argued for a radically empirist[11] theory of spatial perception similar to that of Steinbuch and Herbart. He maintained that the ability to localize objects visually, even in two dimensions, is acquired through experience. Because he allotted an extensive role to experience in spatial perception and compared spatial perception to scientific inference, his position might seem not only "empirist" but also "empiricist." However, although Helmholtz rejected what we should call "rationalism," and although he emphasized that confirmation in experience is the only basis for accepting scientific statements, he was not simply an "empiricist." Indeed, he regarded the philosophical legacy of Locke and Hume to be skepticism (PO 3:32), and he considered his position to be fundamentally Kantian.[12]

Helmholtz wrote in 1884 that he had always been a "faithful Kantian" in fundamental respects but realized that Kant required revision on certain matters (VR 1:viii). The agreement that he envisioned with Kant pertained to problems and solutions in epistemology. Helmholtz distilled the problems of epistemology into the question, "What is true in our intuition and thought?" or, "In what sense do our representations correspond to actuality?"[13] This question asked for an answer in terms of the *kinds* of representations we possess. It is not a question of whether this or that judgment or perception is "true" or "corresponds" with reality, but whether, say, color sensations (or represented temporal relations, or represented spatial relations), considered as a kind of representation, reveal the actual properties of material objects. This question is an accurate statement of one of the chief epistemological problems inherited from the eighteenth century. The question had been posed prior to Kant, and he countenanced its asking (by asking it himself) but forbade answering (if an answer requires determining the degree of correspondence between representations and things in themselves). Helmholtz would seek an answer not by advancing a metaphysical conception of the nature of reality as a standard against which to compare the dimensions of perceptual experience, but by pooling the resources of physics, physiology, and psychology in order to determine the epistemic status of the various kinds of perceptual representations.

Helmholtz agreed with Kant's antimetaphysical stance, expressed in terms of an attitude of epistemic modesty with respect to "things in themselves." However, he had to reconcile this attitude with Kant's claim to derive a broad range of what he termed "metaphysical" knowledge from synthetic a priori principles. And indeed, despite Helmholtz's empiristic stance, he considered himself in agreement with Kant's commitment to a priori principles, in two respects. First, he claimed that the fundamental tenet of his epistemology and psychology of the senses—the doctrine that visual sensations of light and color are uninterpreted "signs" (or "symbols") whose meaning must be learned through experience—was in agreement with the Kantian doctrine that our forms of experience are subjective. He portrayed Müller's doctrine of specific nerve energies—according to which the qualitative character of sensations is attributed to the characteristics of nerve fibers, not to the external stimulus (since any of a variety of nonluminous stimuli can evoke sensations of light in the visual system)—as a physiological confirmation of Kant's doctrine of subjectivity. Helmholtz stated his epistemological conclusion from this doctrine as the position that "sensations are only signs for the properties of the external world, whose interpretation must be learned through experience" (VR 1:17; SW, 475).[14] This position was "Kantian" only in the weak sense that it exhibited epistemic modesty regarding "things in themselves." In its treatment of color sensations as an example of Kant's doctrine of the "forms of intuition," it erred doubly as an interpretation of Kant: it treated Kant's doctrine as a version of nativism, and it missed the epistemic significance of Kant's focus on space as opposed to color.

The second point on which Helmholtz claimed to agree with Kant's doctrine of transcendental a priori knowledge pertained to the causal law. In his lecture of 1855, Helmholtz argued that the causal law must be accepted as a presupposition of cognition of an external world, declaring his allegiance to Kant on this point (VR 1:116). He retained a general commitment to this position, repeating it in 1866, 1878, and 1894. Indeed, under the guise of the "principle of sufficient reason," Helmholtz had appealed to the universality of the causal law in the opening arguments to his memoir on the conservation of energy, arguments which he later would affirm to have been Kantian in inspiration.[15] However, I shall argue that, despite Helmholtz's repeated statements that the causal law is an a priori principle of knowledge, his attitude toward that law and its epistemic foundation changed across the decades as he came to realize the full force of the notion that the law cannot be justified empirically. In the end, he refused to grant it an unequivocal a priori status of the type Kant had assigned it. Correlative with this change in his attitude toward the causal law was a change in his solution to the problem of the external world; from an early reliance on the causal law in inferring the existence of the external world, he came to acknowledge that no such proof could be given, and he adopted a position according to which we accept the causal law on faith, as a guiding principle in our comprehension of appearances.

Despite Helmholtz's frequent insistence on his Kantian ancestry, the "orthodox" Kantians of his day denied his claim. They rejected as most un-Kantian his belief that foremost among the "incidental points" requiring correction in Kant's philosophy was the doctrine that Euclid's axioms apply with necessity to physical space.[16] Kant held that the applicability of the axioms of geometry to physical space is assured through the doctrine that space is the a priori form of outer sense. Helmholtz, who read this doctrine as a statement about the psychology of spatial perception, argued that Kant's mistakenly nativistic treatment of spatial perception led to his equally mistaken doctrine that physical space is necessarily Euclidean. But, Helmholtz contended, recent developments in geometry revealed that Euclid's axioms are only contingently applicable to physical space, and recent developments in the physiology of the senses indicated that the particular form of spatial intuition was acquired through experience. Helmholtz thus found himself in a position to correct Kant on both the psychology of spatial perception and the status of Euclid's axioms as a description of physical space. However, as we shall see, while Helmholtz was in a position to argue effectively against Kant's position on the geometry of physical space, the success of his argument was not based upon his psychology of spatial perception in the way he thought it was.

But even apart from the question of geometry, Helmholtz's relation to Kant was problematic. Throughout his writings on the epistemology and psychology of spatial perception there is a permanent tension between his "Humean" project of naturalization and his "Kantian" commitment to the objectivity of scientific knowledge. Beginning in the 1860s Helmholtz sought, on the one hand, to resolve the unconscious inferences of perception into purely associative processes; on the other hand, he remained committed to the paired beliefs that the processes of perception are of a kind with scientific inferences and that through science nature can be "comprehended" according to objective laws. He recognized that the commitment to objective laws could not be justified through the associative inferential processes that he believed underlay both scientific thought and perception; here he appealed to the law of cause as a transcendental principle lying implicit within scientific and perceptual inference. Nonetheless, his commitment to the associative analysis of inference and his doctrine that the law of cause as a transcendental principle is implicitly contained in such inferences were sometimes expressed within pages of one another in the same work.[17] These commitments exhibit the same sort of tension as existed between the accounts of causality adopted by Hume and by Kant.

In assessing Helmholtz's naturalistic program, it is important to attend both to his explicit methodological pronouncements and to his actual argumentative practice. Helmholtz reveals a number of his attitudes toward the foundations of psychology and the relation between psychology and physiology only by way of (nearly) tacit assumptions invoked in the heat of argument. His arguments against nativism particularly reveal underlying assumptions, both about his conception of the relation between physiology and psychology and about his view of the general canons for theory choice in natural science. Not surprisingly, these underlying assumptions sometimes conflict with his explicit methodological pronouncements. Although he often maintained that psychology must look to physiology for its theoretical starting points, Helmholtz himself persistently treated psychology as an autonomous science that could not be reduced to physiology. Moreover, it is important to keep in mind that Helmholtz advanced the goal of applying the techniques of the natural sciences to mental activities as a programmatic ideal. Antidogmatist that he was, he brought a "try and see" attitude to the project of extending natural scientific methods to the analysis of the mind.

Following Helmholtz's own tendencies, I shall approach his work in perception by first characterizing its more strictly empirical portions, and then I shall turn to the philosophical positions that informed it or were developed out of it. Discussion of his theory of spatial perception, including his arguments for empirism and his doctrine of unconscious inference, will be followed by examination of his epistemological position, including his "sign" theory of perception and its connection with the problem of the external world and the status of the causal law. His conception of the status of the axioms of geometry and their connection with the psychology of vision can then be examined, and the force of his response to Kant's teachings on geometry assessed. My final section considers the extent to which Helmholtz's work prescribes and exemplifies limits to a naturalistic approach to the mind.

1 Helmholtz's Psychology of Spatial Perception

Helmholtz devoted more pages of his published work to the psychology of visual spatial perception than to any other single topic. This is somewhat ironic, for he considered himself to be a physicist and physiologist but not a psychologist. He established his early reputation in physiology through the discovery of the speed of the nervous impulse, and he achieved immediate fame in vision-related studies through the discovery of the opthalmascope, which happened almost incidentally, in the course of preparing a lecture demonstration. Nonetheless, from the mid 1850s the psychology of spatial perception took on an interest for him that was unabated until his death.

Although he first broached the topic of spatial perception in his Kant Lecture of 1855, a systematic statement of his psychological theory did not appear until the third part of the *Physiological Optics*, published in 1866. In his introduction to the entire *Optics*, Helmholtz divided the field of physiological optics into three parts: "the dioptrics of the eye," "the theory of the sensations of the nervous mechanism of vision," and "the theory of the interpretation of visual sensations." The first division traces the path of light rays as they enter and traverse the optical media of the eye. The second pertains to the sensations that such light arouses in the nervous system, "without taking account of the possibility they afford for recognizing external objects"; the visual sensations are treated as uninterpreted signs aroused in the nervous system by the action of light. The final division pertains to the "representations of objects disposed in space" that the sensations enable an observer to form (PO 1:35). Later he explained that this topic belongs properly to the "field of psychology," although both physics and physiology can contribute to its investigation: the first, through an analysis of the physical stimulus for spatial vision, and the second, through an analysis of the physiological activity of the sensory nerves and the ocular musculature (PO 3:3).

Considered broadly, Helmholtz's tripartite division is parallel to the three grades of sense described by Descartes. However, Helmholtz's division shares with nineteenth-century authors such as Steinbuch, Tourtual, and Lotze the assumption that the sensations of vision are nonspatial, varying only in quality and intensity. Along with a number of his predecessors and contemporaries, Helmholtz subscribed to the one-fiber, one-sensation doctrine. He readily admitted that the assumption of nonspatial sensations could not be justified through introspection, a method which he considered inadequate to reveal the basic elements of psychological processes (VR 1:111; SW, 142). Rather, he based his posit of such sensations on the physiological assumption that each sensory nerve conveys its sensory activity into the brain independently of other nerve fibers, like an insulated telegraph wire; in the brain, this activity is "brought to sensation," or as he sometimes put it, "to consciousness"— a terminology he adopted even while recognizing that such sensations are not singly available to introspection (PO 2:29-30).[18] Like earlier investigators, he was quite liberal in positing unconscious sensations when they seemed required in order to explain consciously available representations. He adopted the position that the stimulation of a single fiber in the retina yields not only a sensation of color but also a "local sign" peculiar to that fiber. Like Steinbuch, he contended that originally such signs are without spatial meaning, which they attain only as a result of subsequent psychological processes.

Helmholtz considered the psychological portion of physiological optics to be concerned precisely with the processes by which nonspatial sensations are combined to form spatial representations. His account of such processes was radically empirist; he considered all ability for spatial localization, including the representation of visual direction within the two-dimensional visual field, to be the product of experience. Although he readily admitted that the available evidence did not unequivocally support his approach over its nativist rival, he argued with some vigor in favor of his position and against nativism. The third part of the *Optics* was organized around these arguments. Reflecting Helmholtz's own emphasis, I consider his positive theory of spatial perception first, before turning to his arguments against nativism.

Much of Helmholtz's discussion of spatial perception is independent of his arguments for empirism and against nativism. By far the largest portion of the third division of the *Optics* is devoted to what might be termed the "psychophysics" of spatial perception, that is, to the review and analysis of findings pertaining to the relation between physical stimuli and perceptual representations of space.[19] The factual investigations Helmholtz undertook in mapping the monocular field of sight and determining the horopter, in formulating laws of eye movement, in investigating the conditions that affect the perception of visual direction, distance, and stereoscopic depth, and in studying double vision, need not be judged in light of the success or failure of his empiristic theory.

Helmholtz himself underscored the theoretical neutrality[20] of what I have termed the "psychophysical" portion of his investigation. In presenting his fundamental rule for investigating spatial perception, he was careful to state it in a form that would render it neutral with respect to the controversy between nativism and empirism. He formulated it as follows: "We always represent such objects as present in the field of vision as would have to be there in order to produce the same impressions on the nervous apparatus, the eyes being used under ordinary, normal conditions" (PO 3:4). The rule as stated by Helmholtz contains no mention of the processes by which visual representations are formed, nor does it imply a stand on the origin or mode of operation of such processes.[21] It simply states that a regular relation exists between impressions on the sensory apparatus and the resulting spatial representations. An example of the rule is that stimulation of the outer edge of the retina leads to the experience of a "luminous object" in the opposite portion of the field of vision. Helmholtz was aware that visual appearances could be caused under conditions that produced incorrect (illusory) localization, and indeed that visual appearances could occur even in the absence of luminous stimulation (e.g., by mechanical stimulation of the eye). But he maintained that, no matter what the actual source of ocular stimulation, objects are represented as being in the location that would cause the current stimulation under "normal conditions." By the latter phrase, he explained, no more is meant than that the eye has been stimulated by light "coming from the opaque objects in its path that were the last to be encountered, and having reached the eye along rectilinear paths through an uninterrupted layer of air"; reflected and refracted light, as well as nonluminous stimulation, are considered abnormal on the grounds that they are atypical.

Although the bulk of Helmholtz's discussion of spatial perception in the third part of the *Optics* is presented in this neutral manner, the order of presentation and the inclusion of certain topics reveal the overarching influence of the argument for empirism. After the philosophical and theoretical preliminaries of the opening section, Helmholtz treats of eye movements first, explaining that they play an essential role in the development (*Bildung*) of spatial representations. From there, he describes the monocular field of vision as the mobile eye learns to "measure" it, and determines the "mapping" of isolated retinal stimulation into visual directions as it develops through training. He next examines the impressions of distance and of depth produced by various optical stimuli under conditions of monocular and binocular viewing. In each case, he presents purely psychophysical relations separately from their psychogenesis. Nonetheless, the general shape of the discussion is molded by the theory that the relations have been established through experience, and Helmholtz endeavors to show in each case how such learning could take place. Further, he devotes an entire section to the phenomena of "binocular rivalry," which he considered to provide especially compelling support for the empiristic theory, and he concludes the third part with a review summarizing the evidence for empirism and against nativism.

Helmholtz's general strategy was to explain the origin of the observed rules of spatial perception through a process of associative learning. In broad outline, his account of the development of spatial perception runs parallel to that of Steinbuch: from nonspatial beginnings, the individual first maps the visual field through eye movements that sweep the fovea over the field of view. The "impulse of the will" to fixate the eye, which he also described as the "degree of innervation" of the ocular musculature, is associated with a particular direction of fixation.[22] Once the effort of the will in moving the eye has been mapped into visual directions (directions of the foveal "line of sight"), the rest of the retina can be mapped into directions in space for the static eye. This is accomplished, as in previous accounts, by associating the "local sign" of each retinal element with the "spatial meaning" of the muscular movement needed to rotate the eye so as to fixate the source of light stimulating a particular retinal element. Alternatively, the directional "meaning" of the various local signs could be mapped by using the hand, whose location in space is assumed to be known, to cover and uncover various luminous sources, thereby establishing the direction in space associated with a retinal element for a given fixation of the eye. Helmholtz thus allowed the possibility that "touch educates vision" and assumed that the spatial sense of touch is a given, without, however, insisting upon the absolute necessity of tactual education for the development of spatial perception.[23]

Despite such similarities with previous theory, Helmholtz's account of the development of spatial perception differs from that of the authors discussed in chapter 4. Unlike them, Helmholtz did not adopt a firm position on whether the bare spatial "form" of experience—the ability merely to represent spatial relations—is acquired through learning.[24] His account focused on the development of the ability to localize visual sensations. Further, although his account of this ability agreed with

Steinbuch, Tourtual, Lotze, Wundt, and others in emphasizing the role of the ocular muscles in determining the "meaning" of local signs, he differed from them on the role he assigned to feelings of muscular innervation.

The authors just mentioned derived the localization of sensations from the degree of innervation of the muscles. Tourtual and Lotze postulated that this spatial "meaning" of the muscle sense is innate (at least for two-dimensional representation). Helmholtz rejected this position. But his treatment of the origin of visual direction also differed from that of Steinbuch, Wundt, and other empirists. According to Steinbuch, an inner "visual space" is developed out of random eye movements *in utero*; on the assumption of a precise quantitative relation between the contraction of the eye muscles, the position of the eye, and the qualitative character of the resulting *Bewegidee*, he postulated the construction of an inner hemisphere of *Bewegideen*, associated through similarity, into which the sensations produced by stimulation of the eye are mapped through a process of association by simultaneity. Wundt, writing in the early 1860s in a work frequently cited by Helmholtz, adopted a similar conception, according to which spatial representation in two dimensions originates through the successive apprehension of "muscle sensations" (*Muskelempfindungen*) that are quantitatively ordered in correspondence with the direction of fixation of the eye; through the successive apprehension of these representations, a spatial image is constructed in the "reproductive imagination."[25] According to both Wundt and Steinbuch, the qualitative character of the sensations associated with muscle action vary in such a way that, when ordered according to similarity, these sensations yield a spatially correct mapping of the respective external positions of the eye. Helmholtz, however, doubted that the "feelings of contraction" (*Gefühl der Muskelanstrengungen* or *Spannung der Muskeln*) of the muscles could be sensed with the quantitative precision required by such theories; nor did he assume that sensations implicated in the use of the muscles originally form a well-ordered qualitative series.[26] He thus needed to develop his own account.

Instead of speaking of "feelings of contraction," Helmholtz preferred to designate (PO 3:204-7, 209) the mental sign correlated with the position of the eyes as the "impulse of the will" (*Willensimpuls*) or "effort of the will" (*Willensanstrengung*); he also spoke of a feeling of the "degree of innervation" (*Grad der Innervation*) or "feeling of innervation" (*Innervationsgefühl*) accompanying a voluntary motion (3:433, 466). These feelings are not specific to the contraction of the muscles but consist in the "intent of the will" (*Willensintention*) to focus on one part of space rather than another (3:48, 54). In an adult, a specific degree of innervation will correspond to a particular spatial location. Not so for an infant; Helmholtz maintained that the "intention of the will" requisite for a particular movement must be learned. He was a radical empirist with respect to the "spatial meaning" of voluntary motions (3:54, 207); originally, the "intent of the will" is just another qualitatively distinct element that enters into the associative process by which the rules of spatial construction are learned. His account of this process was quite sketchy, but it made clear that the primary role of the impulse of the will in early learning is to allow the developing perceiver to distinguish changes in sensation that depend upon his or her movement from changes that result from alteration in the objects reflecting light. When faced with a stable visual field, the observer comes to recognize the possibility of repeatedly directing the ocular musculature in such a way that a previously experienced constellation of luminous sensations and local signs is experienced once again ("at will"). The possibility of directing the musculature so as to reexperience a constellation of visual sensations is preserved in "sensory memory"; this memory mediates subsequent intuitions of the spatial directions of visual sensations (3:56-7, 135; SW, 215-16). If Helmholtz's account was to work without positing innate spatial meaning to the "impulses of the will," he had to assume that these impulses vary qualitatively with the direction of fixation in an arbitrary manner, and that association through simultaneity is responsible for establishing topologically appropriate connections among them. Once visual direction has been established for the foveal line of sight, the perceiver maps the relation between directions in the field of sight and local signs for points located off the fovea.

Helmholtz considered the difference between his and his predecessors' accounts of the origin of spatial direction to be significant. Steinbuch, Lotze, and Wundt believed that the muscle sense is primary in the development of visual localization (the sensations of the muscle sense are qualitatively well ordered to serve spatial perception). In contrast, Helmholtz maintained that the "spatial meaning" of various *Willensimpulse* was subject to a constant test of experience, in such a way that the relationship between the "impulse of the will" and a particular location in space was constantly reinforced by, and subject to, retraining through experience. He demonstrated this fact through experiments in which prisms were used to cause a misalignment between the field of sight and the felt location of the hand; in such cases, eye-hand coordination was "re-tuned" in favor of felt location (PO 3:206-7; SW, 198-9).

Despite these differences over the operation of the muscle sense, Helmholtz's general analysis of the perception of visual direction agreed with that of his predecessors: in the mature perceiver, the localization of visual sensations depends upon (1) a feeling of the position of the body and eye and (2) the "local sign" of the stimulated retinal element. Together, these two sources of information establish visual direction from a particular standpoint. At least for the monocular field of vision, however, they do not serve to determine the third dimension; additional learning is required for the perception of distance.

Helmholtz's account of the perception of distance and depth again built upon but departed from previous accounts. The differences began with his refusal to posit a representation of the two-dimensional visual field as a discrete element in the visual process. Although he sometimes spoke of a "field of sight" in which objects appear "as if" they were in a plane, he warned that this manner of speaking was a mere expository convenience. As he remarked, even when we experience objects at a distance and in depth, the topological properties of the "field of sight" are preserved, as if the visible world were a sheet of rubber stretched forward and back to yield a three-dimensional scene (PO 3:132-4).[27] And while he considered the ability to localize within the field of sight to be distinct from the ability to determine distance, he did not contend that this two-dimensional localization occurs prior to the perception of distance, either temporally or causally. Indeed, he considered localization in three dimensions to be the basic form of representation in mature perceivers, and he maintained (3:330; SW, 205) that experience of the two-dimensional ordering of the field of view results from a special attitude or interest on the part of an observer (such as an artist, or a sensory physiologist). In forsaking the two-stage theory, Helmholtz departed from a host of previous theorists, including Descartes, Berkeley, Steinbuch, Tourtual, Müller, and Lotze.

Although Helmholtz did not posit a two-dimensional image as a primitive stage of the visual process, he allowed some efficacy to the pictorial cues already discussed (chapters 2 and 4). However, he contended that the pictorial cues allow us merely to form a *representation* of distance, while a second set of cues (*Hilfsmittel*) underlies the actual *perception* of distance. He observed that the so-called pictorial cues—illumination, atmospheric haze, geometrical perspective, overlap, and known size—depend upon prior knowledge of the objects in the field of view. For this reason he considered it beyond doubt that the efficacy of such cues arises through experience (as all previous authors agreed). Helmholtz adopted the usual line on the unconscious efficacy of such cues: the cues become connected with distance through an associative process that is "not conscious and not voluntary, but is as if through a blind force of nature, even if it follows the laws of our own mind" (PO 3:245). The end product of such associative processes takes on the same immediacy that we ascribe to sensation, and so we seem to see objects at a distance immediately, even when the representation is formed through associations based upon prior acquaintance with particular environmental circumstances and types of object.

The second type of cue does not require prior knowledge of specific objects. These cues depend upon the application of general rules to the sensations produced by optical stimulation (including muscle feelings): included here are accommodation, convergence, motion parallax, and binocular stereopsis. These cues vary in their efficacy and importance. At best, accommodation (PO 1:103) could work only for short distances (beyond which focal length goes to infinity), while convergence (3:260) provides a potential cue to absolute distance (by means of implicit triangulation). Neither cue is especially robust, as Helmholtz reported (3:245-6, 260-70) by reviewing some observations performed by Wundt (to which he added his own). In contrast, he considered motion parallax and binocular stereopsis to be the most important and reliable cues for the perception of depth and distance; or rather, he considered them to be two ways of taking advantage of one and the same source of information about distance: both depend upon "comparison of the perspective images that the same object presents from different viewing positions" (3:246).[28]

Helmholtz's treatment of motion parallax was brief and conversational. It was illustrated with examples from ordinary life in which objects at different distances noticeably change their apparent positions with differing velocities as the observer moves: walking along a path, moving in a thick woods, looking out the window of a moving train. His primary theoretical observation indicated that the comparison of differing images in such cases requires the use of memory to preserve a previous image (PO 3:246-8).[29] In contrast, he provided an extensive discussion of the geometry of stereopsis, of the instruments used to gather stereoscopic observations, and of the results of such observations.[30] Interestingly, his initial discussions of both motion parallax and stereoscopic depth focused on the psychophysics of these modes of stimulation. Nevertheless, while acknowledging that these cues do not depend upon knowledge of particular objects in the field of view for their efficacy but upon general rules, he maintained that these rules are learned. His argument for an empiristic explanation of these cues was less direct than in the cases of the pictorial *Hilfsmittel*. Indeed, only in the sections following his psychophysical treatment did Helmholtz provide an argument for his empiristic interpretation, which relied upon inferences drawn from observations of double vision and binocular fusion (in conjunction with a number of supporting assumptions).

The fundamental characteristics of Helmholtz's theory of spatial perception are clear. The theory would be constrained by a precise empirical characterization of the spatial abilities we in fact have; that would determine the object of explanation. According to Helmholtz's own preferred theoretical perspective, explanations of these phenomena would appeal only to simple psychical elements and laws for their combination.

The elements were nonspatial sensations varying in quality and intensity and the laws were those of associative combination. Helmholtz was quite clear on the latter point. In the concluding section of Part Three he summed up his account of spatial perception by observing that "the only mental processes involved therein are the involuntary ones connected with the association of ideas (*Ideenassoziation*) and the involuntary flow of representations" (PO 3:439). The theoretician's task was to show how the application of such laws to the sequences of elements arising in the course of ordinary experience would yield associative habits sufficient to produce, in given circumstances, the spatial representations possessed by adults.

The portrayal of Helmholtz's project thus far describes his account of the psychology of spatial perception as presented in the technical sections of Part Three of the *Optics* (sections 27-33, including Helmholtz's concluding statement of his theory). My account, reflecting Helmholtz's discussion in those sections, has avoided two topics of considerable interest that he broached in the general introduction to Part Three (section 26) and developed more fully in his lectures on vision. First, I have not discussed Helmholtz's characterization of the psychological process of spatial perception as an *unconscious inference*. This silence accords with the fact that Helmholtz did not use that term in his technical account of the psychology of spatial vision; nonetheless, in section 26 and in his 1868 lectures he portrayed unconscious inference as the workhorse of the psychology of perception.[31] Second, I have not mentioned the role of the law of cause in these inferences; yet Helmholtz concluded section 26 by stressing the role of this law in perception of an external world, and he emphasized its importance in previous and subsequent lectures. These doctrines can receive adequate interpretation only in connection with the philosophical dimension of Helmholtz's work on perception, discussed below in sections 3 and 4.

2 Helmholtz's Arguments against Nativism

The question of the psychogenesis of spatial representation and spatial localization received attention from German philosophers, physiologists, and psychologists throughout the nineteenth century. The question was complex, and the answer given by an author might reflect commitments on a number of issues. In the case of Helmholtz, these issues included the relation between nerve-fibers and sensations, the genesis and character of the mental operations that integrate sensations into perceptions, and the relation between space considered as a "form" of such perceptions and the actual spatiality of the world. This latter topic is epistemological and will be addressed in section 4. Here I shall focus on Helmholtz's arguments against nativism as an hypothesis about the mechanisms by which spatial representations are produced. These arguments particularly reveal Helmholtz's methodological and theoretical assumptions.

Although the question of whether we "learn to see" or are innately possessed of our visual abilities had been the subject of intense discussion for more than a century prior to Helmholtz, he first made this controversy the organizing theme of a comprehensive discussion of spatial perception.[32] Even so, he never clearly separated the varieties of nativism against which he responded but simply took aim at the enemy under whatever guise it appeared. Nonetheless, the controversy between nativism and empirism, as he understood it, had two dimensions. The first pertained to the question of whether spatial localization within the two-dimensional visual field results from innately given sensations of direction. When Helmholtz was writing the third portion of the *Optics* the dominant eighteenth-century view—that a two-dimensional spatial representation is primitively given in vision and that the perception of depth and distance depends upon learning—was still prominent, perhaps largely through its endorsement by Helmholtz's eminent teacher, Johannes Müller. Yet this position had been notably challenged. Helmholtz recognized two groups of predecessors for his position that a two-dimensional representation is not given, but arises through learning: a philosophical lineage stemming from Herbart, and a line of sensory physiologists akin to Steinbuch (PO 3:33). Among the philosophers, he included Cornelius and Waitz who followed Herbart in deriving all spatial representation from nonspatial sensations through the "fusion" or "associative blending" of such sensations.[33] Among the physiologists, Helmholtz named Nagel, Classen, and Wundt, even though only the latter two actually produced empiristic accounts of two-dimensional localization.[34] Nagel was a nativist on this score, and Classen joined him in 1863, denying his own published position of the previous year. Both considered the perception of depth to be a product of learning, but then so had Müller, whom Helmholtz classified as a nativist.[35]

That Helmholtz counted Nagel and Classen as empirists can be explained only by turning to the second dimension of the controversy, which pertained to the postulation of innate anatomical mechanisms to explain certain aspects of spatial vision. Everyone assigned nerve fibers the innate power to produce sensations, Müller contending that such sensations are spatial, but many others (including empirists, and nativists of the Tourtualian variety) denying it. Beyond the spatiality of the sensations themselves, a further question arose regarding the role of nervous mechanisms in spatial perception: the problem of binocular single vision. Authors disagreed about whether innate anatomical connections are required to explain the fact that we have two eyes but perceive only one world. What is the mechanism by which two instances of physical stimulation (one in each eye) yield a single sensation localized in space? Some authors, including Müller, explained single vision by positing innate anatomical connections between the two eyes. Others, including Helmholtz, maintained that stimulation of nervous fibers in the two retinas produces two sensations that are united by being referred, through a psychological process, to a single external source. With respect to the controversy between nativism and empirism, the point of contention was whether single vision is the "psychical" product of learning through experience or the result of innate anatomical connections. Helmholtz classified the work of Nagel and Classen as empirist, even though they were committed to innate spatial representations, because they explained binocular single vision as a "psychical" product of learning.[36]

A clear schema of classification into nativists and empirists lay implicit in Helmholtz's discussion. A position could be classified through its implied response to two questions. Does it explain localization in two dimensions through innately given spatial sensations? Does it explain binocular single vision as the result of innate anatomical connections? If the answer to both questions is either affirmative or negative, the classification is obvious. In the case of a mixed response, Helmholtz counted a negative answer to the second question as sufficient to make one an empirist. Thus, the question of innate anatomical connections carried the greater weight in Helmholtz's classification scheme.

Helmholtz's emphasis on anatomical connections becomes easier to understand when we consider that his chief nativist opponent in the *Optics* was Ewald Hering. Beginning with his *Beiträge zur Physiologie* (1861–1864), Hering argued for an extreme form of physiological nativism. I say "physiological nativism" to distinguish his position from the sort of nativism that attributes spatial localization to an innate disposition of a soul or mind (as in Lotze's account of surface-form localization). Hering was concerned especially to argue that vision must be explained by referring to properties of the nervous system, not to properties of a distinct mental substance. Throughout his work he attacked views that referred to the activity of a "*Seele.*" He sharply distinguished his "purely physiological" approach from the psychologically oriented accounts of Wundt, Nagel, Classen, and Volkmann, and he lamented the fact that recent accounts of vision had taken on such a "psychological" coloring.[37] Hering portrayed the difference between his views and those of Wundt or Classen in terms of the type of explanation that should be given for spatial perception: explanations should be physiological, instead of appealing to a bare "unconscious inference." He did not deny that inferential or judgmental processes play a role in vision, but he insisted that such psychological processes be given a physiological foundation.[38] In practice, he tended to minimize the role of learning in perception, but he would have rejected the suggestion that his demand for physiological explanations forced him to do so.

Hering's emphasis on anatomical and physiological explanations, together with his opposition to Wundt and Classen (whom Helmholtz later would explicitly mark as allies) suggests that the controversy between nativism and empirism may be more fundamentally characterized as a controversy over the relation between physiology and psychology, or over the need to reduce psychological explanations to physiological ones. Hering insisted upon anatomical and physiological explanations for spatial perception, whereas Helmholtz, Wundt, and Classen insisted upon psychological explanations, distinct from anatomical and physiological considerations. Indeed, for some authors the controversy marked a divide between materialism and metaphysical dualism. Schleiden and Ulrici, two authors Helmholtz listed in the bibliography to his initial discussion of nativism and empirism (PO 3:33), emphasized the high metaphysical stakes in the dispute. They argued that evidence in favor of the role of learning in perception was also evidence for the existence of a soul distinct from the brain; for, they reasoned, unconscious inference is implicated in learning, and a special mental substance is required to perform such acts of ratiocination.[39] Helmholtz refused to endorse such metaphysical commitments. At the same time, he labeled the appeal to anatomical and physiological explanations as speculative, and he rejected materialism as being no less metaphysical than "spiritualism." He considered psychology to provide a distinct type of explanation, with its own evidential basis independent of physiology (3:432). Accordingly, the choice between nativism and empirism was for Helmholtz a choice between anatomical and physiological explanation on the one hand and psychological explanation on the other.

In assessing Helmholtz's arguments against nativism, we should bear in mind that he frankly admitted the available evidence did not permit an unequivocal decision between nativism and empirism.[40] He defended his preference for empirism by attacking nativism on three grounds: (a) with empirical evidence, (b) with the claim that empirism is a simpler hypothesis, and (c) with the claim that nativism lacks explanatory power.

2.1 Empirical Evidence
Almost without exception, the empirical evidence came from experiments regarding binocular vision. (The exceptions were evidence from newborn animals and from the newly-sighted blind, upon which Helmholtz placed little weight.)[41] The problem of binocular vision had two main

branches, only one of which entered significantly into Helmholtz's brief against nativism. First, there was the problem of binocular single vision, which had been addressed in nearly every treatise on vision since ancient times. Second, Wheatstone's discovery of the stereoscope in the 1830s drew attention to the previously obscure fact that binocularly disparate images are a potent means for depth perception, thereby raising the problem of how binocular disparity leads to the perception of depth. The problem of binocular single vision had already produced a large literature before Wheatstone's work, and it remained the predominant problem regarding binocular vision through the mid-nineteenth century. The preponderance of stereoscopical results reported by Helmholtz pertained to this problem.[42]

Nativist and empirist explanations of single vision had been widely discussed in the century prior to Helmholtz's work. Nativists explained single vision as an innate disposition of the nervous system to perceive as single those features of the visual field imaged upon corresponding elements of the two retinas, where "corresponding" is defined in terms of the distance and direction of each retinal element from the center of the retina. Some writers explained this innate disposition by hypothesizing anatomical connections in the brain between neural fibers originating from similarly situated elements on the two retinas.[43] In the nineteenth century, the hypothesis of innate anatomical connections between the two eyes was strongly advocated by Müller, who explained that single vision occurs when "identical" elements (those occupying corresponding positions) in the two retinas are stimulated because these identical points are connected neurally to a single point in the sensorium. He compared the eyes to "two branches with a single root, of which every minute portion bifurcates so as to send a twig to each eye."[44] Müller allowed that numerous anatomical routes for the fibers leading from a single point in the brain to the two retinas were conceivable and that insufficient evidence was available to determine the precise route. The essence of his position was that, by whatever anatomical pathway, "identical" points in the two retinas are anatomically connected and that these connections are innate. This view, known as the "identity hypothesis," retained advocates through the mid-nineteenth century, including Valentin, Volkmann, and Hering.[45] The empirist account, according to which single vision is a product of habit or learning, was also strongly represented in the eighteenth century. The essence of this view was that stimulation of the two retinas produces two sets of sensations that we learn to refer to a single object.[46] Although Steinbuch developed a second empiristic account of single vision in his *Beytrag*, accounting for single vision in terms of an acquired physiological correspondence between similarly situated elements in each of the retinas, most subsequent empirists defended the "two sensation" version, including Wundt, Nagel, Classen, Cornelius, and Helmholtz.[47]

In the absence of direct anatomical evidence for the existence of binocular neural fibers, the confrontation between the two groups occurred at the level of empirical observations regarding the various phenomena of binocular vision. The empirists sought to show that certain phenomena of binocular vision, taken together with some assumptions about the way the nervous system works, preclude the existence of innate anatomical connections between fibers projecting into the brain from retinal elements in separate eyes. Nativists responded either by questioning the phenomena or the assumptions used in interpreting them.

One area of investigation pertained to clinical cases in which the alignment of the two eyes is permanently altered in such a way that the optically corresponding parts of the retinal images fall on different portions of the retinas than previously. This occurs when the shape of the eye has been altered through injury, in cases of newly acquired or newly cured strabismus, and in cases where an artificial pupil is cut in the iris to one side of the normal pupillary location.[48] Patients typically reported double vision immediately after the alteration occurred; the double vision abated with time until eventually the patients reported single vision once again, even though the altered relation between the two eyes persisted. Empirists argued that the disappearance of the double vision over time indicated that a new set of corresponding retinal elements had been established; they further argued that the acquisition of a new set of corresponding elements indicated that the original set of corresponding or identical retinal elements also had been acquired.

The empirist argument contains an important methodological assumption, which Helmholtz stated explicitly: if a set of relations between stimulation and perceptual experience is alterable, the relations are acquired; if the relations are inalterable, they are innate (PO 3:12-13). At the root of this assumption lies what later became known as the "constancy hypothesis," the thesis that each anatomically distinct portion of the retina, when stimulated in a particular fashion, will always produce the same sensation to consciousness, via the physiological process transmitted along its attendant neural fiber.[49] This assumption is stronger than the doctrine that each neural fiber produces a single sensation: it asserts that the behavior of single fibers in producing sensations is not alterable as a result of experience. With this assumption in place, it can be argued that the alterability of the relation between stimulation and perception is evidence for the operation of habit or learning. The alleged alterability of the set of corresponding retinal elements could then be taken as evidence that the original correspondence was acquired.[50]

Early nativists, such as Reid and Müller, generally countered this argument not by denying the constancy hypothesis but by denying that a new set of correspondences is formed. Reid and Müller both reported that, in the cases of strabismus they examined, the sensations produced by previously corresponding points remained unaltered. The single vision attained by a strabismic individual was explained with the hypothesis that one of the double images is suppressed, so that the patient sees with only one eye at a time. The suppression itself goes unnoticed, so that the individual might think he saw singly while using both eyes (and here is where the empirists go astray), whereas he actually saw singly because one eye dominated the other.[51] Helmholtz met this argument by using an opthalmic prism to separate the two images in cured strabismics with single vision. The result of this procedure was double vision, which showed that both eyes were capable of producing images at the same time (PO 3:334).

But even if the single vision of cured strabismics resulted in certain cases from an alteration in the corresponding elements in the two eyes, why should alterability preclude innateness? Helmholtz responded by accepting the constancy hypothesis, according to which the physiological portion of sensory processes is viewed as inalterable, unlike the "mental" or "psychical" association of sensations. This argument was not, however, effective as a response to Hering, who pointed out that, if we assume all learning is physiologically mediated, then all learning involves some change in the nervous system. This directly contradicts the assumption that our innate anatomical apparatus is inalterable. Hence, in the absence of any special argument pertaining to binocular connections, Hering would have no reason to accept the hypothesis that innate anatomical connections resist all experiential effects unless he could be convinced that learning is not a physiological process. But Hering contended that learning does affect the structure of the brain (or the physiology of "the sensorium").[52] Helmholtz himself recognized the possibility that learning and memory are physiologically mediated, but he still retained his original argument (PO 3:408). This part of the dispute thus ended in a dead heat, with each side expressing opposed assumptions about the physiological basis of learning in the nervous system, assumptions that neither could establish conclusively.

Helmholtz claimed to derive stronger support for his position from a second area of empirical evidence purporting to demonstrate directly that sensations from the two eyes enter consciousness separately and are psychically fused. He drew attention to three groups of phenomena. First, the phenomenon of stereoscopic luster, which occurs when one white and one black surface are viewed in a stereoscope, each being viewed by one eye. The appearance is of a shiny surface. This result led

Helmholtz to conclude that "the excitation of corresponding points of the two retinas is not indistinguishably combined into a single impression."[53] He argued: "If there were a complete combination of the impressions produced upon both retinas, the union of white and black would give gray. The fact that when they are actually combined in the stereoscope, they produce the effect of luster—an effect which cannot be produced by any kind of uniform gray surface—proves that the impressions on the two retinas are not combined into one sensation" (SW, 207-8; VR 1:346). Helmholtz seems to have assumed that, if two nerve fibers unite, the processes transmitted along each of them are fused into a single physiological process that can produce only a single sensation.

The second group of phenomena was binocular rivalry, which occurs when differently articulated fields are presented to each eye. Perpendicularly situated black rectangles that overlap in the binocular field, or sets of lines arranged so that they cross each other perpendicularly, are examples. Often both images are seen superimposed on one another in the binocular field of view, although sometimes portions of one or the other image will predominate, suppressing the other. In the latter case the phenomenal states typically alternate, first one color (or contour) predominating, then the other. From these facts Helmholtz concluded

> that man possesses the faculty of perceiving each field of sight separately, without being disturbed by the other field of sight, provided it is possible for him, by some of the methods above indicated, to concentrate his whole attention on the objects in this one field. This is an important fact, because it signifies that *the content of each separate field comes to consciousness without being fused with that of the other field by means of organic mechanisms; and that, therefore, the fusion of the two fields in one common image, when it does occur, is a psychic act.* [PO 3:407][54]

Rivalry between the sensory states caused by separate stimulation of the two eyes is evidence that sensations from each eye separately enter consciousness—or at least the realm of unconscious mental states. This arousal of separate sensations by each eye, he argued, belies the hypothesis of anatomical connections.

The third set of phenomena is subsidiary to binocular rivalry and pertains to the effect of attention on the alternating suppression of one rival field by the other. Helmholtz made the following observation and conclusion:

> The rivalry between two fields, as it occurs in binocular fusion of the images above mentioned, is analogous to the careless, vacillating, uninterested state of the attention, accustomed to flit from one

impression to another, until the various objects are gradually passed in review. That this variation does not depend on some organic mechanism of the nervous system, as has been conjectured by Panum and E. Hering, at least on nothing more than underlies our mental activities, appears evident to me from the fact of introspection, namely, that by means of concentrating the attention, which is acknowledged to be and has been named above as purely psychic, the variation can be instantly stopped, without producing any noticeable change in the external conditions (such as changing the direction or movement of the eyes, etc.). [PO 3:408]

Helmholtz here takes the fact that attention, a "purely psychic" factor, can influence the vacillation of rival images as evidence that rivalry does not depend on an innate anatomical ("organic") mechanism, or at least, "on nothing more than underlies our mental activities." Apparently Helmholtz depended upon a distinction between innate anatomical mechanisms that mediate sensations on the one hand, and, on the other, the "mental" processes involved in combining sensations into perceptions, which, if they have an anatomical basis, have one that is not specific to binocular vision.

Previous nativists had admitted the phenomenon of binocular rivalry but denied the interpretation. Müller acknowledged that the failure of sensory states coming from separate eyes to mix was indeed a problem for his identity theory, but he compared the situation to two balance pans swinging back and forth through equilibrium. The alternating rival sensations result from the alternating dominance of the stimulus process transmitted along one or the other of the branches of a single binocular neural fiber. The alternation might be explained not as the rivalry of two mental states, each seeking to enter consciousness at once, but as the rivalry of two physiological processes, each seeking to dominate the activity of the unified neural fiber formed by the union of the separate fibers coming from each eye.[55] Indeed, the phenomenon of rivalry provides evidence for only one of the two rival sensations at a time. The hypothesis that there are two separate sensations continuously in existence, one of which is excluded from consciousness while the other dominates, is thus no more supported than is the hypothesis of physiological rivalry. Furthermore, even granting the supposition that two sensations do arise in consciousness, Helmholtz himself described an anatomical hypothesis that could account for the phenomena: fibers coming from corresponding retinal elements each split into two fibers, one joining the fiber from the other eye and one not, allowing for fusion and for distinct impressions (PO 3:438). Helmholtz thereby admitted that the nativists could account for the phenomena. Binocular rivalry thus gave support to neither side but once again pitted opposing assumptions about the operation of the nervous system against one another. Helmholtz himself concluded correctly that the empirical evidence he marshaled to support empirism was inconclusive.

2.2 The Simplicity of Empirism

Perhaps realizing that he could not achieve victory on empirical grounds, Helmholtz adopted a second line of attack, charging the nativists with needlessly multiplying hypotheses.[56] He sketched this argument in his popular lecture of 1868 as follows:

> The hypotheses which have been successively framed by the various supporters of nativistic theories of vision, in order to suit one phenomenon after another, are completely unnecessary. No fact has yet been discovered inconsistent with the empiristic theory, which does not assume any indemonstrable anatomical structures or any unprecedented modes of physiological activity in the nervous substance. It assumes nothing more than the association of intuitions and representations, which is well known from daily experience and its essential laws. [VR 1:353; SW, 213]

Nativists hypothesize anatomical connections for which they have no independent evidence. Empirists, on the other hand, assume only the operation of the known laws of mental life: association among elemental sensations.

Helmholtz admitted that the nature of psychological processes such as the association of ideas itself remained unknown, and that no explanation of their operation was then possible or likely in the near future. But the existence and actual operation of such laws was, he argued, beyond question, as the nativists admitted (VR 1:353). By contrast, the hypothesized mechanisms of the nativists were just hypotheses. He compared the evidential bases of the two sorts of explanations in the concluding section of the *Optics*:

> No matter what view is taken of the psychic activities, and no matter how hard it may be to explain them, there is no doubt as to their actual existence, and to a certain extent we are familiar with their laws from daily experience. It is safer, in my opinion, to connect the phenomena of vision with other processes that, although indeed requiring further explanation themselves, are certainly present and actually effective, as are the simpler psychical activities, instead of trying to base these phenomena on completely unfamiliar hypotheses as to the mechanism of the nervous system and the properties of

the nervous substance, which have been invented ad hoc and are unsupported by analogy of any sort. [PO 3:432]

In the face of ignorance, proceed by trying to assimilate unknown processes to ones that are known, introducing as few hypotheses as possible. Helmholtz here endorses the ideals of conservatism and parsimony in choosing between scientific hypotheses.

Helmholtz argued that the nativists' assumption of innate spatial sensations also rendered their account inherently more complex then the empirist account. This assumption necessitated appeal to both learned and innate factors in explaining our actual spatial perceptions. Even Hering did not suppose that spatial perception is completely innate; even he allowed a role for experience in the fine-tuning of spatial perception.[57] By themselves, innate spatial sensations are inaccurate; hence, they must be overcome and corrected by experience. And surely, Helmholtz contended, "if the factors derived from experience are able to give the correct information as to the relations of space even in spite of opposing direct space-sensations, they must be still better and more easily able to give the correct information about them when there are no such obstacles to be overcome" (PO 3:452). Why introduce the unnecessary hypothesis of innate spatial sensations if such sensations must be replaced by others, as a result of experience? A nativist such as Hering would reject Helmholtz's argument, for it again rested on the notion that what is innate is inalterable, so that innate spatial sensations, if not equivalent to an appropriate spatial perception to begin with, would constitute only so much mental clutter to be suppressed in favor of representations derived from experience. But Hering's position was that innate spatial sensations are elements which, through combinatorial processes, are molded into appropriate spatial representations. Helmholtz allowed his nonspatial sensations to be combined and arranged in various ways to produce spatial representations (why are they not inalterably nonspatial, if innately nonspatial?), so why should not innate spatial sensations also be alterable?

Nativists need not have conceded defeat in the face of Helmholtz's accusation of multiplying hypotheses. Indeed, they could claim that the shoe was on the other foot. Helmholtz could not show that *nothing* was innate but only that he could account for the data by assuming less of an innate endowment: (1) dispositions to produce nonspatial sensations in a range of qualitative characters, and (2) general laws of psychic activity. But the nativists could claim that, since some innate sensations are involved, the best strategy would be to avoid calling on a complicated learning process and try instead to maximize the innate portion of the explanation. Emil du Bois-Reymond raised this objection in his letter to Helmholtz from April of 1868:

> The great objection to the strict empiristic attitude always seems to me to be that it ought to be possible to carry it through consistently, which, as you yourself admit, is not the case; for if it is innate in the calf to go after the smell of the udder, why should not all its faculties be innate? It appears to me that so much nativism which one cannot get rid of is still left, that a handful more or less does not much matter. In regard to motion, for example, there are countless complicated cases in which we cannot get rid of it. You will say that one can at least try to limit it as far as possible, and that I do not deny.[58]

While one might try to limit the amount of nativistic explanation as far as possible, so much nativism is demanded by the facts that a little more might not matter. Du Bois-Reymond seems to suggest that if empirism cannot be carried through consistently, nativism should be.

The argument reduces to opposing intuitions about the simplicity of the assumptions of the two explanatory programs. There is no conclusive outcome.[59]

2.3 Nativism's Alleged Lack of Explanatory Power

Helmholtz had one final line of argument which, if successful, would be devastating: he contended that nativism actually left our spatial abilities unexplained. This attack was his most frequently repeated[60] and most fundamental. He stated the objection as follows: "The nativistic theory provides virtually no explanation of the origin of our perceptual images; for it simply plunges right into the midst of the matter by assuming that certain perceptual images of space would be produced directly by an innate mechanism, provided certain fibers were stimulated" (PO 3:16). He regarded the assumption that spatial images, rather than mere color sensations, are produced by an innate mechanism to be a gratuitous assumption that blocked further investigation into the origin of spatial perception. Two questions must be answered in order to understand whence Helmholtz believed the force of his attack to come. First, why is the nativists' claim that perceptual images of space are produced by stimulation of certain neural fibers (owing to an innate property of those fibers) more objectionable than Helmholtz's admission, which he shared with nativists, that sensations of color are produced by stimulation of certain neural fibers (owing to their innate properties)? Second, why does an appeal to innate factors in the nervous system preclude further investigation into the origin of spatial representation?

A full understanding of Helmholtz's answer to the first question must await discussion of his epistemology, but the essential point may be stated now: his differential treatment of color sensations and spatial representations was based in his conception—shared with many of his contemporaries—of the relation that instances of each type of representation bore toward their external causes. In conformity with the distinction between primary and secondary qualities, it was widely held that color sensations are mere signs of their causes and that there is no intrinsic connection between the phenomenal character of a color sensation and its external cause. Nativists and empirists alike maintained that the phenomenal character of color sensations depends on some currently inexplicable property of the nervous system (a position that was formalized in the doctrine of specific nerve energies). By contrast, spatial representations were assigned a quite different status; they were deemed not mere arbitrary signs of the external world but representations of the actual relations among things. Because spatial representations were considered to be in harmony with the relations found in the external world, an explanation of the origin of this harmony was demanded. And not only by Helmholtz: Herbart, Steinbuch, and Tourtual all agreed that some such explanation was needed. In Helmholtz's view, the nativists could not provide an explanation of this feature of spatial representation—they were forced merely to assume, without argument, a "preestablished harmony" between the world and our ability for generating perceptual representations of it—whereas he could and did provide a detailed explanation of the origin of this harmony. His explanation ascribed the origin of our spatial abilities to the acquisition of rules for generating spatial representations, the acquisition process being guided by causal commerce with external objects.

Helmholtz's repeated claim that nativism utterly fails to provide an explanation of the origin of this harmony at first presents an enigma, for in the *Optics* and the lecture of 1868 he simply asserted that the assumption that spatial localization is an innate capacity "is tantamount to giving up any attempt to explain the phenomena of localization."[61] In these works, the only semblance of an argument against what he characterized as the nativistic assumption of a "preexisting harmony" was rhetorical; he assimilated nativism to the disreputable metaphysical tendency of identifying the mind and the world (thereby assuring a harmony between the two).[62] One would, however, like to be taken beyond mere guilt by association.

Helmholtz only hinted at his reason for repeatedly charging that nativism lacks explanatory power. My interpretation depends as much on what he did not say as on what he did say. His charge reflected the fact that prior to Darwin, there was no generally accepted means for explaining innate adaptive mechanisms (Lamarck's explanation was not widely accepted). Of course, this lack of explanation attended purely anatomical adaptations as well as the psychological adaptation involved in spatial perception. But in the case of spatial perception, an account was available: an empiristic account, explaining the adaptation as something acquired during the lifetime of the individual.

In his 1869 lecture on "The Aim and Progress of Natural Science," which was the opening address before the *Naturforscherversammlung* in Innsbrück, Helmholtz expressed great enthusiasm for Darwin's work. He introduced his discussion with the claim that "Darwin's theory of the development of organic forms... provides an entirely new interpretation of organic purposiveness [*organischen Zweckmässigkeit*]" (VR 1:387; SW, 237). Previously, he contended, only two theories of organic purposiveness were common: the vitalistic assumption of a special "life soul" that directed the vital processes, and the belief in a supernatural creation of each species. According to Helmholtz, Darwin's theory provided a new theoretical possibility by showing how "adaptation in the structure of organisms can result from the blind rule of a law of nature without any intervention of intelligence" (SW, 238; VR 1:388).[63] Since Helmholtz accepted Darwin's theory and gave it high praise, it is puzzling that he did not grant that the nativists might apply a Darwinian explanation to the case of spatial perception, an explanation which, if plausibly developed, would provide them with an acceptable explanation for the origin of an innate capacity for spatial perception.

In the lecture of 1869 Helmholtz surveyed several areas of progress in natural science; he took up developments in sensory physiology just after his laudatory discussion of Darwin's work. He made the transition as follows:

> While Darwinian theory treats exclusively of the gradual modification of a species after a succession of generations, we know that to a certain extent a single individual may adapt itself or become accustomed to the circumstances under which it must live. Thus even during the life of a single individual, a higher degree of organic adaptation [*Zweckmässigkeit*] can be attained. Moreover, it is especially in that area of organic phenomena where purposiveness of structure has reached its highest form and excited the greatest admiration—the area of sense perception—that, as the latest developments in physiology teach us, this individual adaptation [*Anpassung*] has come to play a most important role. [SW, 240; VR 1:390-1]

Helmholtz characterized spatial perception as an acquired adaptation, without even raising the possibility that it could be considered as a species-level adaptation; yet his chief rival, Hering, had suggested that very possibility earlier in the decade, in a work Helmholtz is known to have read with some care.[64] In the lecture, when he raised the possibility of spatial perception's being innate, he spoke only of its being "a preordained product of an organic creative force" (SW, 240). But surely this is just to set up nativism as a straw man, since he had only a few paragraphs before contended that Darwin's theory of evolution had vitiated the explanation of adaptation in terms of organic creative forces.

Helmholtz's arguments against nativism failed on all counts to provide conclusive arguments against his opponent. He was generally prepared to admit that the empirical data were equivocal; his argument that empirism is simpler was subject to being turned around and urged by a nativist; and his claim that nativism is incapable of explaining the origin of our capacity for spatial perception failed to take into account the possibility of a Darwinian interpretation of nativism.

Nevertheless, Helmholtz's arguments are of interest for what they reveal about how he conceived the available theoretical options. Each argument presents a choice between an explanation based on the postulation of innate anatomical structures and physiological processes or one based on learning, between physiological and psychological explanation. That Helmholtz would constantly prefer "psychic" explanations to physiological explanations may seem odd, given his notorious stand against vitalism and his role in establishing the conservation of energy (widely considered to be the bane of dualism). Indeed, it might seem that Helmholtz has revealed dualistic tendencies in pitting *learning* against *physiology*; for unless one were a dualist, it is not apparent why learning should not be regarded as having a basis in physiology, thereby making the choice into one between the physiology of innate abilities and the physiology of acquired abilities. Hering raised this very objection to the way in which Helmholtz's acknowledged allies—Nagel, Wundt, and Classen—described their psychologically based theories; after the publication of the *Optics*, he leveled the same charge against Helmholtz, labeling him a "spiritualist" who invoked a *Seele* or *Geist* under the euphemism of providing a "psychological" explanation of the phenomena.[65]

In discussing binocular rivalry, Helmholtz distinguished between the positing of a psychic as opposed to an organic basis for single vision. He qualified the distinction, remarking that the psychically induced variation in binocular fusion "does not depend on some organic mechanism of the nervous system, . . . at least on nothing more than underlies our mental activities [generally]" (PO 3:408). This qualification is frustratingly indeterminate. Does Helmholtz think there is an organic basis for psychological processes? Examination of additional passages would confirm that he repeatedly contrasted the "organic" with the "psychic," equating the latter with processes that involve learning.[66] Indeed, in criticizing Panum's arguments for preferring to explain the various phenomena of binocular fusion through "physiological, not psychic forces," he writes as if he were a dualist:

> From the synopsis of Mr. Panum's explanations as given above, the reader can see that, at least so far as fusion and rivalry of the images is concerned, they actually provide the mere form of an explanation, in that the facts are summarized in an abstract concept. If they have any bearing on the question of causal relations, it is in a negative way, by insisting against the influence of psychic processes, a stand which is invariably supported through incomplete observation of the facts. Furthermore, these explanations attribute forms of activity to the nervous substance which are well recognized in the sphere of the lower psychic activities [Seelentätigkeiten], but nothing similar to which has ever been discovered in the domain of the material world [Körperwelt].[67]

The contrast between the domain of the material world and that of the lower psychic activities is particularly striking.[68]

In the final analysis it would be incorrect to classify Helmholtz's scientific position as either materialist or dualist. His considered position was to avoid choosing between these alternatives, on the grounds that such a choice would amount to adopting a metaphysical hypothesis that could not be supported scientifically. He rejected both "spiritualism" and materialism as equally metaphysical.

> I acknowledge that we are still far from a real scientific understanding of psychic phenomena. We may agree with the spiritualists that such understanding is absolutely impossible, or we may take precisely the contrary view along with the materialists, according as we are inclined toward one speculation or the other. For the natural philosopher, who must stick to factual relations and seek their laws, this is a question for which he possesses no basis for choice. It must not be forgotten that materialism is just as much a metaphysical speculation or hypothesis as is spiritualism, and that it therefore does not provide one with the right to choose between factual relations in natural science without a factual basis. [PO 3:432][69]

Helmholtz wrote these words in opposition to the tendency to posit unobserved physiological mechanisms to explain psychical phenomena. His own explanatory preference, as we have seen, was to appeal to the known laws of psychology, even if they are and remain unexplained physiologically. We shall find that this preference was deeply consonant with his general conception of good methodology in natural science.

To say that Helmholtz failed to refute nativism is not to say that he lacked auxiliary reasons for preferring empirism. No small part of his preference for empirism is accounted for by the unity this position allowed him to posit between the processes underlying visual perception and the processes by which scientific hypotheses are formulated and tested. Helmholtz maintained that both processes yield inductive conclusions, even if some of these conclusions are formed unconsciously.

3 The Theory of Unconscious Inference

The classical statement of Helmholtz's theory of unconscious inference occurred in section 26 of the *Optics*, which he summarized in his lectures on vision of 1868. In Helmholtz's own words, the psychological theory of the *Optics* (and, by implication, the theory of unconscious inference) was designed to provide an account of the operations by which we come "to form representations of the existence, form and position of external objects" on the basis of the sensations that arise from stimulation of the retina, together with ocular muscle feelings (PO 3:3). Thus, the primary function of the theory was to give an account of the process by which sensations are integrated into perceptions. At the same time, in section 26 and in the lectures Helmholtz connected the theory with issues in metaphysics and the theory of knowledge, and specifically with the problem of the external world. This connection had at its root an analogy between perceptual inference and scientific inference: he drew attention to the role experience and experiment plays in the testing of both perceptual and scientific hypotheses about the existence and character of natural objects.[70]

Although Helmholtz was committed to the doctrine of unconscious inference (under one name or another) throughout his career, his expression of the doctrine did not remain stable across the decades. The doctrine was intended to speak to distinct problems, some philosophical, some psychological, and it changed as his epistemology and psychology developed. Helmholtz was at first, in his lecture of 1855, uncertain about the psychology of unconscious inference; but once he presented an account of this psychology in 1866, the content of the theory remained virtually unchanged through his final essay in 1894, even if he developed reservations about the term "unconscious inference" (EW, 132; SW, 508). The philosophical side of the doctrine also underwent development as a result of Helmholtz's changing conception of the relationship between the doctrine of unconscious inference and the Kantian postulation of the causal law as an a priori given.

I shall first look briefly at the development of Helmholtz's account of the psychology of unconscious inference before turning to his mature presentation of the theory itself, including his identification of the processes underlying perceptual inference with those underlying scientific inference.

3.1 The Development of Helmholtz's Theory

Although Helmholtz used the term *unbewusster Schluss* in a small portion of section 26 of the *Optics* he published in 1860, the first full presentation of the theory of unconscious inference came in the portion of section 26 he published in 1866 (most likely written in 1865).[71] In his lecture of 1855 and in the second part of the *Optics*, he had indicated his belief that the fundamental processes of perception are inferences but had left open the question of the precise nature of the psychological processes involved.[72]

The lecture of 1855 contained an extended discussion of the psychology of vision. Helmholtz was clear on several points about the general characteristics of perception: that sensation is not perception, that true visual perception (*das Sehen*) results from the "understanding" of sensations and is therefore a psychological process, that we learn to see, and that vision results from judgments, conclusions, and deliberations of which we are not conscious (VR 1:101-2, 110). He illustrated the judgmental nature of vision by appeal to visual illusion, arguing that the falsehood of such illusions should not be attributed to the processes in the nervous fibers but to the judgment that is applied to the resulting sensation. With respect to the nature of the processes underlying such judgments, he frankly admitted the difficulty of determining their nature and lamented the lack of help from the discipline of psychology.

> The determination of the nature of the psychical processes that transform light sensations into a perception of the external world is a difficult assignment. Unfortunately we receive no help from the psychologists, since for psychology introspection has previously been the sole path to knowledge, while we have to do with mental activities about which introspection gives us no knowledge, indeed, whose essence we can first infer from the physiological investigation of the sense organs. [VR 1:111][73]

Physiology, as Helmholtz argued repeatedly, establishes that the elements of perception are single sensations produced by the stimulation of single nerve fibers. But that discovery, in itself, does not reveal the character of the processes by which such sensations are molded into perceptions.

Helmholtz's statement of the great difficulty he found in determining the character of those processes was not merely rhetorical, and he did not achieve a clear resolution of his question in the course of his lecture. But he did make an attempt. At one point, he characterized visual inference as "a mechanically drilled [inference], which has arisen through the series of involuntary associations of ideas" (VR 1:112). Later on, however, he asked whether characterization of the psychological process of vision as associational would result in the implication that "that which I have earlier called the thinking and inferring of representations in actuality is not thinking and inferring, but nothing more than a mechanically drilled association of ideas" (115). His question allows the possibility that inference and association are fundamentally different kinds of process. And at least for one type of perceptual judgment, that by which we posit the existence of an external world, he found a compelling need to posit a true (nonassociative) inference. Inference is necessary to bridge the gap between the direct perception of our own sensory states and the external objects that are their causes:

> We never perceive the objects of the external world immediately, rather we perceive only the effects of these objects upon our nervous apparatus, and this has been so from the first moment of our life. By what means then have we been brought from the world of the sensations of our nerves to the world of reality? Obviously only through an inference; we must posit the presence of external objects as causes of our nervous excitation; for there can be no effect without a cause. [VR 1:115-16]

Helmholtz went on to announce his agreement with Kant that the rule "no effect without a cause" is "a law of our thought given prior to all experience." At present, however, I wish to stress the fact that in 1855 Helmholtz distinguished inferences using this law from associational processes, suggesting that he considered associational and judgmental processes to differ in kind.

Helmholtz discussed the notion of unconscious inference on several occasions between 1855 and the publication of the third part of the *Optics* in 1866, but only in a lecture of 1862 did he add much to his earlier discussion. In a letter to his father in March of 1857 he expressed the need for a deeper philosophical understanding of what he termed the "unconscious arguments from analogy by which we pass from sensations to sense-perceptions."[74] In the material from Parts Two and Three of the *Optics* published in 1860 Helmholtz characterized the process as an unconscious inference from analogy, but gave no description of the process.[75] In the lecture of 1862, he added to this picture the claim that these judgments are inductive inferences of the same sort that underlie inductive generalization in natural science; he referred both sorts of inference to a "psychological instinct" for forming general rules as a consequence of myriad experiences of instances of the rule.[76] Still, he gave no precise characterization of the psychological processes in inductive inference, of the sort that he would provide in sections 26 and 33 of the *Optics*.

Between 1858 and 1862 Helmholtz's junior colleague Wilhelm Wundt published a series of articles on sense perception, anticipating the form and content of Helmholtz's theory of 1866 in several ways.[77] Wundt contended that perception must be understood through an analysis of the mental processes that underlie it and that a science of psychology distinct from physiology was needed to carry out this analysis.[78] He drew a rigorous distinction between *sensational* elements and the *perceptual* representations constructed from them through elementary psychical processes; he listed color sensations, local signs, and muscle feelings as the elementary sensations of vision. Wundt denied that spatial representation is a priori in the Kantian sense but admitted that specific senses (viz. touch and vision) have a "disposition" for acquiring spatial intuition.[79] He explained the development of spatial intuition in terms of mental operations on the elements of sensation. These mental operations he repeatedly described as unconscious inferences.[80]

Wundt developed his views on unconscious inference most fully in the sixth article in his series, which had been completed by January of 1862.[81] He characterized the process of unconscious inference as one of induction, in which laws relating the simplest of cognitive elements—sensations—are formulated through experience. He was not content merely to label the process as one of induction and rest his case, but sought to analyze in detail the psychology of inductive inference, citing Mill's *System of Logic* and Whewell's *Philosophy of the Inductive Sciences*.[82]

According to Wundt's analysis, there are three processes involved in perceptual inference. The first is the association of sensations through constant conjunction; he labeled this process the "colligation" of sensations. The second process is the "synthesis" or "fusion" of the previously associated elements into a unified whole which differs from the elements. The third process is that of inference from analogy: from the presence of some of the properties usually found to belong to a certain object, the presence of other characteristic properties is inferred. Conclusion by analogy is not an essential part of perception, but serves as a short-cut process in everyday perception where familiar objects are involved. Colligation and synthesis are essential.[83] Synthesis plays the central role in spatial perception, fusing previously associated nonspatial sensations into a spatial representation. The creation of spatial representations occurs through the development of series of associations between muscle feelings and local signs,[84] presumably in accordance with the special "disposition" of the optical system to form spatial representations.[85] This synthesis, he maintained, is a constructive process possessed of its own "creative activity," although, he advised, this creative activity follows from necessary laws.

The primary processes identified by Wundt, association and synthesis, are reminiscent of Herbart's hypothesized combinations and fusions and Steinbuch's association by simultaneity and similarity. However, whereas Steinbuch saw these processes as belonging solely to sensibility, Wundt was closer to Herbart in conceiving them as fundamental to mental life in general. But Wundt portrayed mental life in a manner that owed as much to Mill and Whewell as to Herbart. He assimilated the fundamental operations of thought to inductive inference of the sort exemplified in the natural sciences: inference from finite samples of data to universal statements.[86] Given Helmholtz's interest in the psychology of inference, he must have found this aspect of his younger colleague's work particularly intriguing.

3.2 The Psychology of Unconscious Inference

On first reading the third part of the *Optics*, the reader may be struck by the apparent lack of connection between Helmholtz's discussion of unconscious inference in the introductory section and his treatment of specific groups of perceptual phenomena—such as the development of the monocular field of vision, or the perception of visual direction—in the succeeding sections. In his description of unconscious inference in section 26 Helmholtz attributed an active role to the perceiver in the formation of judgments regarding the causes of sensations, whereas in the succeeding sections he seems to refer only to passive associational processes. This associational account is at work in Helmholtz's explanation of the associational mapping of local signs into visual directions; as he put it, "the ability to recognize and to judge the arrangement of objects in the field of view without moving the eyes can be acquired, as is espoused by the empiristic theory," which meant that "without the hypothesis of innate knowledge of the arrangement of retinal points, the facts can be explained by the known capacities of sensory memory" (PO 3:134-5). The process of matching local sign with visual direction would seem to consist of a passive collection of associations in memory rather than an active process of judgment; for although movement of the eyes and body might be required in order to produce the paired sensations requisite for mapping local signs into directions in the visual field, the process by which the learning takes place is ascribed to sensory memory.

Although there exists a tension between the active inferential account of section 26 and the subsequent reference to bare association, Helmholtz himself did not believe that the two accounts were in conflict; he considered them identical. Helmholtz thought that the psychology of unconscious inference could be wholly explicated in terms of the ordinary functioning of memory and association, a view he attributed to Mill (PO 3:23-4). The general rules for visual localization are formed using the same kind of inductive process as underlies the discovery of scientific generalizations. In each case, the general law arises out of particular experiences as a result of a tendency for the irregular and the variable to be "washed out" in preserving experiences in memory. No active process of inference is required in the subsequent application of an associatively based rule; given a representation of the antecedent, the associated consequent comes to life. In Helmholtz's account the active role of the perceiver is restricted to the "testing" of associatively formed rules of association through motor activity. But what is the basis of such testing? Could it be associative, too?

In order to see how association and active testing fit together in Helmholtz's theory, it will be necessary to delve more deeply into his analysis of unconscious inferences and of the circumstances in which they are formed, an analysis which allowed him to identify the integration of perceptual images with the formation of inductive conclusions in natural science.

Following Mill, Helmholtz remarked that seemingly deductive syllogisms often are no more than inductive inferences. If we argue from "All men are mortal" and "Caius is a man" to "Caius is mortal," this is a simple case of deduction in which the conclusion is implied by the major premise (*all* men would not be mortal unless Caius, who is man, was mortal). According to Helmholtz, Mill's insight was that the major premise in such inferences is really nothing more than a mnemonic device for summarizing the numerous observed cases of mortality that serve as the (incomplete) evidence for the general statement; in principle, the universal major premise could be eliminated in favor of a statement of its evidential basis. In other words, the chain of reasoning in such (actually inductive) inferences may proceed directly from the body of previously observed cases of type A that have had result B to the conclusion (based on analogy) that this new case of A will have result B. In ordinary life, this sort of inference goes on unconsciously, by virtue of the previously encountered connection among instances of types A and B, as imprinted in memory. Reasoning from a consciously formulated general principle does nothing but introduce an abbreviated representation for the evidential and causal basis of the inference.

Helmholtz contended that the same sort of reasoning by analogy occurs in perception:

When those nervous mechanisms whose terminals lie on the right-hand portions of the retinas of the two eyes have been stimulated, our usual experience, repeated a million times all through life, has been that a luminous object was over there in front of us on our left. We had to lift the hand toward the left to hide the light or to grasp the luminous object; or we had to move toward the left to get closer to it. Thus while in these cases no actual conscious inference is present, yet the essential and original office of such an inference has been performed, and the result of it has been attained; simply, of course, by the unconscious processes of the association of ideas going on in the dark background of our memory. [PO 3:24]

On the basis of experience we learn to relate retinal location to visual direction. The process of learning is associative. No conscious reasoning is involved, yet the process is effectively the same as an inference. Indeed, Helmholtz would claim, it is an inference.

A striking fact about the conclusion of perceptual inferences is that they result in the *appearance* of a luminous object in the field of vision. The conclusion of such inferences is a spatial perception, not a linguistic expression. Helmholtz was very much aware of the ostensible difference between perceptual inferences and linguistic inferences, but he did not consider the psychological processes underlying them to differ:

There appears to me in reality only a superficial difference between the inferences of logicians and those inductive inferences whose results we recognize in the intuitions of the outer world we attain through our sensations. The chief difference is that the former inferences are capable of expression in words, while the latter are not, because instead of words they deal only with sensations and memory-images of sensations. Indeed, it is precisely the fact that such sensations cannot be described in words that makes it so difficult to discuss this area of mental activity at all. [VR 1:358; SW, 217]

The chief difference between paradigm cases of inferences expressed in words and perceptual inferences are the materials by which the inferences are expressed.

Helmholtz insisted on the parallel between linguistic inferences and inferences involving perceptual images, asserting that "it is clearly possible, by using the sensible images of memory instead of words, to produce the same kind of combination which, when expressed in words, would be called a proposition or judgment" (SW, 218). This happens for individual propositions—such as the identification of a connection between two characteristics of an object through perceptual acquaintance (in the form of perceptual images), as when we hear a voice and recall its owner's face—and also for universal propositions. Indeed, the rules of inference that Helmholtz supposed perceivers to acquire as they learned the meaning of, for example, local signs, would be rules applying to that local sign universally, in all perceptual situations. Here is his example of such a universal perceptual "proposition":

> If I know that a particular way of seeing, for which I have learned how to employ exactly the right kind of innervation, is necessary in order to bring into direct vision a point two feet away and so many feet to the right, this also is a universal proposition, which applies to every case in which I have fixed a given point at that distance before or may do so hereafter. It is a piece of knowledge which cannot be expressed in words but which sums up my previous successful experience. It may at any moment become the major premise of an inference—whenever, in fact, I fix on a point in this position and feel that I am doing so by looking as that major premise states. This perception of my fixation is then my minor premise, and the conclusion is that the object I am looking at is to be found at the location in question. [VR 1:360; SW, 219]

The universal in this case is the relation between (A) a particular degree of innervation, together with the local sign for foveal vision, and (B) a location in (viewer-relative) space; the major premise is "All A's are B's," the occurrence of the sensations mentioned under A serves as the minor premise, and the phenomenally experienced location is the conclusion. It is difficult to make sense of Helmholtz's dictum that this perceptual universal "cannot be expressed in words," since he has just expressed it in words. But what he may have meant is that the phenomenal character of something appearing in a particular location cannot be simulated by words, though it can be described. In any case, the thrust of his point is that the processes by which perceptual images are formed are of a kind with the processes underlying scientific and other verbally-expressed inferences.

Helmholtz's account of imagistic reasoning extended to inferences among the perspective projections that may be attained from a given object. He explained that connections among such projections help account for our ability to perceive a known object as three-dimensional, even though we receive only a single perspective image. These connections can also account for our ability to predict what further perspective images we would receive if we changed our position with respect to the object. Indeed, Helmholtz believed that this sort of inference among perspective images exhausts the content of our idea of an object in three dimensions; it constitutes our concept of the object. He developed this point in the *Optics* but explained it most fully in the paper of 1894.

The memory images of pure sense impressions can also be used as elements in combinations of representations, where it is not necessary or even possible to describe those impressions in words and to grasp them conceptually through such description. A large part of our empirical knowledge of the natural relations among the objects around us obviously originates in this way. The blending of the many perspective images of an object into the representation of a three-dimensional form seems to me an especially clear example of this kind of combination of sensory intuitions which corresponds to an inference. . . . Indeed, the representation of a three-dimensional figure has no content other than representations of the series of visual images which can be obtained from it, including those which can be produced by cross-sectional cuts. [WA 3:544-5][87]

The idea of a body's corporeal form constitutes a function among all possible sense-impressions of the object; Helmholtz's account bears an apparent kinship with Kant's doctrine that concepts are functions among intuitions. Whether he modeled his account after Kant's or not, he equated sequences of interderivable images with the concept of a body.[88]

Helmholtz believed that his theory of unconscious inference, which conceived the psychological processes responsible for the generation of spatial perceptions as mechanisms of inference, was a step forward over previous theories in which intuition (conceived as belonging to the Kantian faculty of sensibility) was distinguished from thought (conceived as concept-mediated cognition). He repeatedly returned to the point that intuition should be resolved into thought, and thought in turn should be resolved into associative inferences.

I believe the resolution of the concept of intuition into the elementary processes of thought is the most essential advance in the recent period. This resolution is still absent in Kant, which is something that then also conditions his conception of the axioms of geometry as transcendental propositions. Here it was especially the physiological investigations on sense perceptions which led us to the ultimate elementary process of cognition. These processes had to remain still unformulable in words, and unknown and inaccessible to philosophy, as long as the latter investigated only cognitions finding their expression in language.[89]

Whereas, he believed, for Kant spatial intuitions were unanalyzable into more primitive components and arose without the activity of the understanding, he had shown how they are in fact psychologically derivative, produced by processes that underlie all judgment and belief.

Helmholtz regarded the dissolution of the sensibility-understanding distinction as one of the chief advantages of his theory of unconscious inference, equivalent to breaking down the distinction between *Kennen* and *Wissen*, that is, between phenomenal acquaintance with objects as opposed to propositional knowledge. The solvent performing these reductions was the notion that underlying the supposedly distinct operations that result in sensory perception on the one hand and judgments on the other is one kind of psychological operation—the association of ideas. Mental life is unified. Having something look some way to you is not a different kind of achievement, and does not rely on different psychological operations, than does subsuming an object under a concept or drawing an inference.

As an account of the production of perceptual images, Helmholtz's theory did not differ essentially from previous associational accounts. But identifying the associational processes underlying perception with the fundamental laws of thought distinguishes his (and Wundt's) theory from previous theories of perception. Helmholtz appealed to the authority of Mill, who had identified inference with association in his *System of Logic* (1843):

> If reasoning be from particulars to particulars, and if it consists in recognizing one fact as a mark of another, nothing is required to render reasoning possible except senses and association; senses to perceive that two facts are conjoined, association as the law by which one of those two facts raises up the idea of the other.[90]

We have seen an echo of this conception of inductive reasoning in Helmholtz's own account of reasoning "from particulars to particulars."

Despite Helmholtz's enthusiasm, there are a number of reasons for being skeptical about the attempt to resolve inference into association, only some of which he ultimately addressed. The most general philosophical objection, that inductive inferences presuppose the law of causation or some other principle of the uniformity of nature, was widely discussed throughout the nineteenth century, and Helmholtz dealt with it on several occasions. But two other problems arise within the mechanics of the associative account itself.

The mechanical difficulties connected with the associational account arise in connection with the account of visual direction. They involve the question of the role of active testing in the formation of associative connections. According to Helmholtz, when a given retinal location is stimulated the resulting sensation has two components: a sensation of light, and a local sign unique to the retinal location. Local signs initially have no spatial meaning (they merely label the retinal element), but through a process of association—say, between visual sensations and the locations of objects as discerned through the sense of touch—a rule is learned associating each local sign with a direction in space. Careful consideration of the conditions under which such learning must take place invites the question of how the perceiver chooses a sound rule for assigning a particular visual direction to a local sign from among the mass of possible directions. Within the constraints of Helmholtz's theory, an answer to this question requires positing a mechanical process of induction. Although in the *Optics* Helmholtz was content simply to refer to the abilities of "sensory memory," in the lecture of 1878 he provided a more extensive response:

> When the traces of like kind which are left behind in our memory by often repeated perceptions reinforce one another, it is precisely the lawlike which repeats itself most regularly in like manner, while the incidental fluctuation is erased away. For the devoted and attentive observer, there grows up in this way an intuitive image of the typical behavior of the objects which have interested him, and he knows as little afterwards how it arose as the child can give an account of the examples whereby it became acquainted with the meanings of words. [EW, 131][91]

Apparently, the regularity of the lawlike is enough to bring it to the foreground of thought, while the merely incidental is washed out.

This appeal to mechanical induction may seem somewhat quick. Imagine the situation of the learner, faced with the flux of sensory stimulation and armed only with some general laws of association—e.g., associative laws of "similarity" and "spatiotemporal contiguity." The situation appears hopeless. If the representations to be associated are images with complex elements, then what is to count as similarity? An indefinite number of similarities might be defined, but to give the associative laws a "head start" in selecting certain patterns would be to abandon the empiristic account.

German physiologists, psychologists, and philosophers in the nineteenth century made an heroic effort to overcome this problem. The outcome was the doctrine of local signs. The effort consisted in regarding the primary elements of psychological analysis as nonspatial or punctiform sensations defined by variation in quality and intensity alone (e.g., EW, 119-25). With such simple elements, the associative laws might have but a single dimension of similarity over which to associate. The problem of picking out the "relevant" dimension of similarity might thereby be avoided.

Once the problem had been reduced to the association of elements varying along only one or two dimensions, the theoretician was still faced with giving an account of how the percipient gets from the flow of punctiform sensations to the construction of a spatially ordered world. Helmholtz assigned the perception of our own movements and their consequences an important role in the acquisition and maintenance of the rules of spatial localization:

> We do not merely have alternating sense impressions which come upon us without our doing anything about it. We rather observe during our own continuing activity, and thereby attain an acquaintance with the enduring existence of a lawlike relationship between our innervations and the becoming present of the various impressions from the current range of presentables. Each of our voluntary movements, whereby we modify the manner of appearance of the objects, is to be regarded as an experiment through which we test whether we have correctly apprehended the lawlike behavior of the appearance before us, i.e., correctly apprehended the latter's presupposed enduring existence in a specific spatial arrangement. [EW, 135-6]

The operative notions are that through voluntary movements we "test" whether we have "correctly apprehended the lawlike." Such testing might indeed serve to sort out the possible associations by allowing expectations to be tested against empirically obtained results. However, in the positing of such testing there has been a shift in the analysis, from pure association to the assumption of certain cognitive abilities, such as the ability to match prediction with result. A purely associative account would simply take each movement as the production of a further constellation of sensations to be associated.

At this point Helmholtz's radical empirism and his associative account of inference come into conflict with his comparison of motor activity to an "experiment" or active test. The role of the feelings of innervation in forming general rules for spatial location is initially that of distinguishing between motions caused by the percipient and changes resulting from an alteration in the field of view. Feelings of innervation originally can have no "spatial meaning," or else Helmholtz would be admitting an innate mechanism, in the form of the "impulse of the will" or "feeling of innervation," for localizing objects in space. However, according to the analogy between unconscious inference and scientific experimentation, the impulses of the will are an active element that allow the perceiver to test the relation between perceptions formed in accordance with the rules of perceptual inference and the sensations that are obtained through actions guided by the those perceptions. Let us grant that such testing might occur in the developed perceiver. The real question concerns how we are to imagine it in the visually naive. In Helmholtz's analysis of the origin of spatial representation, the impulse of the will must be denied the innate "meaning" that a particular change in the pattern of stimulation has resulted from the action of the observer; the "feeling" associated with muscle movement is just one more element that enters into the mix of nonspatial sensations, out of which rules of spatial localization emerge. If the empiristic account is carried through consistently, it precludes the kind of innate "meaning" that would be presupposed in an active testing of spatial relations through voluntary motions that carry as part of their content an expectation of a particular change, or any change at all, in the pattern of sensation. The initial help that was promised from the metaphor of "active testing" can be secured only by a violation of the project of resolving thought into associative learning.

A similar theoretical tension arises in Helmholtz's discussion of the case of the associative basis of the discovery of scientific laws. Initially, he describes the position of the scientist as one in which the lawlike is simply given:

> What we can find unambiguously, and as a fact without anything being insinuated hypothetically, is the lawlike in the phenomena. From the first step onwards, when we perceive before us the objects distributed in space, this perception is the acknowledgement of a lawlike connection between our movements and the therewith occurring sensations. Thus even the first elementary representations contain intrinsically some thinking, and proceed according to the laws of thought. [EW, 138]

The lawlike presents itself immediately in perception, without conscious hypothesis. If the empiristic account of perception is carried through consistently, the laws of thought in question can be only associative laws operating over meaningless sensations. Yet in comparing the cognitive processes of the scientist with those of perception, Helmholtz needed more; he needed to characterize the process of perception as a grasping of the objectively lawful, not as a mere mechanical filter for regularities. His solution was to read an additional law of thought into the perceptual process itself. Retrospectively describing his associative account of the inductive inferences of perception, he asserted that they depend not merely on the laws of association but also on the Kantian "law of causality": "The law of causality is an a priori given, a transcendental law. A proof of it from experience is not possible, since the first steps of experience, as we have seen, are not possible without employing inductive inferences, i.e., without the law of causality" (EW, 142). The attempt to resolve thought into mechanical processes of association has ended with the statement that at least one cognitive principle must be presupposed at the outset. The attempt to resolve thought into blind laws for combining simple representations has ended with thought still resisting the analysis.

Here we arrive at perhaps the fundamental tension in Helmholtz's thought, that between his antagonism toward innate ideas and principles of thought, on the one hand, and his appreciation of the problem of induction, on the other. This tension became particularly acute in connection with his desire to derive from psychology an account of both the causal origin and justificatory ground of inferences from particular instances to general laws.

4 Epistemology of "Signs" and the Causal Law

The fundamental tenet of Helmholtz's epistemology was that "sensations are only signs for the properties of the external world, whose interpretation must be learned through experience" (VR 1:17; SW, 475). In the context of his epistemology, sensations act as signs, and perception is their interpretation. This was also the fundamental tenet of his sensory physiology and psychology, inasmuch as he treated nonspatial sensations as the elements from which spatial perceptions are formed.[92] There was indeed a thoroughgoing continuity between Helmholtz's epistemology and psychology of perception: the uninterpreted signs of epistemology are identical with psychologically primitive sensations, and the formation of spatial representations ascribes a "meaning" to or "interprets" those sensations. Psychologically derivative spatial perceptions are also epistemologically derivative—they result from combining elemental sensations according to psychical laws of inference. Thus his psychological account of spatial perception as resulting from inferences based on uninterpreted "signs" merges completely with his epistemological distinction between uninterpreted signs and their epistemological interpretation.

Helmholtz formulated the fundamental problem of epistemology in terms of the representational relation between the various dimensions of sensory experience and the external world. He asked, "What is true in our sensation and thought?" As Schlick observed, the word "true" is used nonstandardly in posing this question: we usually think of judgments as the bearers of truth. But Helmholtz was not posing a question about judgments. Rather, he was asking, in a manner reminiscent of earlier discussions pertaining to the distinction between primary and secondary qualities, about the representational accuracy of the various dimensions of sensory experience. He came closer to his intent in rephrasing the question as, "In what sense do our representations correspond to actuality?" However, he was not asking whether the various particular representations we form of the size and color of objects are accurate, but rather, to what extent and in what manner color or shape as types of representation correspond to real properties of objects.[93] In the context of answering this question, he classified sensations as signs for the properties of objects.

Helmholtz counted among "sensations" mental states varying only in quality and intensity, such as color sensations, and he counted spatial representations as perceptions. He was aware that historically these two groups, under the names "primary" and "secondary" qualities, had been thought to differ markedly in the degree of "correspondence" between an individual instance of a given type of representation and its external cause. In characterizing sensations as mere *signs* of their causes, Helmholtz considered himself to agree with earlier authors such as Locke.[94] He regarded color sensations as having a regular connection with external objects, which he expressed as follows: "Similar light produces under like conditions a like sensation of color. Light, which under like conditions excites unlike sensations, is dissimilar" (VR 1:319; SW, 185). He considered the existence of such a relation to be a sufficient condition for color sensations to be signs of their external causes.

In their status as *signs,* sensations are to be contrasted with representations that are *images* of their causes:

> ... our sensations are, insofar as their quality is concerned, only *signs* [*Zeichen*] of external objects, not *images* [*Abbilder*] with any sort of resemblance to them. An image [*Bild*] must be similar in some respect to an object. A statue, for example, has the same bodily form as the human being after which it is modeled; a painting has the same color and perspective projection.[95]

Note that the two examples of the imaging relation are spatial isomorphism in two or three dimensions and similarity of color. But since color, along with other sensations, has been classified as a sign, that leaves only spatial isomorphism to exemplify the imaging relation for the purposes of epistemology.

In his popular lectures on vision of 1868, Helmholtz held out the hope that spatial representations might be "images" of the external world.[96] He thus adopted a position similar to the traditional view of the ideas of primary qualities, as found in Locke and others. But he had tempered such epistemological optimism in his *Optics,* where he contended that "the only respect in which there can be a real agreement between our perceptions and reality is the time-sequence of the events" (PO 3:21).[97] He denied that spatial representations provide images of external spatial configurations. Indeed, he maintained that our ideas can possess no more than a "practical" correspondence with the world, and that the only comparison possible between ideas and things is in terms of the guidance our representations provide us for engaging in actions that bring about a desired result (in the form of an expected group of new sensations). He regarded any other sort of comparison as "unthinkable" and "nonsensical" (3:18). Thus spatial representations may guide our actions as we interact with things in themselves, but that is no guarantee that they picture to us the actual properties of things in themselves. The request for some further kind of comparison between spatial perception and the external world, other than the practical test of moving or bringing touch to bear, is meaningless: "To ask whether the idea I have of a table, its form, strength, color, weight, etc., is true apart from any practical use I can make of the idea, and whether it corresponds with the real thing, or is false and owing to an illusion, has just as much sense as to ask whether a certain musical note is red, yellow, or blue" (3:18). Spatial representations are to be viewed as "correct" or "in correspondence" with external objects just insofar as they allow the perceiver to predict what sensations he will receive upon further sensory interaction with the table. Thus, spatial representations, too, are symbols, ones that can be translated by voluntary movement into new sets of spatial representations.[98]

Even though in the *Optics* Helmholtz classified both color sensations and spatial representations as signs or symbols, he nonetheless posited a great difference between them in their epistemological and psychological status. Sensations stand as signs for various (nonspatial) properties, and spatial perceptions serve to represent the relations among these properties. Although it may be too much to hope that spatial representations are *images* of the relations among things in themselves, they nonetheless signify the actual relations that obtain among the objects that interact with the organ of vision. Inasmuch as they represent these relations well enough for practical purposes (guiding of voluntary motion), they exhibit "correspondence" or "harmony" with external objects, and thus correspond with reality. And so although we are allowed to posit an external world with which we are in causal interaction and to distinguish various properties in that world, we are not allowed to ascribe actual spatial relations to it. There are real relations in the external world that are causally implicated in our spatial representations, but these external relations cannot be known to be spatial. In a general way, this position corresponds to that of Herbart, and to a reading of Kant that accepts a real causal role for the thing in itself.

But this position was only one in a sequence of positions Helmholtz adopted with respect to the problem of the external world. In the late 1840s and 1850s he had been even more optimistic about our knowledge of that world, treating the physicist's description in terms of forces among particles as a description of how things actually are (WA 2:608-9).

However, by the time he wrote the "Facts in Perception," he had given up the notion that the existence of the external world could be established through argument or reasoning; he regarded the postulation of such a world as one hypothesis among others, which, if it is accepted, must be accepted on faith. Which is not to say that he gave up the idea that experience reveals reality; rather, he retreated to the position that the "lawlike" regularities manifest in the changes among sensational signs is constitutive of the real. He ended up by answering the question "What is true in our experience?" by pointing to the lawlike itself. The most general of such laws is the causal law—in Helmholtz's interpretation, the law of lawfulness that must be presupposed in the search for other laws by means of experience. His changing interpretation of the status of this fundamental law is the key to the other changes in his position.

Helmholtz was committed to treating the law of cause as the fundamental principle of natural science from very early on. In the philosophical introduction to his memoir on the conservation of energy,[99] he characterized the theoretical portion of natural science as follows:

> The theoretical part . . . seeks to ascertain from their visible effects the unknown causes of natural processes; it seeks to comprehend them according to the law of causality. We are compelled to and justified in this undertaking by the fundamental principle that every change in nature must have a sufficient cause. [WA 1:13; SW, 3-4]

He went on to characterize the goal of natural science as the successful tracing of natural phenomena to "inalterable ultimate causes," and at one point used the law of cause interchangeably with the principle of sufficient reason. He regarded the causal law as the basic principle of theoretical natural science; the only limitation he mentioned on its applicability was the possibility that the "spontaneity of freedom" would break its necessity, a possibility he chose not to address (SW, 5, 4).

Helmholtz invoked the absolute a priori validity of the law of cause in his lecture of 1855 on vision, in which he accorded it a fundamental epistemic role in knowledge of the external world. In that lecture he repeated the position he had expressed in 1852, that "we never perceive the objects of external world immediately, but perceive only the effects of these objects on our nervous apparatus." He did not find this a hopeless position as regards knowledge of the causes of those effects, for he was prepared to invoke the causal law as a "law of our thought given prior to all experience" (VR 1:116).[100] Far from being a proposition based on experience, Helmholtz regarded the causal law as a requisite to experience of an external world of objects in space: "we need this proposition, before we have any knowledge of things in the external world; we need it, in order merely to reach the knowledge that there are objects around us in space at all, between which a relation of cause and effect can obtain." Here the causal law serves to support a robust realism about the external world, a world in which objects are arranged in space and causally interact with other objects, including the sense organs of the perceiver. We infer the existence of this world as the cause of our own sensory experience.

From this realism about the spatial world Helmholtz made a two-stage retreat. The first stage was expressed in the third part of the *Optics*, where he rejected the notion that our spatial representations of the world have any more than "practical" reality; spatial representations are merely a system of symbols that designate subsequent patterns of sensations. But this position did not call into question the reality of the external world, just our ability to "picture" that reality, and he continued to invoke the causal law as the basis for inferring an external world.

> We can never emerge from the world of our sensations to the representation of an external world, except through an inference from the changing sensations to external objects as the cause of this change. . . . Accordingly, the law of causation, by virtue of which we infer the cause from the effect, has to be considered also as being a law of our thinking which is prior to all experience. Generally, we can have no experience of natural objects unless the law of causation is already active in us, and so it cannot first of all be derived from experiences which we have had with natural objects. [PO 3:29-30]

The effect of his injunction about the merely "practical" reality of spatial intuitions was to alter the epistemic informativeness of inferences to the external world based upon the causal law: that law ceased to serve as a bridge out to the external world, and acted more like a telegraph line into a closed room. For, as Helmholtz had admitted earlier, in the discussion of sensations as signs, the informativeness of such inferences to the external world is limited by the fact that we are directly acquainted only with the effect side of this equation; but since such effects are conditioned both by the cause and by the affected subject, the subject is in no position to attain a knowledge of the cause as it is in itself.

But what about the law of cause itself? Does it not provide objective knowledge of the course of events in the external world? At first it seems so. Helmholtz went on to repeat the idea, expressed in the introduction to his famous memoir, that the law of cause amounts to the supposition of the "comprehensibility" of all natural phenomena. As such, it could not in principle be refuted by experience: "For if we founder anywhere in applying the law of cause, we do not conclude that it is false, but simply that we do not yet completely understand the complex of causes mutually interacting in the given phenomenon" (PO 3:30).[101] Nonetheless, his position in the *Optics* did represent this second retreat from his earlier causal realism. For although he regarded the causal law as irrefutable, he did not consider it to be a law of nature; rather, he regarded it as a subjective condition on our apprehension of nature. Once again equating the law of cause with the law of sufficient reason, he wrote:

> The law of sufficient reason is really nothing more than the *urge* of our intellect to bring all our perceptions under its own control. It is not a law of nature. Our intellect is the faculty of forming general conceptions. . . . Just as it is the characteristic function of the eye to have light-sensations, so that we can *see* the world only as a *luminous phenomenon*, so likewise it is the characteristic function of the intellect to form general conceptions, that is, to search causes; and it can *conceive* the world only as being *causally* connected. [PO 3:31]

In attributing the causal law to the intellect, he did not mean to suggest that the discovery of regularities among phenomena had merely subjective significance; if we discover that the phenomena follow certain laws, this is to be counted as an objectively valid fact. But we should not suppose that our ascription of a causal relation to such regularities is anything more than our way of expressing that the phenomena are lawful; the causal connection can come to nothing more than this.[102]

In comparing the relation between the causal law and the intellect with the relation between sensations and the eye, Helmholtz robbed that law of its ability to underwrite inferences to an external world. The causal law was merely our way of expressing the lawfulness of nature; it could provide no real leverage in getting from sensations to an independent world.

Although Helmholtz did not explicitly draw this conclusion in the *Optics*, he discussed it at some length in the "Facts in Perception." In that work he repeated his previous conclusions regarding the status of sensations as signs of their causes. But he went still further in considering the status of both the realist and the idealist hypotheses about the character of those causes. By the idealist hypothesis, he meant that the world is the product of the knowing mind, like a consistent dream; the realist hypothesis, by contrast, posits a mind-independent reality. Although he found the idealist hypothesis "improbable" and "unsatisfying," he was forced to admit that it could not be refuted. He thus had to admit that "we cannot acknowledge the realist hypothesis to be more than an excellently serviceable and precise hypothesis"; we cannot ascribe it truth, because the competing idealist hypothesis cannot be ruled out (EW, 137-8).

Furthermore, even if the realist hypothesis were accepted, according to Helmholtz that would not involve positing an "imagistic" relation between perception and the external world. He continued to characterize the relation between spatial perceptions and the world as parallel to the sign relation of color sensations. Reviewing once again his "*Willensimpuls*" theory of spatial intuition, according to which we "call the relationship which we alter in an immediate manner by the impulses of our will a *spatial one*," he continued:

> In this case space will also appear to us—imbued with the qualities of our sensations of movement—in a sensory manner, as that through which we move, through which we gaze forth. Spatial intuition would therefore be in this sense a subjective *form of intuition*, like the sensory qualities red, sweet, cold. Naturally, the sense of this would be just as little for the former as for the latter, that the place specified for a specific individual object is a mere illusion [*Schein*]. [VR 2:224-5; EW, 124]

Just as color sensations are not illusory by virtue of being mere signs of their causes, neither are spatial perceptions. On Helmholtz's theory of space, the outer relations that are represented spatially in our experience may be "quite unknown to us," but as long as spatial representations are found adequate for the guidance of action, that is, for predicting subsequent patterns of sensations, they are serving their essential epistemological function.

In attempting to answer his initial question regarding the "true" in our thought and intuition, Helmholtz was not content to stop with denying both the realist and idealist hypotheses. He found an opportunity to proceed a step further by abstracting from those two hypotheses and reflecting upon the regularities in the appearances that they both must concede. He contended that "what we can find unambiguously, and as a fact without anything being insinuated hypothetically, is the lawlike in the phenomena" (EW, 138). Throughout the paper, he distinguished between "the real" (*das Reelle*) and "the actual" (*das Wirkliche*); the actual, for Helmholtz, is the lawlike pattern of relations among phenomena. Although he allowed that scientists could discuss even the metaphysically "real" under the guise of hypothesis, he believed that such investigations must remain inconclusive. Invoking Kant, he maintained that knowledge of the "real," or the "thing in itself," is denied us. But that limitation does not foreclose our epistemic options: "we can attain knowledge [*Kenntniss*] of the lawlike order in the realm of the actual, admittedly only as presented in the sign system of our sensory impressions." The domain of science lies in the investigation of the actual. It can "seek for the laws of the actual," leaving aside the "soaring schemes of the metaphysicians" as expressed in the realist and idealist hypotheses (VR 2:241-2, 245; EW, 140-1, 143).

But what is the status of this knowledge of law? How does Helmholtz's limitation of our knowledge to regularities among impressions differ from Hume's similar conception, a conception that Helmholtz had earlier portrayed as leading to the denial of the possibility of objective knowledge? In the *Optics*, he had asserted that laws unifying a range of the phenomena may *ipso facto* be accepted as objective. His discussion of the lawlike in 1878 has much the same tone. At the same time, he found himself confronted with the problem of induction, with little room to maneuver. Helmholtz was torn. He considered the tendency to respond to the lawlike in phenomena to be a fundamental fact of our psychology, and one that had merely to be refined in the service of science. But he recognized that this fundamental law of thought could not provide unconditional justification for the inferences among the phenomena that we form in accordance with it. He hankered for just such a justification, but he was unwilling to pay the price of ascribing more structure to the mind than the primitive operations of association.

This tension in Helmholtz's thought is amply revealed in the essay of 1878. Within a few paragraphs he makes the following points:

> 1. Every inductive inference is based on trusting that an item of lawlike behavior, which has been observed up to now, will also prove true in all cases which have not yet come under observation. This is a trust in the lawlikeness of everything that happens.
>
> 2. Trust in lawlikeness is at the same time trust in the comprehensibility of the appearances of nature.... we call the regulative principle of our thought that thus impels us the *law of cause.*
>
> 3. We have no further guarantee for the applicability of the law of cause than this law's success.
>
> 4. The law of cause is an a priori given, a transcendental law.
>
> 5. A proof of it from experience is not possible, since the first steps of experience, as we have seen, are not possible without employing inductive inferences, i.e., without the law of cause.... Here the only valid advice is, have trust and act! [VR 2:243-4; EW, 141-2]

The law of cause is an unprovable presupposition that must be accepted on faith, a regulative principle that drives our thought, and an a priori given. In standard Kantian usage, if the law of cause is an a priori given, then it is not an unprovable presupposition that must be accepted on faith; its necessary and universal validity is established through its deduction as a condition on the possibility of experience. Helmholtz seems to hint at this last point at (5), but does not use it to make an assertion of the law's universal validity. The final portion of (5) places us back at points (1) and (3).

In the end, Helmholtz adopted the position that we are driven to accept the causal law, but we remain unable to assert unconditionally its universal applicability. Contrary to the satisfaction that he took from the "Kantian" justification of the law in 1847 and 1855, in his later writings Helmholtz was no longer able to accept that this "law" of thought gives us purchase on objective reality. And indeed, in an appendix to the "Conservation of Force" composed in 1881, Helmholtz explicitly rejected his earlier Kantian position: "The philosophical discussion in the Introduction is more strongly influenced by Kant's epistemological views than I would now think correct. It was only later that it became clear to me that the principle of causality is, indeed, nothing more than the presupposition of the lawfulness of all natural phenomena"; he followed this statement with some remarks on law, power, and force that are strongly reminiscent of both section 26 of the *Optics* and the 1878 lecture.[103]

What would it have meant for Helmholtz truly to attribute objective validity to the causal law, or guarantee its universal applicability? One possibility is for him to have actually accepted the Kantian point that the law of cause is a priori, and to have accepted it in the Kantian sense. But Helmholtz never revealed an appreciation of the Kantian notion of a transcendental deduction of the law. For him, there could be only one interpretation for the claim that the law of cause was both "a priori" and of guaranteed universal applicability: to treat it as innate and as innately "corresponding" to an independent, objective world. But he could not accept the law on these terms, for that would amount to postulating a "preestablished harmony" between a principle of thought and a mind-independent world of the very sort that he roundly rejected in his arguments against nativism.

However, Helmholtz might have understood the claim that the law of cause is an a priori principle in two senses that he would have found acceptable. First, it could be an innate mechanism for discerning the lawlike in the phenomena. This indeed was one component of Helmholtz's position. In his 1878 paper, the tendency to accept the law of cause was just the tendency to form generalizations through association: the mind is a filter for the lawlike because it has an associative memory. When he called the law of cause a "law of thought," he was not postulating it as a principle of the understanding in the strict Kantian sense. For although he often remarked that his work allowed for the dissolution of Kantian "intuition" into laws of thought, he resolved the laws of thought themselves into the operations of the "sensory memory," that is, into the laws of association.

Second, the law could be "a priori" in the sense of being a presupposition of the scientific activity of searching for laws. As in (2) above, the law of cause would now be treated as equivalent to the lawfulness of phenomena, and the search for the lawful in the phenomena would be treated as the aim of science. To treat past observation as evidence of lawfulness is to "trust" in the lawfulness of phenomena and the comprehensibility of nature, which, according to Helmholtz, is to trust in the law of cause. But to "trust" in the comprehensibility of nature is not to establish the a priori certainty that nature is comprehensible; to presuppose the law of cause is merely to "have trust and act."

This dual reading of Helmholtz's attitude toward the causal law (as innate tendency of thought and as pragmatic presupposition) is confirmed by a passage from his unpublished papers. The causal law, he wrote, is distinguished from the particular laws of nature by the following points:

1. It is the presupposition for the validity of all the others.

2. It provides the only possibility for us to know anything non-observational.

3. It is the necessary basis for purposeful behavior.

4. We are driven to it by the natural mechanics of our association of representations.[104]

Recognizing that these points do not establish the law's validity, he suggested that we are strongly driven "*to want* it to be be correct, for it is the basis of all thought and behavior." But whether we are *entitled* to consider ourselves as thinkers can be answered only by acting in a way that presupposes the law. And it can never be answered finally. Even though our mind is such as to search for the lawful, we are unable to provide an absolute justification for this search. Again, the solution is to have trust and act.

It is Helmholtz's attitude toward scientific practice that distinguishes his position from Humean skepticism. Perhaps because Hume was so earnestly engaged in curbing the claim of reason to metaphysical knowledge through the causal law, he stressed the negative conclusion of his attack on causal reasoning: induction and causal reasoning admit of no rational justification. Even in his positive program of developing a "science of human nature," he simply concluded that we have the animal habit of associating constantly conjoined ideas. There was only the slightest suggestion that it would be well to cultivate such habits. And there was absolutely no mention of whether, or why, it would be reasonable to allow present observations to bear on subsequent practice. Hume left it as a mere fact of our nature that we do.

By contrast, for the mature Helmholtz the important point was to acknowledge early on the limitations on the metaphysical claims of reason, and to stress the fact that the practice of science itself presupposes the search for ever more general laws. The causal law becomes a kind of practical, or pragmatic, a priori principle. Not only are we to recognize that our mind is a filter for associative regularities; we should explicitly set the task of comprehending nature—that is, of referring the phenomena to ever more general laws—as our goal. Current observations are counted as evidence for universal laws, even if no absolute justification can be attained; the laws of nature are granted provisional acceptance. The justification for such laws is not cumulative but resides in the daily application of previously gained generalizations to the phenomena as they present themselves. Without being sure of our title to be thinkers, we proceed in the project of making nature comprehensible. The discovery of ever more general connections among observed phenomena is its own reward.

Helmholtz used the same strategy of ascertaining the assumptions of a particular scientific practice and seeking to give them a practical justification in his work on the applicability of geometry to physical space.

5 Helmholtz and Kant on Geometry and Perception

In papers from the 1860s and 1870s Helmholtz sought to revise Kant's conclusions about the status of Euclid's geometry, both as a description of physical space and as a description of visual space. He contended that the applicability of Euclid's geometry to physical and to visual space is an empirical question, and concluded that the attempt to establish the necessity of Euclid's axioms through a Kantian appeal to space as a transcendental form of intuition must fail. Helmholtz attributed his great predecessor's errors regarding the necessity of Euclid's axioms to the underdeveloped state of both mathematics and sensory physiology at the time he wrote.

Helmholtz's argument for the contingency of Euclid's geometry proceeded in two stages. First he questioned the idea that Euclid's geometry is the only conceivable geometry by invoking the work of Riemann and Beltrami as well as his own work on n-dimensional manifolds (EW, 11-17).[105] He had become interested in n-dimensional manifolds as a means of describing the relations among primary colors. Such manifolds define a function among some number of dimensions of variation, in this case, a function from possible values for three primary colors to the color resulting from their combination. A number of authors (including Herbart and Grassmann, who were modifying Newton)[106] had observed that the color function (among others) can be displayed spatially, in this case, as a "color circle." But it was the insight of Riemann and Helmholtz that no spatiality is implied by the n-dimensional function itself, as is now familiar from applying a three-dimensional function to the relations among colors, or creating a two-dimensional manifold relating auditory pitch and intensity. Helmholtz considered space (regarded as the object of geometry) to be a special case among continuously differentiable manifolds. In geometric space a metric exists assigning a definite numeric value to the distance between any pair of points. He went on to describe "spatial manifolds" that do not satisfy Euclid's axioms but that nonetheless yield a consistent geometry, including manifolds with both positive and negative curvature, in Riemann's sense.[107] Helmholtz concluded that other consistent geometries besides Euclid's are indeed conceivable, in the sense that they can be described within the resources of analytic geometry. He carried out this part of his argument by considering only analytic geometry (rather than diagrams or spatial images) in order to avoid any appeal to intuition in Kant's sense.[108] In effect, he was using analytic geometry to provide a means for attaining results about continuous magnitudes[109] without beginning with a spatial conception of such magnitudes (as did geometry proper, which was understood to have an inherently spatial subject-matter).

In the second stage of his argument, Helmholtz contended that empirical measurements could conceivably be obtained that would provide evidence that physical space is non-Euclidean. Helmholtz, like Kant and Riemann, considered geometry proper to be the science of physical space. The claim that the mathematics of manifolds allowed for consistent geometries other than Euclid's would not have been particularly interesting unless it could be applied to physical space as an object of measurement (or to the spatially extended objects defining a physical space). In order to establish his conclusion that the axioms of geometry have an empirically contingent status as descriptions of space, Helmholtz needed to show that it was possible to describe a set of measurements that would confirm the hypothesis that physical space is non-Euclidean. He would then hold that our acceptance of Euclid's geometry is based on contingent experience with actual space, and so it does not admit of a priori justification.

Helmholtz limited his argument to conditions of measurement consistent with a space of constant curvature (whether Euclidean, or possessed of positive or negative curvature). He considered the basic geometrical operation to be the determination of congruence through the procedure of translating one body into a position such that it coincides with a second body.[110] The fundamental assumption of his argument was that rigid bodies exist that can be moved freely in any direction and rotated about any axis without distortion.[111] This assumption having been granted, one can imagine a sequence of measurements with a rigid instrument in which, say, the three angles of a triangle are determined to sum to more or less than the sum of two right angles; or, assuming that light rays are "straightest" lines (whether Euclidean or not), one can imagine a measure of parallax for distant stars that would yield a value consistent with space of negative curvature but not with flat or positively curved space (EW, 18, 154). If the assumption about rigid bodies is granted, the arguments are correct.

Helmholtz further contended that non-Euclidean geometries are conceivable in the robust sense that one can imagine a series of spatial intuitions an observer would have in a space that deviated significantly from zero curvature. He appealed especially to the work of the Italian mathematician Eugenio Beltrami, who had provided an account of the spatial intuitions that would be available to an observer moving through a space of negative curvature.[112] Using Beltrami's device of mapping this "pseudospherical space" into the interior of a Euclidean sphere, Helmholtz provided an account of what it would be like for a person with a body, eyes, and habits such as ours (i.e., a person whose sensory apparatus was accustomed to Euclidean space) to move about in a non-Euclidean space (EW, 21-2). He showed that the successive experiences one would receive upon moving about would contradict Euclidean expectations but would conform with what is expected in a pseudospherical space of constant negative curvature (again on the assumption that there are "fixed" or rigid bodies—perhaps our own—and that these remain rigid as they move about: 15-17, 24). For example, it might appear to an observer that two lines converge at the horizon, say, some 100 feet away (that is, the lines might appear the way actually converging lines appear in Euclidean space). Advancing toward the point of convergence, the observer would find that the lines were separate at that position but appeared to converge farther on, and this experience would be repeated for as far as he followed the lines. In such a space, lines that appeared to be parallel would diverge as an observer followed them (21-2). Helmholtz also provided an account of the experiences of an observer who is imagined to enter a space with constant positive curvature; as he suggested, "the strangest visual feature of the spherical world would consist of the back of our own head" filling the distant background (22-3).

Helmholtz therefore concluded that Kant was wrong in holding that the axioms of Euclid's geometry apply to physical and visual space with necessity. He maintained that the applicability of the axioms must be decided empirically; the belief that they apply to physical space with necessity is a result of the (surmountable) limits placed upon our imagination by the structure of our sense organs and the character of the perceptions we actually have. In arriving at these conclusions, Helmholtz did not argue that the geometry of n-dimensional manifolds is itself empirically derived or supported. Rather, he invoked Riemann's distinction between mathematically defined geometrical structures on the one hand and their fit to physical space on the other (EW, 12). Not the axioms of geometry considered purely mathematically, but the axioms of geometry as descriptions of space must be subject to empirical test, according to Helmholtz. The question of which axioms describe physical space thus cannot be decided with apodictic certainty. It takes on the status of other empirical claims in physics (18, 152-63).

In Helmholtz's view, the status in physics of empirical claims with any generality was not simple factuality. Helmholtz did not believe that the Euclidean character of physical space could be determined as a fact of observation. Although he had a tendency to speak of the "facts" underlying geometry, he seems to have meant not that the geometry of space can be factually determined in a straightforward manner, but that any attempt to assign a geometry to space must rely on empirical considerations. Indeed, he did not claim that space is in fact Euclidean but only that the measure of curvature "is at least not noticeably distinct from zero" (EW, 14). Thus, he was aware that limitations on the precision of measurement precluded a determination of the precise geometry of space. He suggested that the assignment of a geometry to physical space has the same epistemic status as does a natural law supported by induction. As empirical generalizations, the axioms "would participate in the merely approximate provability of all laws of nature through induction."[113]

Nonetheless, his claim that the geometry of physical space could be shown, on empirical grounds, to be "nearly Euclidean" was not without its problems. He acknowledged that the "factual" determination of the curvature of space depended upon the assumption that rigid bodies exist, or at least bodies that are affected only by local forces such as temperature. He realized that a Kantian opponent could contend that space was flat, but that measuring rods were deformed by forces that operated upon them depending on their location in space. Indeed, he recognized the complete observational equivalence of a situation in which space has a negative curvature and rods remain rigid when translated, and one in which space is Euclidean and measuring rods change in length as they are transported. Thus, even if the measurements he envisioned as supporting the conclusion that space is non-Euclidean should occur, one could retain a commitment to Euclidean space by making suitable adjustments in mechanics. If one stipulated that space is Euclidean, these measurements would then be taken as evidence that rigid bodies deform when translated through space (EW, 19-20, 25, 156). But, as he observed, such a position would violate Kant's stricture that the axioms are synthetic; the envisioned position would make the axioms "analytic," indeed, the analytic consequence of a stipulative definition.[114]

On first appearance his own position is no better off, for it required positing the existence of rigid bodies. As he admitted, the existence of such bodies cannot be determined simply by measuring the bodies at hand, for the measurements would require some bodies already known to be rigid. Helmholtz struggled to provide a satisfactory solution to this problem. Initially, he argued that we must simply assume, as a principle of mechanics, that physical objects can be translated without distortion through space. He classified as methodological the assumption that the same physical processes operate in the same manner independent of spatial location (EW, 25). This assumption, however, would parallel the Kantian stipulation of Euclidean geometry, unless one provided a means of determining "same manner" independently of measurements presupposing that one's instruments undergo no alteration when translated through space.[115] His solution to this problem was to argue that certain mechanical processes could be used to determine whether one region of space was "physically equivalent" to another: he defined as "physically equivalent" those regions of space in which "under like conditions, and in like periods of time, like physical processes can exist and run their course." Unfortunately, this proposal was not decisive. As Schlick later observed, the time-determination involved must ultimately rest upon spatial measurements.[116] Effectively, Helmholtz's argument remained in the position of asserting the principle that the possibility of translation without distortion is a presupposition that accords with experience and provides for a satisfactory mechanics, but that cannot be established absolutely.

Despite his inability to solve this problem decisively, Helmholtz was on solid ground in his criticisms of Kant for accepting as universal and necessary what can receive only empirical support. Even failing a criterion for picking out rigid bodies, we can imagine empirical evidence that would force us to choose between revising mechanics and abandoning the Euclidean structure of space. Since Kant had considered both the structure of space and the fundamental principles of mechanics to rest on a priori foundations, Helmholtz's argument showed that the Kantian picture required revision. Interestingly, Helmholtz himself did not precisely understand how his arguments entailed a revision in Kant's position, for he did not appreciate the notions of "transcendental" and "synthetic a priori" knowledge as Kant employed them. Helmholtz equated these notions with the positing of innate knowledge or cognitive mechanisms. He gave them a physiological and psychological interpretation that he did not distinguish from Kant's quite different philosophical understanding of these notions.

Consider how Helmholtz described his agreements and disagreements with Kant. First, he indicated he agreed with Kant that space is a "subjective," "necessary," and "given" form of intuition that is "possessed prior to all experience." By "subjective" he meant that our representation of space depends on our nervous apparatus in a manner similar to sensory qualities such as red, sweet, or cold. By "necessary" he seems to have meant that it is analytic that we equate the "external world" with that which we perceive as having spatial properties. By "given prior to experience" he meant that our motor and perceptual apparatus must "be given to us, by our makeup, before we could have spatial intuition" (EW, 124). While agreeing with Kant (as he interpreted him) on these points, he disagreed with Kant's belief that space is "transcendentally determined" to be Euclidean. He chalked up Kant's belief in the necessity of Euclid's axioms to the simple fact that the development of non-Euclidean geometry occurred after Kant's death; since Euclid's was the only available geometry when Kant wrote, its applicability to space might well have seemed to be guaranteed.[117] But he also suggested that Kant's position derived in part from his belief that the form of spatial intuition is innate:

> In his conclusion that spatial relationships contradicting the axioms of Euclid could never in any way be represented, just as in his whole conception of intuition in general as a simple psychic process, incapable of further resolution, Kant was influenced by the contemporary states of development of mathematics and the physiology of the senses.[118]

Helmholtz considered himself to be in a position not only to correct Kant's mathematical conceptions but also to correct Kant's view that, from the standpoint of the physiology of the senses, the space of intuition is a primitive given.

The notion that correcting Kant's physiology of the senses would help salvage his position on geometry was doubly in error. First, as Helmholtz himself had shown, even if one assumed that the visual system has the habits (or, we might add, the innate disposition) appropriate to Euclidean space, measurements could be attained to indicate that physical space was not flat (given the assumption of rigid bodies). The question of the "geometry" of visual space could be separated from the question of the geometry of physical space; a nativistic theory of vision was compatible with the position that the geometry of physical space is only contingently Euclidean. Thus, even if Kant had held a nativistic theory of vision, that alone could not explain the error of the Kantian position on the geometry of physical space. Furthermore, Helmholtz's assertion that we could come to have a visual system attuned to non-Euclidean space,[119] far from supporting his claim that the structure of physical space is only contingently Euclidean, itself presupposed that claim. It was as a result of showing that physical space could be non-Euclidean that Helmholtz was able to contend that if spatial perception is indeed an acquired ability, the character of the ability we acquire contingently depends on the actual character of physical space. On the supposition that our visual habits are formed through experience, it may be supposed that they would be different in a non-Euclidean space (EW, 22).

Second, and more fundamentally, the notion that Kant's point about the status of Euclid's axioms could be distilled into a thesis of psychological nativism revealed Helmholtz's misunderstanding of the notions of "transcendental knowledge" and "a priori form." We have met with this misunderstanding above, when Helmholtz equated the doctrine of specific nerve energy with the Kantian notion of a subjective form of perception. This mistaken equation, which was made prior to Helmholtz and has been repeated since, suggests where the misunderstanding took place. Helmholtz seems to have believed Kant's doctrine of the subjective character of space was exhausted by the mere "subjectivity" itself—that is, by the doctrine that spatial representations have the same status Locke assigned to the secondary qualities. Helmholtz understood the term "transcendental," as applied to spatial representations, to be exhausted by the assertion that, upon stimulation, our visual system in fact will come to develop spatial representations (representations characterized by "next-to-ness").[120] Thus, the point that the character of color sensations depends upon the characteristics of the sensory nerves could seem to be equivalent to the claim that space is a subjective form of intuition.

Helmholtz failed to appreciate both the content and the basis of Kant's doctrine of space. First, he did not see, or refused to accept, that Kant had asserted that the a priori form of intuition could condition the actual content of experience. He argued that Kant's position did not require that his (alleged) nativism condition the character of physical space, as opposed to the character of visual space. Helmholtz thus failed to see that Kant intended the conclusions of his "Transcendental Aesthetic" to set a condition on, and to provide synthetic a priori knowledge of, all possible experience of space, including measurements undertaken in physics. Second, he did not appreciate that Kant was not attempting to establish a thesis about the psychology of vision, and that, far from Kant's intending his claims to be a contribution to empirical psychology, he held them to be established independently of that discipline, through transcendental argumentation.

Helmholtz's assimilation of Kantian subjectivity of space to the accepted notion of the subjectivity of color overlooked the special epistemic character Kant ascribed to our knowledge of space and to Euclid's axioms as a description of the space we know. Kant described his theory of spatial intuition in the "Transcendental Aesthetic" as a "deduction"; as such, it could not be based upon experience but must appeal to the conditions on any possible experience. *Prima facie*, Helmholtz's claim that our sensory apparatus "conditions" our sensory representations may seem equivalent to Kant's claims. But in Kant's case, part of the "experience" to be accounted for was the necessity of the intuitive "fact" that the shortest distance between two points is a straight line (and other Euclidean "intuitions"). Necessity could not be ascribed to such statements if they were supported empirically; hence, Kant's appeal to space as a transcendental "form" of intuition that conditions all experience, and (allegedly) guarantees the applicability of Euclid's geometry to all experiences of the physical world.

The notion of a transcendental argument based upon the conditions for the possibility of experience—where "experience" itself must claim Kantian objectivity, that is, universality and necessity—was utterly foreign to Helmholtz's way of thinking. By his lights all universal claims about the physical world are "merely probable." The law of cause was the only exception, and Helmholtz came to regard it not as a law of nature— and certainly not as a principle of knowledge that had been proven to apply to all experience with necessity—but as a methodological principle whose acceptance constituted a presupposition of empirical investigation.

To what extent, then, were Helmholtz's arguments against Kant's teaching on the status of Euclid's geometry as a description of physical space effective? Although their effectiveness did not derive from his own empirically based psychological theories, they worked precisely because Kant's doctrine of space contained empirical content in claiming that all sensory experience must conform to Euclid's geometry. I argued earlier that this part of Kant's doctrine should be regarded as the assertion, on transcendental grounds, of a "brute fact" about the form of outer sense. Helmholtz, drawing upon Riemann and others, was able to show that this brute fact could not have the necessity Kant claimed for it. The impact of Helmholtz's arguments against Kant derives from the plausibility of our being able to have experiences that could show the structure of objects in space to be non-Euclidean—that non-Euclidean geometries are not only mathematically possible but capable of receiving empirical support as the geometry of physical space (allowing the methodological assumptions pertaining to mechanics).

Helmholtz was not thereby doing psychology: he was investigating the epistemological status of claims about the geometrical structure of physical space. The imagined sequences of experience are significant as potential evidence for a claim in physics. What Kant had regarded as a brute but necessary fact of transcendental psychology (pertaining to the form of outer sense), Helmholtz showed to be more plausibly regarded as a contingent fact about physical space. Kant left himself open to this argument because he had himself argued that the form of human outer sense determines the spatial character of physical space, thereby supporting his claim that we can have knowledge of space in an a priori manner. Helmholtz effectively denied that a priori knowledge of the structure of physical space is possible. This conclusion was not based on his work in psychology. The plausibility of his claim in psychology that spatial perception might be non-Euclidean was instead based on an examination of the sorts of evidence that might be brought to bear on certain claims in physics.

Although Helmholtz could not fully appreciate his own achievement because he did not attribute a commitment to synthetic a priori principles to Kant, he effectively showed that Kant had adopted an untenable conception of the relation between the "structure of the human mind"—whether conceived transcendentally or as an object of empirical psychology—and the range of epistemically possible theories. The moral of Helmholtz's arguments is not merely that an "innate" space of human intuition could not decide the structure of physical space. By implication, he revealed that any attempt to set substantive a priori constraints on the curvature of physical space was doomed to fail.

6 The Limits of Natural Science

Helmholtz's work in sensory perception and cognition was an attempt to bring the mind, or at least a certain set of mental phenomena, within the domain of natural science. But he did not limit his application of natural-scientific explanation to sense perception; under the banner of the theory of unconscious inference, he extended natural-scientific modes of explanation to the mind of the natural scientist itself. According to his analysis of the psychology of inference, the mind of the scientist is a filter for the lawlike; he singled out the search for lawfulness within the fluctuations of the phenomena as the defining characteristic of scientific cognition. At the same time, his own theory of perceptual and scientific cognition was itself (at least by his lights) an instance of the perception of the lawful in the flux of phenomena, for it brought unity to the phenomena of the mind by an appeal to general laws; in this case, the laws of association.

Helmholtz was in fact prepared to extend his naturalistic approach to yet other domains of mental phenomena, and especially to those associated with the arts. Perhaps because his first academic position involved teaching anatomy to artists, he produced in 1871 a lecture on the relation of optics to painting. He brought physiology and psychology to bear on painting in several ways, explaining aspects of the experience of painting through the theory of color contrast, seeking scientific insight into certain established artistic practices, and proferring "tips" to the working artist on the use of aerial perspective.[121] The paradigmatic example of his application of sensory physiology to the arts, however, was his earlier treatment of the consonance and dissonance of musical tones. His explication of consonance and dissonance, which served as the centerpiece for the *Sensations of Tone,* was based upon physics, physiology, and psychology, all three. Starting from the wave-analysis of sound and the Fourier analysis of waves, Helmholtz sought to account for the interactions between musical notes that lead us to experience the pleasant and unpleasant sensations associated with consonance and dissonance. His account focused on the interactions among the upper partial tones that accompany the sounding of the "fundamental" pitch of a note played on a musical instrument. According to his analysis, consonance and dissonance stand at the foundation of musical scales, and the history of musical tonality may be seen as a slow process of discovering the consonances and dissonances associated with various musical intervals and various tunings and formulations of the scale.[122]

By no means did Helmholtz consider his natural scientific approach to be exhaustively adequate as a means of investigating art. Especially when discussing the expressive quality of art and the arousal of feelings in the spectator, he spoke of reaching the limits of natural science. A particularly striking example occurred in a lecture from 1857 on the physiology of harmony. Near the end, he compared the rhythmic crashing of ocean waves, an enthralling sound in its own right, with the flow of a musical composition.

> Whereas in the sea, blind physical forces alone are at work and the final impression on the spectator's mind is nothing but solitude, in a musical work of art the movement follows the outflow of the artist's own emotions. Now gently gliding, now gracefully leaping, now violently stirred, penetrated by or laboriously contending with the natural expression of passion, the stream of sound, in primitive vivacity, bears over into the hearer's soul unimagined moods which the artist has overheard from his own, and finally raises him to that repose of everlasting beauty of which God has allowed but few of his elect favorites to be the heralds. [VR 1:155; SW, 107-8]

Helmholtz's audience must have thought that their speaker would now bring the same insight to these matters that he had brought to the experience of consonant and dissonant tones. But Helmholtz raised the question of artistic expression only to place it firmly beyond his reach. He closed his lecture with a simple statement: "Here however are the boundaries of natural science, and I am commanded to stop."

Two questions are posed by this last turn of phrase. What are the boundaries of the application of natural science to art, that is, how are they to be recognized so that we know when we've reached them? And does the fact that Helmholtz says he was commanded to stop suggest that he wanted to go on?

In the *Sensations of Tone,* as he passed from examining the physiological and psychological effects of isolated or simultaneous tones to discussing musical scales and tonic relations, Helmholtz realized that he would need to address questions of musical style. He noted that the character of his investigation must change because he was crossing the boundary just mentioned.

> Up to this point our investigation has been of a purely natural-scientific type. We have analyzed the sensations of hearing, and investigated the physical and physiological bases for the phenomena discovered—partial tones, combinational tones, and beats. In the whole of this research we have dealt solely with natural phenomena, which present themselves mechanically, without any choice, to all living beings whose ears are on the same anatomical plan as our own. In such a field, where mechanical necessity reigns and all free will is excluded, science is rightfully called upon to establish constant laws of phenomena and to demonstrate strictly a strict connection between cause and effect.[123]

This description asserts Helmholtz's conception that natural-scientific investigation is essentially a search for laws. Now he contrasts this type of investigation with the one at hand:

> Because in this third part of our inquiry we turn primarily to music and wish to furnish a satisfactory foundation for the elementary rules of musical composition, here we tread on new ground, which is no longer purely natural-scientific, although the knowledge which we have gained of the nature of hearing will still find numerous applications. We pass on to a problem which by its very nature belongs to the domain of aesthetics. When we spoke previously, in the theory of consonance, of the agreeable and the disagreeable, we referred solely to the immediate impression made on the senses when an isolated combination of sounds strikes the ear, and paid no attention to artistic contrasts and means of expression; we thought only of sensuous pleasure, not of aesthetic beauty. The two must be kept strictly apart, although the first is an important means for attaining the second.[124]

Sensuous pleasure may be fixed by natural laws governing the physiology of sensation, but aesthetic pleasure is not. As Helmholtz put it for the aesthetics of music, "the system of scales, modes, and harmonic tissues does not rest upon inalterable natural laws, but is the result of aesthetical principles, which have already changed, and will still further change, with the progressive development of humanity."[125] Taste is historically and culturally conditioned. The method of natural science, with its search for strict causal laws, is limited to the material elements of the arts. Aesthetical inquiry can discover principles of style only through historical investigation of the various periods of art and national or cultural artistic traditions.

This looks like the distinction between *Naturwissenschaft* and *Geisteswissenschaft*—and in fact it is. In a lecture entitled "The Relation of the Natural Sciences to the Sciences in General," presented before the assembled faculty of the University of Heidelberg in 1862, Helmholtz mapped out the relations between the *Naturwissenschaften* and the *Geisteswissenschaften* in a parallel manner. He contrasted the natural and moral sciences in terms of both their subject-matter and their characteristic modes of thought. The natural sciences seek universal categories and laws, which may be taxonomic, as in natural history, but which will tend to be exceptionless nonetheless; physics and mathematics are the paradigms. The moral sciences, by contrast, deal with human institutions and actions; they cannot achieve the certainty of the natural sciences (SW, 128-38).

Helmholtz's discussion of the methods of the natural and moral sciences is of special interest. In contrasting their respective methods he also contrasted their characteristic modes of cognition. In his 1862 lecture, he said that the method of the natural sciences involves especially the conscious formation of general categories and laws. When a natural scientist arrives at a generalization from a mass of facts, such as that all mammals have lungs, he makes what Helmholtz (SW, 130-2) termed a "logical" induction (which Helmholtz knew could not be reduced to a strict logical principle). Once such an induction has been established and codified it can serve as the basis for a deduction—e.g., from the observation that something is a mammal, to the conclusion that it must have lungs. Precise, consciously formulable inferences characterize the natural sciences.

By contrast, Helmholtz characterized the intellectual activities of the moral sciences by their lack of clearly formulable principles. The jurist, the philologist, and the historian must all be prepared to bring enormous amounts of information to bear on single cases: to see what fits and what does not, which generalizations should be applied and which should not. This requires a certain "psychological instinct" or "feel"—"*Tact*" and "*Gefühl*" in Helmholtz's words—that results from a process of unconscious reasoning. This feel for what is relevant, this ability to grasp what is essential in a large body of information or in the flux of everyday experience, Helmholtz termed "artistic induction":

> In opposition to *logical induction,* which operates only with clearly defined universal propositions, we shall call this kind of reasoning *artistic induction,* since it is most conspicuous in the most exceptional works of art. It is an important part of artistic talent to be able to reproduce—by words, forms, colors, or musical tones—the characteristic external indications of a person's character, and to grasp by a kind of instinctive intuition, uncontrolled by any definable rule, the steps by which we pass from one mood to another. [SW, 132]

The moral sciences, dealing as they do with human actions and motives, make particular use of this "artistic" induction.

In 1862, Helmholtz assimilated the characteristic thought processes of the moral sciences to those of the arts, thereby setting both apart from the natural sciences. In subsequent years, his conception of the relations among the thought processes of the scientists and artists changed. Although there is no evidence that his attitude toward the distinction between the natural sciences and the *Geisteswissenschaften* altered, his attitude toward the relation between the psychology of scientific inference and "artistic induction" underwent a complete turnabout in the space of a few years. By the end of the decade he stressed the similarity between the intellectual activity of the natural scientist and that of the artist, emphasizing the role of what he had earlier called "artistic induction" in the thought processes of the natural scientist.[126] After about 1870, he downplayed the importance of deduction in natural science, confining it to the presentation of results already obtained. Correspondingly, he emphasized the importance of inductive intuitions in the work of active natural scientists. One might speculate that this resulted from his own deeper involvement in physical research after being called to the Physics chair at Berlin in 1871. A perhaps more fundamental factor was the development of his doctrine of unconscious inference. For the mature Helmholtz, the intellectual processes involved in both science and art are one with the intellectual processes that underlie perception.

In order to see how the psychology of inference could yield a unified conception of the psychology of science and art, we need to examine Helmholtz's understanding of what he termed "the highest aims of art." While he thought of art as expressing the intuitions of the artist, he conceived the artist, like the scientist, to be seeking truth. The truth he seeks is that of the ideal type. He does not slavishly imitate nature; art involves translation. The artist sifts through ordinary experience to bring together a set of features that may never occur together in nature. This translation results in an "ennobled vision" of nature. A work of art "should present us with a vivid, unified perception of all the features of an ideal type, the separate fragments of which normally lie scattered chaotically in our minds."[127]

In his conception of the artist as grasping the ideal, the invariant, the lawlike in nature, Helmholtz sealed the bond between artist and scientist. The development of this aspect of Helmholtz's thought is apparent in his two papers on Goethe, written forty years apart. In the first, delivered in 1852, Helmholtz attempted to explain Goethe's achievements as a natural historian by showing how his abilities as a poet could carry over to his grasp of natural phenomena.[128] Helmholtz emphasized the dividing line in Goethe's abilities. The poet was successful in those areas, such as plant and animal morphology, in which he could treat nature as a work of art and put his sensibilities to work in grasping manifest though subtle kinships in form or shape. He failed in his theory of color, because that area of investigation requires the ability to abstract precise mathematical laws under controlled conditions, a requirement that conflicted with the poet's direct, intuitive approach to natural phenomena. In the terminology of the 1862 paper, the poet succeeded in the first area through the use of artistic induction; he failed in the second because it required the abstraction and precision of logical induction.

In his second paper on Goethe, delivered in 1892, Helmholtz revised his treatment of Goethe. He now presented art as a second way, besides science, of grasping the lawful in phenomena. He rejected what he termed the common philosophical conception that artistic intuition stands in opposition to thought proper. Artistic intuition acts like a filter for the invariant in nature: the lawful or the ideal is captured in memory, allowing the variable to fall by the wayside. In its intuitive grasp of lawful regularity, art is like sense perception and scientific inference. In all three cases the mind retains only what is regular and invariant in the flux of experience.[129] The artist's imagination grasps the lawful in nature through the associative laws of thought. This grasp informs the artist's productions. The audience to the artist's works then can find in it a truth they recognize in their own experience. The artist has distilled it out and presented it in a clear and forceful manner. In this sense artist and scientist have the same aim: each seeks to make comprehensible what is lawful in nature.

The unifying feature of Helmholtz's accounts of cognition in art, perception, and science is the reduction of mental activity to the associative filter of memory. This unity had its cost. Helmholtz recognized that an associational account of scientific inference could not explain why the laws so inferred are justified. His reduction of mental processes to natural law was in tension with his desire to justify the search for laws, perhaps by securing the validity of the causal law itself. Given that desire, why did he reject his earlier tack of ascribing objective validity to the causal law on Kantian grounds? One reason was that it would amount to positing a "preestablished harmony" between thought and the world.

But there was an even more powerful consideration for appealing to the laws of association in explaining various acts of cognition. In reducing thought to associational processes, he subsumed it under a common form of explanation in the natural sciences: that of the interaction of simple elements in accordance with law. A striking parallel may be drawn between Helmholtz's use of this form of explanation in physics and psychology. In his memoir of 1847, he characterized the ideal of physical explanation thusly: "natural phenomena should be traced back to the movements of material objects which possess inalterable motive forces that are dependent only on spatial relations" (SW, 5). With only a few substitutions of terms, this ideal fits his preferred mode of explanation in psychology as well. Associational laws are laws for relating sensations in accordance with their elementary properties, including their qualitative character and intensity and their temporal relations. Helmholtz's "unconscious inferences" occur merely through the lawful interaction of simple sensations in accordance with their qualitative characteristics and strict laws of association. The same scheme of explanation applies in both physics and psychology.

The point is not that Helmholtz sought to explain psychological processes through physical ones. Helmholtz was not a physicalist or a materialist, if one takes these terms to imply a commitment to the project of reducing the laws of psychology to physics. The all-too-common description of the biophysics movement of the 1840s as "materialist" is simply mistaken. Although Helmholtz, du-Bois Reymond, and their friends were united in opposition to the postulation of "vital forces" to explain physiological processes in the organism, and they all maintained that only physics and chemistry are needed to account for physiological processes such as digestion and respiration, they did not extend their search for physical and chemical explanations of organic processes to include the processes of thought.[130] What is common to Helmholtz's explanations in physics and psychology does not derive from a reduction to physical laws, but from a common style of explanation, a common explanatory schema.

Helmholtz himself invoked the unity found in *psychological* explanations as grounds for preferring them to *physiological* explanations of spatial perception. Having characterized both "materialism" and "spiritualism" as metaphysical hypotheses in the concluding section of the *Optics*, he thereby precluded an appeal to materialism to undergird a preference for physiological over psychological explanations. He provided a methodological argument for his own psychological bias:

The only justification I can see for [having recourse to physiological hypotheses] would be after all attempts had failed to explain the phenomena by the known factual relations. But, in my judgment, this is by no means the case with the psychological explanation of visual perception. On the contrary, the more attentively I have studied the phenomena, the more the interplay of the psychic processes has revealed its universal uniformity and harmony, and the more consistent and coherent this whole region of phenomena has appeared to me. And so I have had no scruples in connecting and unifying the facts in the preceding sections by explanations which were founded essentially on the simpler psychic processes of the association of ideas. [PO 3:432-3]

Because the facts of perception could be better unified by appeal to the "known laws" of psychology, the laws of association, Helmholtz rejected physiological reduction. In this he certainly acted in accordance with the epistemological precept of his 1878 paper to reject metaphysical speculation in favor of explanation through the lawful. It would indeed have been yet another metaphysical speculation to assert that all phenomena can be explained by appeal to the laws of physics, in advance of producing instances of physical explanations for those domains of phenomena in which other laws were available.

In the final analysis, however, Helmholtz was forced to admit that he had not resolved thought itself into the law-governed naturalistic processes of psychology. He recognized that in subsuming the search for lawfulness under a set of mental laws he did not speak to the question of justifying the search and its outcome (whether that outcome was the mental laws themselves or laws in another domain). Postulating psychological laws to explain the processes of inference does not justify the inferences drawn in accordance with those laws: the postulation of *laws* of thought does not guarantee that they are laws of *thought*. Helmholtz refused to abandon the concept of the scientific investigator as thinker as he would have done, by his own lights, had he simply equated the natural mechanism of inference formation with the norm for sound thought. He chose instead to embrace a conception of the investigator as thinker in an act that he came close to characterizing as a leap of faith.[131]

At this point Helmholtz left the domain of psychological investigation for the field of philosophical reflection, switching from charting laws to analyzing the presuppositions of scientific practice. Instead of merely asking how the mind may be described as governed by certain natural laws, he also sought to determine the presuppositions of scientific thought and then to embrace them as presuppositions. Without fully reflecting upon the boundary he had crossed, his shift in analysis exemplified the limits to his natural scientific approach and marked the point of entry into the question of the epistemic standing of the enterprise of science itself. Thus, not only at the boundary between natural science and artistic expression did he recognize a limit to naturalistic modes of explanation: he recognized such a limit in the understanding of scientific thought itself. The normative practice of natural science resisted complete natural-scientific analysis.

Chapter 6
Summary

A central topic for modern philosophy developed at the intersection of theories of mind, theories of knowledge, and theories of spatial perception. Because of the importance of spatial properties in early modern science, conceptions of the mind's ability to know space became central to attempts to establish (or to undermine) metaphysical foundations for the new science. The mind's ability to apprehend spatial properties—including both the mind's grasp of the properties of matter and space through geometry and its determination of specific spatial configurations through visual localization—was a central topic for philosophers from Descartes to Hume. Kant's theory of geometry and his doctrine of transcendental idealism recast discussions about space into discussions about the spatial form of intuition. The Kantian problematic provided a sighting point for subsequent discussions within the German scientific and philosophical community, discussions that culminated in the work of Helmholtz.

During the early modern period, theories of the mind's apprehension of space were intimately connected with theories of the mind itself as a perceiving and knowing power. By the eighteenth century two distinct attitudes toward investigating the mind's perceptual and epistemic abilities became manifest: naturalistic and normative. Some authors explicitly applied the attitude of the natural scientist to the study of the mind, hoping to explain the mind's activities in a "Newtonian" manner by appealing to simple mental elements and to laws for combining them, the laws of association. Hume was an early advocate. This natural-scientific attitude toward the mind was directed against the normative conception of Descartes and others, according to which the mind possesses a special power of intellectual intuition allowing it to determine substantive truths regarding the constitution of the material world, a view represented in eighteenth-century Germany by Wolff, Crusius, and Tetens. These authors implicitly or explicitly denied that the mind's knowing power could be adequately explained associationistically. They contended that notions such as objective validity and intellectual apprehension of the truth could not be eliminated from explications of the mind's knowing abilities and that these notions could not be reduced to causal laws of association.

Kant made explicit the contrast between naturalistic and normative approaches to the mind. Moreover, he affirmed the legitimacy of each, albeit with qualifications that would have been rejected by the proponents of each position. While he affirmed the universal applicability of natural law to mental states, he denied the adequacy of the resulting explanations when applied to the mind's knowing abilities, where he found that notions of objective validity were required. In the case of geometry he maintained that the intellect has a primitive ability to recognize the truth. He denied, however, that the intellect has the power of grasping the properties of the world of things as they are in themselves. In the case of spatial perception, he denied the mind's ability to determine that the fundamental properties of mind-independent objects are geometrical, as proponents of the new science had claimed. He expressed this position in his doctrine of transcendental idealism.

In early nineteenth-century Germany sensory physiologists and psychologists as well as philosophers assessed with differing results the prospects for using psychology to respond to Kant's doctrine of transcendental idealism, including its application to spatial properties. Some authors attempted to use psychology to refute, support, or extend Kant's work; others argued that psychology was irrelevant to determining the grounds for valid knowledge.

These discussions formed the background to the work of Helmholtz, whose adjudication of theories of spatial perception reflected his advocacy of a particular approach to the investigation of the mind. In his *Physiological Optics*, Helmholtz developed a highly articulated theory of spatial perception, arguing that both the ability to represent space at all and the ability to localize objects within space are acquired through learning. He characterized the process of learning as inferential and developed an account of the unconscious inferences of perception modeled after his explicitly stated ideal of natural-scientific explanation, itself a continuation of the "Newtonian" form of explanation. Helmholtz further contended that unconscious inferences in perception are of a kind with mental processes characteristic of the natural scientist, thus adopting parallel natural-scientific accounts for both sorts of mental activity. Helmholtz also found that his theory of spatial perception provided a basis for refuting Kant's doctrine that Euclid's geometry applies with necessity to physical space, predicating his belief on the further belief, shared with several previous writers, that the doctrine was an instance of nativism.

I have argued that Kant's doctrine of spatial intuition was not an instance of nativism. His argument that physical space is Euclidean depended upon a "psychological" thesis of sorts, that a single faculty of sensibility conditions both visual perception and geometrical intuition. The unity of the faculty of sensibility provided a basis for asserting that the properties of space known through geometric intuition must condition empirically attained sensory experience of the physical world. In any event, Helmholtz had no conclusive argument against the thesis that the faculty of sensibility has innate spatial abilities, but he used the recent discovery of non-Euclidean geometries to undermine Kant's claim that the structure of physical space is necessarily Euclidean. He also argued plausibly that, in principle, the metric of space could be discovered through empirical measurements to differ from the metric of ordinary perceptual experience.

Naturalistic and normative accounts of thought and the mental underwent considerable change during the eighteenth and nineteenth centuries. The three sections of this chapter examine these changes by summarizing chapters 2 through 5. The concluding chapter considers whether the naturalistic and normative approaches constituted an historical divide between psychological and philosophical discussions of mind, and it brings my historical case study to bear on current discussions of naturalistic and normative approaches to the mental.

1 Modern Naturalistic Theories of Mind: Naturalism without Materialism

Chapter 1 distinguished between "metaphysical" and "methodological" naturalism. As applied to theories of mind, metaphysical naturalism asserts that the mind is part of the natural order because materialism is true: the mind is identified with organized matter and is thereby rendered truly a "part" of nature. Methodological naturalism views the mental as subject to the form of explanation characteristic of the natural sciences. In its original version, methodological naturalism expressed the thesis that mental phenomena can be explained by appeal to primitive elements and laws for the interaction of such elements; the laws were understood by analogy with laws in mechanics, thereby forsaking the traditional mentalistic vocabulary of knowledge, truth, validity, and justification.

Materialistic naturalism played virtually no role in fully articulated naturalistic accounts of the mind such as those developed by Hume, Hartley, Lossius, Steinbuch, Tourtual, and Helmholtz. In the hands of other authors, materialism played the negative role of precluding certain types of nonnaturalistic accounts of mind—those that invoked supernatural agency and immaterial substances; in this manner materialism served as a kind of cosmological posture in the writings of such influential naturalists as Lucretius, Hobbes, Priestley, or d'Holbach. Part of their influence arose from their vilification; because of the reproach heaped upon them, materialism became a position to be avoided, and it was explicitly disavowed by numerous eighteenth- and nineteenth-century authors whose positions were seemingly otherwise compatible with the scorned materialistic position. Notwithstanding, the naturalistic authors who rejected or remained agnostic about materialism were able to formulate reasonable grounds for their naturalism. Indeed, avowal of materialism would have contributed little or nothing to the positive explanatory endeavors of authors who, unlike Hobbes, Priestley, and d'Holbach (and later materialists such as Vogt and Moleschott), put forward detailed accounts of the workings of the mind.[1]

For early modern naturalistic writers actually engaged in producing theories of the mind, the vocabulary of matter and motion offered little by way of an ultimate explanatory foundation. Hartley and Lossius made use of the doctrine of vibrations to explain the physiological basis of the laws of association, but vibratory physics as an ultimate account of the mind required a satisfactory account of how vibrating matter can be identical with phenomenally present sensations. None of the authors who produced articulated theories of the mind felt adequate to this task; indeed, they cited as reasons not to adopt materialism its inability to explain the phenomenal character of sensations or the unity of consciousness.[2] Moreover, none of those who employed the doctrine of vibrations as a physiological explanation of association presented it as a "deeper" physiological grounding for their commitment to psychological laws of association. Typically, they treated the psychological laws of association as better known than physiological speculations about vibrating fibers. Even Hartley and Lossius treated the laws of association as autonomous laws of psychology that could be justified by appeal to psychological phenomena of habit formation.[3] Authors as diverse as Steinbuch, Tourtual, Kant, and Helmholtz all agreed that, while a strict correspondence was to be expected between psychological and physiological phenomena, psychology was nonetheless autonomous, under no sway even from physiology. A naturalistic connection between "inner" states of mind and the "outer" world of matter was secured not by affirming materialism but through the weaker claim that mental processes have physical correlates in the brain.

Materialism aside, then, the projects of various eighteenth- and nineteenth-century authors may be conceived as "naturalistic" because they applied a mode of explanation, characteristic of Newtonian natural science, that employed the vocabulary of elements and laws. Merely adopting a Newtonian vocabulary was not in itself sufficient to qualify a theory of the mind as naturalistic, for Wolff, Tetens, and Eberhard each did so while retaining a conception of the mind according to which "recognition of the truth" is a primitive capacity or "law" of thought. Nor was the use of a "causal" vocabulary sufficient to establish a position as naturalistic. The language of causality was found in Cartesian accounts of the mind as a knowing power, and Wolff and his followers retained it, but none of their writings was thoroughly naturalistic. Ultimately, the mental theories of avowed naturalists such as Hume were distinguished from their nonnaturalistic opponents because their explanations appealed to causal laws cast in the truth-neutral, nonepistemic vocabulary of mental elements and their associative interactions.

Oftentimes the naturalistic thesis was expressed in the vocabulary of mental faculties. In the eighteenth century, theories of mind could be distinguished as naturalistic through their attempted reduction of the understanding to sense and imagination; these theories provided a counter to the claim that the human intellect was different in kind from the faculties possessed by animals, and so to the claim that humankind was separate from nature. The common means for effecting such a reduction was to analyze formation of beliefs in terms of associative habits. The reduction of all mental faculties, including the understanding, to associative laws became the hallmark of psychologically articulate naturalism in the eighteenth century. Indeed, as the century progressed, talk of mental faculties itself began to give way to associative reduction; early in the nineteenth century, Herbart proposed eliminating all talk of faculties in favor of a psychology in which associative processes were taken to be basic or fundamental. As a theory of the senses, the associationist approach received its most thorough application in Steinbuch's account of spatial perception. Helmholtz combined the associative account of belief formation with the associative theory of spatial perception in his doctrine of unconscious inference, thereby attempting to provide a naturalistic explanation of both perceptual and scientific inference. The associationist brand of naturalism remained the most common type of naturalistic account of the mind well into the twentieth century. However, as the case of Herbart reveals, it would be a mistake simply to equate associationism with naturalism, for despite his associationist reductionism in psychology, Herbart adopted a methodologically normative account of the mind.

Well in advance of Helmholtz's culminating statement of the associationist psychology of perception, a second strand of methodological naturalism arose within German physiology and psychology of the senses. Proponents of this brand of naturalism did not adopt the explanatory model of elements and laws. Instead, functional explanation became the leading component in their analysis of the mind. When Tourtual and Müller embraced the functional approach they could not, of course, appeal to the theory of natural selection to explain the origin of functional organization. Nevertheless, they counted functional organization among the manifest properties of organisms, ascribing it not only to anatomical structure but also to psychological process. Finding that attempts to explicate functional organization in terms of physics and chemistry alone had failed, they admitted it into their explanations as a primitive, leaving open the question of its further explication. In so doing, they did not deny that the laws of physics and chemistry govern all physiological processes but only that they were adequate to explain functional organization.

The situation changed with the publication of Darwin's *Origin* and its favorable reception by a number of prominent figures in German physiology. Attributions of functional organization could be given a clear foundation, even if each ascription of a specific functional adaptation to the members of a species required its own justification. Within a few years of the publication of Darwin's *Origin*, Hering appealed to natural selection as the means for explaining the etiology of the functional organization of the visual system, including the mechanisms underlying spatial perception. He thus used Darwin's theory as a means of validating functional analysis against those who regarded it as inherently connected with vitalism or creationism. Hering's appeal to natural selection did not, however, cause Helmholtz to reconsider his oft-repeated charge that nativist accounts of spatial perception tacitly appeal to vitalism or creationism to explain their posit of innate spatial abilities—even though Helmholtz recognized the theory of natural selection as a generally valid ground for explaining innate adaptive structures.

In sum, although materialistic naturalism as an attitude toward the mind has been available within the tradition of Western philosophy from ancient times, articulated natural-scientific accounts of the mind first arose in the eighteenth and nineteenth centuries. Such accounts were associationistic. Although the phenomena of association had been long recognized and were ascribed some explanatory role by Hobbes as well as by Descartes and Spinoza, the idea that the laws of association form the fundamental laws of a general science of human belief-formation first received explicit articulation in the eighteenth century. Similarly, although the notion that mental as well as physical structures submit to functional analysis is ancient, the adoption of a physiologist's attitude toward the functional organization of the mind became entrenched only in the nineteenth century. Natural-scientific theories of the mind are a modern invention.

2 From Metaphysical to Methodological Normativity

The content of "normative" conceptions of thought and the mental underwent comparable change and development in the eighteenth and nineteenth centuries. Initially, normative accounts were built around the notion of the intellect as a truth-recognizing faculty. Descartes and Leibniz in the seventeenth century, reflecting the assimilation of Greek philosophy into Christian (and Jewish, and Islamic) theologically conditioned metaphysics, explained this remarkable mental power by appeal to the divine origin of the mind. For Descartes, the mind is constituted in such a way that the will is compelled to assent to the clear and distinct perceptions of the intellect; the human judgmental faculty was treated as essentially a truth-detecting faculty, divinely warranted. Clear and distinct perception on the part of the intellect, and the associated compulsion of the will, were taken to be primitive operations, admitting of no explanation in terms of psychological laws, processes, or mechanisms. Which is not to say that Descartes did not apply the language of causality to the will—he was ready to speak of the will as "compelled" to choose. Rather, he distinguished between the *mental* causation of the will in the face of clear perception of the truth and *natural* causation pertaining to either mind or body. Habit is an example of a natural cause in the mental sphere; the impact of a moving body upon a second body is an example of material causation. In addition, Descartes was, at least nominally, committed to mind-body interaction as a third species of natural causation. These instances of causation may be assigned to the ordinary course of "nature" and thereby distinguished from the metaphysically normative causal force of clear and distinct perception.

The Cartesian conception of a special truth-recognizing ability provides a standard for what I have termed the "metaphysically normative" account of judgment. Various positions from the eighteenth century can be classified through their responses to metaphysically normative accounts of the mind: some reject the normative approach, others adopt and support it, while still others recast the notions involved. According to one dominant train of thought in the eighteenth century, the fundamental, rock-bottom explanatory notion in accounts of judgment was "the ability to discern the truth," and the only widely accepted explanation of this ability attributed it to a divinely instituted immaterial substance. At least two means of attacking this position were available. One might directly deny the existence of immaterial substance, in order to undercut the accepted explication of the immediate mental intuition of substantive truths about the immaterial and material worlds. This strategy, metaphysical naturalism, is d'Holbach's. It asserts, on materialistic grounds, that human beings lack a faculty of reason distinct from the senses and the imagination and capable of grasping the essences of things (whether in an a priori or a posteriori manner); human epistemology becomes a matter of sorting through sensory experiences to find regularities, without the guidance of a special truth-discerning faculty.[4] However, the more theoretically articulated denials of the power of reason came from methodological naturalists who, as in the case of Hume, attacked the power of reason from within by means of arguments directed against the notion that human reason possesses the special intellectual abilities that had seemed to require an immaterial substance for their explanation. Such attacks on the existence of an immaterial intellectual agency did not require establishing a materialistic counter-ontology. Having denied we have the abilities whose alleged existence had motivated the positing of such agency, they could remain neutral about the ontology of the remaining mental faculties. Methodological naturalists, having denied that rational insight undergirds beliefs in matters of fact, attempted to provide an explanation of belief-formation based upon the laws of association, the laws of sense and imagination. Hartley and Lossius attempted to apply this form of naturalism across the board, to include not only judgments of matters of fact (in Humean terminology) but also arithmetic judgments (to which they assigned the status of factual knowledge).

A variety of authors in mid-eighteenth-century Germany implicitly or explicitly denied this form of naturalism. While generally accepting the law of association as the law of imagination, they also posited a distinct sphere of mental causation proper to the intellect or understanding, separating the causation of truth recognition from the causation of habit or association. Although Wolff, Baumgarten, Tetens, Eberhard, and Crusius differed in their accounts of the basis of knowledge of the mind and to some extent in the vocabulary they used to describe the mind, all agreed in granting some scope to associative laws of the imagination and in rejecting the idea that the deliverances of the intellect can be reduced to associative habit. Tetens was quite explicit in denying that association could provide a complete analysis of the judgments of the understanding. He, along with the other authors mentioned, remained committed to a metaphysical basis for the normativity of human knowledge, founded on the judgmental powers of a special immaterial substance whose operations could be reduced neither to material processes nor to the mental naturalism of association and habit.

The work of Kant reoriented the relationship between naturalistic and normative accounts of thought and the mental. Like most of his contemporaries, he denied that metaphysical naturalism, in the form of materialism, could be established; unlike many of his contemporaries, he provided an argument for this conclusion in his brief against the claims of traditional metaphysics to know "things in themselves." At the same time, he affirmed the general applicability of methodological naturalism to all mental phenomena; he granted that the causal law is applicable to the succession of mental phenomena, in effect endorsing the universality of the laws of association. But, notoriously, he denied the negative conclusions about the possibility of objective knowledge Hume had drawn from naturalism. For whereas Hume, Lossius, and others put forward their naturalistic account of belief formation to replace normative conceptions of the mind, Kant allowed naturalistic and normative descriptions of the mind to coexist. He followed Hume and others in forsaking the metaphysically normative conception of the mind as an immaterial substance possessed of special powers for grasping the transcendently real, and he explicitly denied the legitimacy of an appeal to divine institution to explain the origin of such powers. Yet he affirmed the legitimacy of the normative conception of thought by providing it with an allegedly nonmetaphysical basis within "transcendental" philosophy.

In the case of Kant, the crucial difference between naturalistic and normative conceptions of the mind consisted in the way in which he conceived the object of explanation—the act of epistemic judgment—not in the explanatory vocabulary he adopted. It would not have been sufficient for Kant merely to consider "judgment" a primitive and unanalyzable mental ability; earlier naturalistic accounts of vision had done the same. Rather, he contended that the thing to be accounted for in cognitive judgments was inexplicable from a naturalistic viewpoint. Famously, he cast the distinction between questions that would be asked of mental processes from normative and from naturalistic points of view as a distinction between questions of fact and questions of right. His own answer to the question of right appealed to mental faculties; but conceived as "transcendental" faculties that must be investigated transcendentally, not as natural entities subject to naturalistic investigation.

The status of transcendental knowledge of the mind is problematic in Kant. Some interpreters judge that such knowledge smacks of metaphysical claims about the noumenal self. The synthesis that underlies the transcendental unity of apperception seems to be an activity of the noumenal mind, for it stands outside the time-determinations of inner sense and thus is not part of the causally-determined chain of mental representations. To other critics, Kant's description of the mind's activity in knowing has seemed to involve an implicit psychology. And it must be granted that much of the mentalistic vocabulary in the first *Critique* is continuous with Kant's own empirical-psychological vocabulary.

I argued in chapter 3 that Kant's transcendental psychology should be assimilated neither to latent claims about the noumenal mind nor to empirical psychology. Kant affirmed methodological naturalism and empirical psychology, but not as adequate descriptions of the subject as knower. And he retained the language of mentalistic accounts of the knower, but not as a metaphysical description of the mind as it is in itself. To distinguish his nonnaturalism from a metaphysically normative approach, I have labeled Kant's approach "methodologically normative." Just as I characterized methodologically naturalistic explanation both by a certain content (appeal to elements and laws expressed in nonepistemic vocabulary) and by a certain mode of justification (natural scientific or empirical), so I characterized methodologically normative accounts of the mind by both content and mode of justification. The content was the vocabulary of objective judgment, universal validity, and necessity. The mode of justification was transcendental, and it appealed neither to the divine origin of the understanding nor to the special properties of an immaterial substance. In his transcendental arguments, Kant argued that his conception of the knower provides the only way to make sense of our ability to possess the knowledge that we do in fact have. Twentieth century students would say that it provided Kant with the only means he could envision for explaining the possibility of knowledge he expected his contemporaries to agree that they had.

The method of transcendental philosophy was exemplified through Kant's transcendental analysis of the possibility of geometrical knowledge. Although the example pertained not to the transcendental analysis of the understanding but of sensibility, the central methodological points transfer to the case of the understanding. Kant proceeded by eliminating alternative explanations for the possibility of a priori knowledge until only one explanation remained. With hindsight, we can think of other available possibilities. We also reject Kant's conception of what requires explanation, if for no other reason than because our conception of the status of Euclid's geometry and Newton's physics differs from his. We thus reject the letter of Kant's arguments for methodological normativity; the spirit may well live on.

3 Naturalistic and Normative Responses to Kant

The historical presence of an author is constituted not only by what he or she writes, but by the way successors read his or her work. In chapter 4 I adopted the strategy of examining various readings of Kant as a way of revealing his legacy for the history of theories of mind and of perception, thereby also providing a set of actually existent positions against which to locate Kant's position. Seeing how a Steinbuch or a Fries gets Kant right or wrong may give added definition to Kant's own position.

Steinbuch and Tourtual were physiologists of the senses who applied physiological thinking to both bodily structure and mental process. As readers of Kant, they agreed in interpreting his claim that space is a form of intuition as an innateness thesis. They then responded to this allegedly physiological and psychological thesis in quite different ways, one attempting to refute it and the other to extend it.

Steinbuch sought to refute Kant's "innateness thesis" by establishing a theory according to which the spatial form of intuition is acquired through associative processes operating *in utero*. He believed that his theory would reveal a possibility Kant had denied, that of providing a valid explanation of the correspondence between the spatial properties of objects considered in themselves and the spatial form of intuition. His basic approach was to naturalize the mind entirely; he pursued the associationist approach to spatial perception in as radically pure a manner as did anyone. But, in so doing, he failed to meet Kant's normative arguments for his theory. Reading Kant as an adherent of nativism, he did not refute Kant's theory: he changed the subject. We may therefore be tempted to see Steinbuch's position as psychologistic. However, he did not formulate the truly psychologistic project of solving philosophical problems by means of psychological theory. Instead, he attempted to use psychological theory to speak to problems that he failed to recognize as philosophical. It is more accurate to say that he responded to Kant psychologically rather than psychologistically.

Whereas Steinbuch tried to refute what he took to be Kant's nativism, Tourtual attempted to elaborate Kant's allegedly nativist account of geometry into a full-fledged theory of spatial perception. He characterized his work as a "physiological" extension of Kant's "transcendental" theory of spatial intuition. Like Steinbuch, he took exception to Kant's transcendental idealism. Unlike Steinbuch, however, he did not attempt to provide a psychological account of the relation between spatial perception and the spatial properties of things, for he realized that a psychological theory of spatial perception could not speak to Kant's grounds for adopting transcendental idealism. Instead, he adopted an attitude of "pragmatic realism," observing that despite our inability to "prove" the existence of independently existing objects in space we in fact accept their existence. Tourtual realized that this position did not "disprove" transcendental idealism; he offered it as an alternative more in line with "common sense" than Kant's. He was surely correct. Moreover, his pragmatic attitude was characteristic of a good many German natural scientists in the middle years of the nineteenth century, including Müller and Helmholtz.

Müller's response to Kant provides an interesting contrast with that of Tourtual. Although his attitude toward transcendental idealism was consistent with Tourtual's, he was even more insistent than Tourtual in separating questions pertaining directly to the theory of the senses from questions pertaining to the status of scientific and geometrical knowledge. Despite the fact that Müller has often been read as himself engaging in a kind of "physiological Kantianism," his *Handbuch* quite explicitly divorced questions about the physiology and psychology of the senses and the understanding from questions about the status of scientific knowledge of the external world. It is fair to say that Müller adopted neither a psychological nor a psychologistic reading of Kant, and that he agreed with Kant, Herbart and others that natural-scientific theories of the senses could not resolve philosophical or epistemological questions. Philosophically, he adopted a form of Herbartian realism.

Recall that Fries's response to Kant stands opposed to Steinbuch's, for it is psychologistic rather than psychological. Indeed, Fries's interpretation has been taken as the paradigm of a psychologistic reaction to Kant: he expressly set out to provide a psychological foundation for Kant's critical philosophy. Yet his psychologism is more refined and interesting than the usual stereotype, which construes psychologism as the attempt to apply the findings or theories of empirical psychology to philosophical problems upon which they have (allegedly on principle) no bearing.[5] However, Fries did not attempt to extend a naturalistic approach to the mind (of the sort developed by Hartley, Hume, and Lossius) to the problems that Kant had addressed in his first *Critique*. Rather, he attempted to extend the mode of justification found in the natural sciences to central Kantian doctrines connected with the synthetic unity of apperception. Fries understood transcendental idealism to be a strictly subjectivist position, in the sense that it limited knowledge to the subject's awareness of his or her own mental states; any notion of intersubjective agreement could pertain only to agreement among subjects about what each could be expected to ascertain from this subjective perspective. Fries then treated the notion of the "unity of the world," and the idea that our knowledge of the world constitutes "one truth," as basic facts that are ascertainable from, or at least accepted within, the subjective standpoint. Similarly, the fact that the principle of contradiction applies to our thought implies for him that all a subject's representations must be brought to unity. These "subjective" facts of consciousness were "data" to be explained by a theory of the unity of apperception.

Fries's psychologism thus differs from other versions by virtue of the "data" it makes primary: the conceptions of "one world" and "one truth." Nevertheless, the theory was "psychologistic" in two senses. First, it drew heavily upon a conception of synthesis as a causal process.

Second, it treated the justification of this causal process as analogous to the justification provided for universal natural laws in the other sciences: through their explanatory adequacy for all known facts. For Fries, the causal process of synthesis does the real work in explicating the possibility of geometry and the applicability of the concept of cause. He interprets the unity of apperception in concrete causal terms, as though a single "spray-gun" of formal apperception spreads the matter of sensation on an inner screen (a subjective space of intuition) in accordance with laws of spatial intuition and causal connection that constrain and guide the operation of the gun.

Herbart agreed with Fries that Kant's position contained an implicit appeal to the concrete causal operations of the mind, but he differed from Fries on the implication. For, although Herbart had some sympathy with Kant's conception of philosophy proper as the analysis of concepts, he rejected Kant's notion that "transcendental philosophy" stands prior to philosophy proper: he did not agree that we gain a better understanding of philosophical analysis by beginning, as in "transcendental analysis," with the subject itself, or with the subjective contribution to the justification of knowledge claims. Herbart implied that the Kantian analysis of the subject would inevitably devolve into a natural history of the mind, thus rendering the analysis irrelevant to the distinctively philosophical task of capturing normatively "ideal" considerations. His charge that there were psychological tendencies latent in Kant's program, and that they rendered Kant's philosophy inadequate to its task, constituted the invention of the charge of psychologism. Herbart's was a pure version of a methodologically normative approach to the mind, banning all talk of cognitive faculties such as judgment or understanding from the analysis of the grounds for knowledge. He pursued a direction in philosophy not found in the period of modern philosophy that extended from Descartes to Kant. He thought it neither possible nor appropriate to begin philosophy by investigating the mind or the understanding or the faculty of reason. For him, such an investigation could neither produce nor ground the principles of philosophy or of natural science.

Herbart defined the aim of philosophy as the analysis of concepts. He found it possible to practice philosophy as an historically situated activity of reflecting upon our concepts and bringing them into consistency by reflecting upon the role they are expected to play in the intellectual economy of our thought, where "thought" includes the demands both of natural science and of the metaphysical urge to grasp "the real." Thus, Herbart attempted to carry on metaphysics after Kant, without, however, claiming a transcendent intuition of the real. He adopted the attitude that, once we have developed the most consistent theory of reality we can construct from all the intellectual materials available at a given time, it is pessimistic niggling to refuse to regard this theory as a description of reality.

Herbart's metaphysics led him to the view that all psychological states and processes must be built up from the mutual interaction of nonspatial "Reals." Like Leibniz's monads, the Reals are characterized by internal states of perception, but unlike monads, Reals are initially incapable of spatial representation. They acquire the ability only through causal interaction. However, unlike Steinbuch, who also believed the ability to represent space is acquired, he considered the relation between spatial perception and the external world, and the psychogenesis of spatial representation, to be two different problems requiring different sorts of consideration. Psychology, as Herbart understood it, pertains to the natural history of soul. Its fundamental framework for explanation derives from metaphysics; its theories of the causal processes of the soul must, however, be both mathematically expressed and adequate to the "data" of inner experience. Not psychology but metaphysics adjudicates the epistemological status of spatial representation: it assesses the metaphysical status of space itself considered as a putative property of "Reals." For Herbart, space has a foundation in the real but is an idealization of the causal processes that occur among nonspatial Reals. Geometry itself is an idealized science that describes an "intelligible space" within which causal interactions take place.

Lotze shared something of Herbart's attitude toward both metaphysics and psychology. In advancing claims about "metaphysical reality," he adopted a reflective attitude toward science and toward what may be termed "value concerns"; like Herbart, he eschewed the claim that there is a "real use" of the intellect. He embraced mechanism in all departments of natural science, and he endorsed dualism in psychology. When he put these positions together into a metaphysics, he found himself positing "two-faced" reals, mechanical on the outside, Leibnizian on the inside. This attempt at reconciling the material and the immaterial has strong affinities with Herbart's posit of "Reals"; the similarities between the two authors perhaps reflected their separate efforts to accommodate the various "pulls" in the metaphysical tradition. Both authors sought simultaneously to encourage scientific psychology and to insulate philosophy from it. Lotze in particular reprised and developed the analysis of spatial perception in connection with the ocular motor apparatus Steinbuch and Tourtual bequeathed the nineteenth century. However, he considered the psychological theory of the basis of spatial perception irrelevant to the philosophical question of the foundations of geometry, a position he expressed in his epistemological writings.

That there was an interplay between Kant's theory of geometrical intuition and theories of spatial perception during the first half of the nineteenth century is apparent. Some authors attempted a psychological refutation or, alternatively, an extension of Kant's theory; others contended that the theory of spatial perception is irrelevant to geometry and to the description of physical space. But despite disagreements over the implications of psychology for Kant's theory of the foundation of geometry in spatial intuition, all the authors accepted Kant's notion that geometrical demonstration proceeds by appealing to spatial intuition. They also assumed the universality and necessity of Euclid's geometry as applied to physical space. The seemingly unproblematic character of this assumption changed after mid-century when the uniqueness of Euclidean geometry was successfully challenged, thereby reopening the question of the relation between geometry and spatial perception, and between geometry and physical space. Perhaps no one was in a better position to do so than Helmholtz, who combined mathematical competence with a deep knowledge of physiological and psychological theories of spatial perception.

In both physiology and psychology Helmholtz was a programmatic naturalizer. His naturalism was methodological, extending as far as it would go a form of explanation modeled after his understanding of explanation in physics. The explanatory model he adopted in the psychology of spatial perception exemplifies his application of methodological naturalism to mental life. For him, the psychological processes of spatial perception resolve into nonspatial sensations and laws of association. Helmholtz did not claim, however, that the naturalistic mode of explanation was adequate for all intellectual subject-matters. The *Geisteswissenschaften* pursue their own subject-matters with their own methods: they use the particularistic and historical mode of research characteristic of legal and historical studies. In his writing on science and the arts, Helmholtz recognized that a natural-scientific approach encountered limits in explaining the products of the human mind. Thus, although he offered a physiological account of the basis of harmony and dissonance in the sounding of chords and in simple melodic progressions, he denied that natural science could give a complete account of musical experience, or even of the structure of the scale. As he saw it, musical scales and musical experience are partly the outcome of an historical process that acts within the limits set by physiological considerations. These limits allow multiple possibilities; the structure of the scale could be explicated only by examining the choices made throughout the history of music. Physiology in itself was not adequate to explain musical experience.

While Helmholtz explicitly drew attention to the boundaries of the naturalistic approach in the case of musical perception, his accounts of scientific inference and of the foundations of the geometry of space exemplified rather than drew those limits. In the case of geometry he claimed to have responded to Kant on the basis of naturalistic psychology, but in fact his response gained its force by appealing to measurements of the sort that would count as evidence in physics. As a refutation of Kant, his empiristic psychological theory was beside the point. His effective argument against Kant was the envisioning of a scenario in which, assuming the straightness of light rays and the rigidity of rods, certain measurements would count as evidence that space has a non-zero curvature. Here, an allegedly naturalistic argument actually requires evaluating physical evidence normatively.

Helmholtz also provided a naturalistic account of scientific inference, patterned after the inductive inferences of perception. But Helmholtz's reduction of unconscious inference to associative processes broke down when applied to scientific inference. Although he attempted to explain how the mind might discern general laws in nature—the law of cause being the most general—by a kind of associative memory filter, he came to recognize that such an explanation did not ground the application of the law of cause or justify appeal to the law. His resolution of the problem of induction required positing the "law of cause" as basic. The scientific practice of accepting universal laws on the basis of finite data samples resisted naturalization.

Chapter 7
Conclusions

Helmholtz and his neo-Kantian opponents disputed about the relevance of psychological considerations to philosophical theory, and in particular over whether the innate or acquired status of spatial perception bore upon Kant's theory of geometry. Opponents who charged Helmholtz with confusing a psychological question with an epistemological one distinguished between questions of causal origin, which admitted of empirical investigation, and questions of justification, which only epistemological reflection could decide. The case of Helmholtz can be juxtaposed with Steinbuch's early misreading of Kant's position, mistaking a normative, philosophical question for a natural-scientific, psychological one. If the division between philosophy and psychology drawn by Herbart, Tourtual, Müller, and Lotze is also considered, then the following general historical thesis suggests itself: the distinction between psychological and philosophical theories of the mind runs precisely parallel to the distinction between naturalistic and normative theories.

This thesis has much to recommend it as a description of the division between philosophy and psychology advocated by the early proponents of a natural science of the mind. Eighteenth-century authors such as Tetens and Kant invoked this division (even if the latter envisioned strict methodological limitations to naturalistic psychology), as did nineteenth-century proponents of a scientific psychology such as Herbart and Lotze. The thesis is further confirmed when those who would dissolve epistemology into scientific psychology confront those who would defend a separate domain of epistemological inquiry that is immune to the methods of natural science. Such confirmation covers a good bit of the twentieth century: the antipsychologistic stance of Frege rests on this understanding of the relation between philosophy and psychology. The idea that thought itself is subject to a special nonempirical investigation has also been advanced by C. I. Lewis, Ryle, Sellars, and D. W. Hamlyn, who equate psychology with causal investigation and preserve a separate domain of the "logical" or "epistemological" for philosophy.

Nonetheless, the thesis pairing naturalism with psychology and normativity with philosophy offers an inadequate picture of both the conceptual and the institutional division between philosophy and psychology in the nineteenth as well as the twentieth centuries. I want to focus first on the conventionally accepted claim that philosophy and psychology became separate disciplines at the end of the nineteenth century when psychology allegedly was founded as an experimental, natural-scientific discipline separate from or in place of philosophy.

1 A New Purchase on the History of Psychological Approaches to the Mind

The claim that psychology was institutionalized as a natural-scientific discipline at the end of the nineteenth century seriously distorts historical reality. There are in fact two problems: one concerns the precise dating of the institutionalization of psychology; the other concerns the character of the psychology thus institutionalized. The conventionally prescribed dates are 1860, 1879, or a time in between. But one can easily make a case that the discipline of psychology was institutionalized in ancient, medieval, and early modern times, as were the other classical "disciplines" (in the sense of subjects taught in school) of arithmetic, geometry, astronomy, and physics. Psychology regarded as the science of the soul is as old as Aristotle's *De anima*. It was taught in ancient, medieval, and early modern schools along with the other "disciplines" in Aristotle's corpus. In the mid-eighteenth century, the philosophical system of Christian Wolff institutionalized "psychology," so conceived, in the German curriculum under that very name. In the late eighteenth century Kant and others conceived "empirical psychology" as constituting a special domain of investigation, distinct from philosophical concerns with epistemic normativity.

One might claim that this early psychology is not truly a scientific psychology on the grounds that Wolff's version was "metaphysical," not natural-scientific, and that Kant refused to promote his version to the rank of a true science. So let us ask when "natural-scientific" psychology was first taught in school. This occurred in the early nineteenth century, through the efforts of Herbart and his students. But here again, although this psychology was natural scientific in intent, it remained "metaphysical"; Herbart allowed that psychology could learn from metaphysics. We are thus led to identify the first regular teaching of a nonmetaphysical, experimental psychology in school by looking to the second half of the nineteenth century, to courses based on Wundt's textbooks.

Still, it is overly hasty to allow course-headings in schools of higher learning to determine the conditions for the existence of disciplines. Subject-matters and research methodologies more properly define disciplines. The presence of practitioners and their intellectual products thereby determine the beginning of a discipline's existence. On this understanding of "discipline," experimental, nonmetaphysical psychology existed as a discipline long before Wundt wrote his textbooks. It existed in the proposals of Bonnet, Kant, and others that the mind could be conceived as a part of nature and could be studied through experience. It existed in the practice of Steinbuch and Tourtual as empirically-minded physiologists who recognized sensory psychology as a branch of sensory physiology distinct from the physiology of bodily processes, a branch they explicitly labeled "psychological." It is thus fair to say that psychology as a subject taught in school is very old (even if the name is modern), but that it was not conceived as having a naturalistic subject-matter and as accessible to the techniques of natural science until the eighteenth century. It is also fair to say that psychology as an empirical research discipline was first conceived in the eighteenth century and first practiced in the first half of the nineteenth.

These reservations about the conventional dating of the founding of psychology would matter little if the discipline that became widely institutionalized in the second half of the nineteenth century had been experimental psychology pure and simple. Then, apart from some remarks about the sense in which the "institutionalization" of all disciplines was transformed in the decades spanning the turn of the twentieth century, one would simply modify the usual claim by saying that experimental psychology first became "significantly" institutionalized at the conventionally recognized time. But the psychology that became institutionalized in the closing decades of the nineteenth century was not experimental psychology pure and simple. It included a burgeoning experimental, natural-scientific psychology, but it also included psychology as a "philosophical" or humanistic subject-matter that was to be studied with methods appropriate to philosophical or humanistic research.

Recent writers on the history of psychology, and particularly historians who have taken a second look at Wilhelm Wundt as the "founder" of experimental psychology, recognize this point. Especially in his early years Wundt proselytized widely for experimental psychology, but by the 1870s he made it apparent that he considered the social or humanistic side of psychology to be of equal or greater importance than the naturalistic, experimental side. It also became clear that he considered the subject-matter of social and humanistic psychology to require research methods distinct from those of the natural sciences.[1] Thus, the alleged "founder" of experimental psychology was actually a persistent voice urging psychology to be more than an experimental natural science. Wundt's case was by no means unique. Others in Germany who occupied chairs of psychology also envisioned a psychology that was more than, or at least other than, an experimental natural science.[2] Psychology grew most rapidly in the United States, of course, not in Germany, and Wundt's students in this country emphasized the natural-scientific side of their teacher's work. But in the United States—and in England, too—formidable scholars were institutionally recognized as "psychologists" who did not regard psychology as an experimental, natural scientific discipline pure and simple. John Dewey, Hugo Munsterberg, and G. F. Stout were among them.[3]

Thus, when psychology became widely institutionalized in the latter half of the nineteenth century, there was no general agreement that it should be regarded exclusively as an experimental natural science. And although the question of whether it should be so regarded remains open to some scholars even today, eventually there came to be widespread agreement among academic psychologists that it should be so regarded. The circumstances in which this happened would no doubt make for a fascinating study, but now it is sufficient to consider this development as it was reflected in histories of psychology written earlier in this century. Disciplinary histories often reflect as much or more about a discipline's self-conception at the time the history was written as they do about earlier historical developments. Otto Klemm's history, published in 1911, traced the antecedents of psychology to antiquity, dated the beginnings of psychology as a natural science from the eighteenth century, and placed its origin as an exact experimental science in the mid-nineteenth.[4] J. C. Flugel's history, published in 1933, told the story of two foundings of psychology: one by Herbart in the early nineteenth century, and one through the adoption of experimental methods in the work of Fechner, Helmholtz, and Wundt, between 1860 and 1879.[5] Finally, Boring's *History of Experimental Psychology*, published in 1929, ascribed the founding of experimental psychology around 1860 to Fechner, Helmholtz, and Wundt, relegating Herbart's work to "the preparation for experimental psychology within philosophical psychology."[6] These histories reveal that the process of consolidating academic psychology into a natural-scientific discipline extended well into the twentieth century.[7]

In sum, the conception of psychology as a natural-scientific discipline employing natural-scientific methods arose in the eighteenth century and was put into practice in the early decades of the nineteenth, while the institutional consolidation of academic psychology as an experimental natural-scientific discipline occurred in the twentieth century.

2 The Natural and the Normative Now

How, then, do matters stand now? A large number of philosophers, many of whom have Kantian tendencies, remains satisfied with a picture according to which psychology and philosophy divide the subject matter of the mind along the lines of the natural and the normative, psychology studying causal circumstances and philosophy examining questions of justification. Psychologists have been willing to accept this picture, leaving "normative" questions of how we should think to philosophers while they pursue the empirical question of how we do think; although a growing number of studies, inspired by the work of Kahneman and Tversky, examine the relationship between behavior and normative standards of rationality, these studies aim to discover the circumstances under which norms are met, not what the norms themselves should be.[8] These groups might peacefully coexist for many years, each taking a detached interest in the efforts of the other. But this easy division of the territory has not been accepted by all; imperialists have arisen on both sides. Some philosophers argue that there cannot be an empirical science of thought, because thought is essentially normative and so must be studied philosophically.[9] And some psychologists and cognitive scientists claim that the methods of the natural sciences are exhaustively adequate for the study of thought, leaving no aspect of the mental for philosophical investigation.[10] Interestingly, some philosophers, working under the banner of naturalism, have presented arguments to undergird the claims of the natural scientific approach to exhaustive adequacy.[11]

Naturalism means many things. If it is stated as the thesis that human beings are part of the natural order and that our cognitive capacities arise from naturally evolved mechanisms realized in the central nervous system, then contemporary adherents of naturalism form a broad group. A moderate form of naturalism might deny that all interesting questions pertaining to our "normative" practices can (or should) be decided by appeal to natural science. Thus, one could adopt a moderate naturalism while maintaining that all of the known facts of human biology and psychology are compatible with a wide variety of ethical and political systems (even while granting that such facts are not irrelevant to ethics and politics). Further, one might hold that mutually exclusive conceptions of the correct norms of scientific method, or of the standards for knowledge, are compatible with that same set of facts, and that no discovery of further facts will settle the question of which conception to choose. If naturalism is merely the doctrine that human beings are "natural" beings, naturalistic description may underdetermine our normative self-conceptions. This moderate form of naturalism has not stirred the current controversy.

The ongoing controversy over the tenability of the naturalistic point of view was sparked by the claim that the facts of biology and psychology provide an adequate basis for understanding all areas of human activity, including thought or cognition. Disagreement arises somewhere between the benign assertion that human beings are part of the natural order, so that human mentation is subject to empirical investigation, and the bold claim that such investigation is wholly sufficient for understanding the thinker *qua* thinker. The interesting disagreement, by my lights, is located inside the extremes set by the reductive naturalist and the antinaturalist. The position that natural-scientific study cannot shed any light whatsoever on the mental is not worth taking seriously; sense-perception, if nothing more, clearly is amenable to natural-scientific study. The interesting disagreement pertains to the boundary of legitimate natural-scientific study of the mental.

Naturalistic studies of the mind—or at least of a range of mental processes and abilities—have blossomed in the course of the twentieth century. Neurobiology and physiological psychology may seem the most obviously "natural scientific" areas of progress, but the psychology of spatial perception, for which Helmholtz's *Handbuch* is a standard source, should also be included among the successful natural-scientific pursuits of our century and of recent decades. Advances have been made in understanding the optical basis for and the psychological processes underlying the perception of size, distance, depth, and motion, even if theoretical convergence of the sort experienced in the classical period of the physical sciences has not emerged.[12] Moreover, just as Helmholtz believed that the study of spatial perception could illuminate cognitive abilities such as object recognition, so too in recent years investigators have envisioned that an account of the perception of spatial configurations will yield insight into the processes underlying object recognition and classification.[13] Beyond these efforts, cognitive psychologists have continued to treat a variety of cognitive abilities—including memory, learning, problem solving, and language use—as objects of natural scientific investigation.[14]

Despite the partial success of such endeavors, the question of the extent to which "the mind" is properly an object of natural-scientific investigation rightly remains open, and for reasons that parallel those in the period from Kant to Helmholtz. The exercise of judgment in determining distances or classifying local fauna into natural kinds, in articulating natural-scientific theories, and in rendering legal opinions or historical and literary interpretations is, in each case, called "mental." Whether all aspects of such mental activities can be exhaustively treated through natural-scientific investigation remains open. A thoroughgoing naturalistic account of a particular type of cognitive judgment must explain such judgments completely while omitting nothing essential.

Eighteenth-century theories of the structure of matter supposed that laws describing fundamental forces are adequate for knowledge of all of the properties of matter, and that no information other than location, mass, and motion are required to explain exhaustively the properties and states of an isolated system of mass points. Consider these theories as ideals of "exhaustive" adequacy. The correlative question, then, is whether theories of cognitive structures and mechanisms can be expected to provide analogously exhaustive explanations of cognitive phenomena. In assessing the applicability of natural-scientific explanation to the mind, this question must be posed not only with respect to whether theories of spatial perception can be extended to all dimensions of the cognition of space, but with respect to all cognitive practices whatsoever.

A good many philosophers writing in the mid-twentieth century denied that natural scientific accounts of the mind could settle issues about one area of cognition, viz., epistemology. They tended to regard psychological and epistemological domains of description as irrelevant to one another, assimilating epistemological descriptions to "the logical space of reasons" and psychological descriptions to "the logical space of causes."[15] Normative epistemology deals with reasons—with evidence, justification, validity—whereas naturalistic psychology deals with causes described in the objective language of physics, physiology, or mechanistic psychology. This recent distinction between naturalistic psychology and normative epistemology shows both continuities and discontinuities with its historical counterpart.

One difference between contemporary and previous naturalism pertains to the notion of causal processes in the mind. By some contemporary accounts, the appeal to "causal processes" is sufficient to render a description of the mind's activities "naturalistic." Accordingly, a concern with "causal processes" as opposed to "justificatory relations" distinguishes the naturalistic psychologist from the epistemologist. Such a distinction would not have worked in the seventeenth and eighteenth centuries, when "causal processes of the mind" were spoken of within the wholly normative context of explicating the mind's tendency, say, to affirm the truth. Hence, categorizing the theories of Descartes and others as "naturalistic" or "psychological" because they spoke of mental causation would impose our present conceptual vocabulary upon a domain it does not fit. Because we no longer regard the metaphysically normative account of the mind as a serious philosophical position, we are likely to suppose that any causal vocabulary used to describe the "processes" of the mind must be naturalistic. Descartes conceived the primary basis of human knowledge as lying in the influence of "pure intuition" on the will, compelling assent; this "causal" account of knowledge is not psychological, either in content or intent.[16]

For similar reasons, contemporary talk of "normative epistemology" is in many ways discontinuous with pre-Kantian epistemology. Theories of knowledge in the seventeenth and eighteenth centuries were theories of the cognitive faculties—of the senses, imagination, and understanding, and their relative contributions to human knowledge. However, such faculties were not conceived as natural organs continuous with the rest of nature or with the mechanical causality of the body; they were conceived as noetic powers. The notion of "intellectual intuition" played a central role in the epistemologies of both Descartes and Locke, even if they differed over the extensiveness of the power of intuition. Contemporary epistemologists, however, cannot appeal to the "faculty of perceiving the truth" as the primary object of their analysis. Not only has the metaphysical basis for such a faculty been rejected along with the conception of the mind as an immaterial substance having special noetic powers, but the notion that the human mind, whatever it is made of, has the power of directly intuiting substantive truths about the world has faded from the active list of philosophical positions. Whether some other conception of the human cognitive system as a "truth detector" can be defended is a question to which I shall return.

What is the status of the vocabulary of normative epistemology today? At best, it is problematic; at worst, it is a historical artifact of the earlier belief in the noetic powers of an immaterial soul. The best case scenario commits us to two irreconcilable vocabularies for describing ourselves during the stretches of time when we make epistemic evaluations: one is epistemic, the other is causal and psychological. Our talk of justification, evidence, objectivity, and other elements of epistemic evaluation reveals the extent to which normative epistemic notions are embedded in our conception of ourselves as knowers. We are unable to dispense with this vocabulary and still retain our understanding of ourselves as beings who reason on the basis of evidence. At the same time, we willingly countenance the natural scientific description of ourselves as organisms responding to the causal flux of surface irritations and internal homeostatic mechanisms. It is widely accepted that cognitive processes should be covered by Helmholtz's dictum that vital processes do not violate the laws of physics and chemistry. This suggests that the mechanisms in which human psychology is realized are wholly subject to natural law. And yet, a description of the knower in the language of physics and physiology leaves out the normative vocabulary of epistemology. Like

Kant, we seem committed to separate domains of discourse for psychology and epistemology. Which need not imply that psychology is irrelevant to epistemology, only that it is not sufficient for epistemology.

This "two vocabularies" approach has recently been the object of ever more vociferous attack from those who adopt the view that the terms of traditional epistemic appraisal belong to an outmoded form of description destined to go the way of the "philosopher's stone" of the alchemist or the "dephlogisticated air" of a more recent, but nonetheless defunct, chemical vocabulary. Such attacks are not new; we have met with their ancestors in Hartley and Lossius. Antimentalistic naturalism has been a constant current in twentieth-century American philosophy, but it has only recently approached a position of dominance.

Although many recent naturalists characterize their position as "materialist," the defining feature of American naturalism at mid-century required no reference to materialism.[17] Naturalism meant simply the extension of the methods of natural science to the study of all subject-matters; or, better, the belief that the natural scientific approach was exhaustively adequate for all serious intellectual pursuits. Here, "the methods of natural science" implied the use of publicly available empirical evidence and the application of "scientific" standards of reasoning. This was not thought to entail "the unity of science" in the reductionist sense. The naturalistic approach could include autonomous disciplines of psychology, anthropology, sociology, and history, in addition to the traditional natural sciences of physics, chemistry, and biology.[18]

This "naturalism" set itself apart from two opponents. By far its most enthusiastic attacks were aimed at those who were nonnaturalistic owing to their belief in supernatural agency and their attribution of a "spiritual nature" to humankind. The primary attack on nonnaturalists was that supernaturalist claims had been investigated and found wanting by the canons of natural-scientific investigation. Naturalists formulated the dispute as a disagreement about whether human beings are "merely" evolved animals. They answered yes, thereby precluding (by their lights) a supernatural origin (without implying physicalistic materialism).[19] The second set of opponents were the "nontheological antinaturalists." In this case the dispute concerned the foundations of the normative standards of ethics. Naturalists accused nontheological antinaturalists of a "lack of respect for scientific method," attributing to them the fear that, if a wholly naturalistic account of the human organism were accepted, there could be no absolute foundations for ethics.[20] This claim reveals the characteristic attitude of American naturalism: the view that the methods of natural science wholly exhaust the domain of intellectual inquiry.[21]

More recent versions of naturalism have naturalized the mind using two strategies that correspond to two strategies for making the mind an object of empirical investigation. The first is the eliminativist strategy, which seeks to eliminate normative mentalistic discourse from the domain of the cognitively significant; the second is the reductionist approach which seeks to reduce normative mentalist discourse to naturalistic discourse without eliminating it.

The father of eliminativism is W. V. Quine, the most eminent of American naturalists. Quine's naturalism is motivated by the methodological naturalism that extends the standards of natural science to all subject-matters, but it differs in the conclusion it draws. Whereas the earlier naturalists envisioned psychology, anthropology, sociology, and history as retaining their autonomy while meeting the rigorous standards of natural science, Quine contends that applying such standards will result in the elimination of "mentalistic" vocabulary from these disciplines and their ultimate reduction, through behavioristic psychology, to physiology and finally to physical mechanisms.[22] Quine's materialistic ontology and his antimentalism are presented as the results of a strict application of the standards of natural science, not as presuppositions of naturalism.[23]

Quinean naturalism holds high the banner of "naturalized epistemology."[24] In Quine's own version of this project, "naturalization" has two phases, a negative and a positive one. The negative phase belittles the "old epistemology" for trying to provide a foundation for knowledge of the external world, and it promotes the project of beginning one's epistemological investigations from a standpoint within natural science, rather than (foundationally) beneath natural science.[25] In terms of my contrast, the negative phase attacks the normative conception of the mental as in Descartes' attributing special a priori epistemic powers to the mind (the workhorse of "old epistemologists"). The attack might proceed by cataloguing the recognized failures of such alleged powers in their putative application to various subject matters, from physics to logic and mathematics. It would thus become apparent that the "conditions for knowledge" cannot be investigated by surveying a common fund of rational principles rooted in the mind of the knowing subject. Investigation must turn instead to the "natural products" of knowledge, that is, to the principles and propositions that receive community-wide acceptance within the various divisions and departments of knowledge. The negative phase of naturalization paves the way for the second, or positive, aspect of the program, which brings the knowing subject directly within the domain of natural science. The aim of this latter project is to provide an explanation of the knower that need not in principle use mentalistic vocabulary. The goal is to achieve a physiological or even physical explanation of such epistemic activities as learning and perception. Truth itself is explicated in terms of interactions among the laws of conditioning (as understood in behaviorist learning theory), including especially the drive for simplicity.[26] The payoff of such a reduction is an enlivened hope that once the mentalistic idiom has been replaced with naturalistic description and explanation, activities such as learning, which used to be relegated to the "mental," will be amenable to empirical investigation that meets the standards of the natural sciences.[27]

More recent eliminative materialists echo Quine's sentiment that a strict application of the standards of natural science will lead to the elimination of mentalistic talk rather than merely reducing it to physiology or physics. They agree with Quine that the vocabulary of mentalism is too far from the truth to admit reduction to physiology or physics, just as alchemy was too far from the truth to be reduced to chemistry proper.[28] The eliminativists are thus in accord with the "two vocabularies" approach that naturalistic and normative modes of description are irreconcilable. However, whereas the latter approach proposes peaceful coexistence, eliminativists advocate dropping the vocabulary of normative mentalism altogether. Although the eliminativist conclusion is more stringent than that of earlier American naturalism, the motivation is the same: strong affirmation of the belief that human beings are part of nature, and strong advocacy of the natural sciences as the only legitimate cognitive practice.

In his Woodbridge lectures, P. F. Strawson contends that "naturalism" need not lead to eliminativism. He does so by drawing a distinction between "hard" and "soft" naturalism. Hard naturalism is the naturalism of the physicalist; it accepts only the vocabulary of physics and physiology in its description of the human organism, ruling out the "soft" vocabulary of "subjective experience," "phenomenal qualities," and "moral and personal reactive attitudes."[29] Strawson himself advocates soft naturalism, contending that we are "naturally" disposed to adopt the soft vocabularies. He advances a kind of social naturalism, which he associates with Hume and Wittgenstein.[30] Such a position is naturalistic in giving up the notion that there are transcendent truths and values that might serve to guide our epistemic and moral practices. But it does not try to reduce our description of ourselves to the purely physical vocabulary of the hard naturalist, accepting instead a framework for description that is implicit in the culture to which we belong. Strawson observes that we "naturally" acquire the normative vocabulary of our social group or culture, and that we are generally unable to make sense of ourselves and our interactions with others without recourse to this vocabulary.[31] The acceptance of this vocabulary is "natural" inasmuch as it is "natural" for humans to be acculturated.

As portrayed by Strawson, hard naturalism conflicts with the language of "subjective experience" and "values," but he presents no parallel conflict with epistemic evaluation—i.e., with the language of normative epistemology. Most likely, he found no conflict between hard naturalism and epistemology because he conceived epistemology primarily in terms of the problem of the external world; and he considered that hard naturalism, steeped as it is in the vocabulary of physics, is committed to the existence of an external world.[32] In Strawson's view, the hard naturalist does not question the status of epistemic evaluation, not counting it among the "soft" vocabularies; rather, she questions our knowledge of the "internal world," including the "phenomenal experience" attendant upon stimulation of the sense organs.

Strawson's discussion notwithstanding, eliminativists are hard naturalists about knowledge. The "hard naturalist" elimination of normative epistemology has been advocated from as far back as Lossius, who "redescribed" the perception of truth as a feeling of consonance attendant upon the vibration of brain fibers. Today, the hard naturalist of the eliminativist variety describes the knower in terms of the truth- and knowledge-neutral vocabulary of physical or physiological causal processes. Epistemic description as applied to the organism drops out under a hard naturalistic description; hence, the *eliminativism* in eliminative materialism. Of course, the eliminativist does not wish to give up the ability to distinguish cases in which an organism is properly tuned to its environment from cases in which it is not. But such cases will be distinguished in terms of the relation between states of the nervous system and states of the environment under physical description. Perhaps the distinction will be built up from the vocabulary used to describe "fly detectors" in the frog, or "hawk detectors" in the field mouse.[33]

The hard naturalists' endeavor to eliminate the vocabulary of epistemic evaluation has invited attempts to turn it against itself. One attempt at turnabout recalls the fact that natural science itself adopts the vocabulary of epistemic evaluation as part of its conception of its own practices, the very practices underlying the physical theories that fund the eliminativist description of the knower. Natural scientists judge arguments, evaluate theories against data, assert the rationality of a particular method, and so on. So the position of the hard naturalist seems to undercut itself: by precluding the epistemic vocabulary of science itself, it invalidates its own claims.[34] Another such attempt appeals to the fact that the language of epistemic evaluation is deeply embedded in our conception of ourselves as thinkers and knowers; to adopt the eliminativist position is a kind of mental or cognitive suicide.[35]

Eliminativists of course concede that we cannot at present make sense of ourselves as well-tuned cognizers within an eliminativist vocabulary; however, they contend that this fact simply reveals the nascent state of the eliminativist science that will explain what we now call "knowledge."

The eliminativist puts her faith in the ongoing development of scientific description, maintaining that brain science will one day provide the proper understanding of the human organism as something that learns, accepts theories, etc. Included in this faith is the signal belief that such an account will not need to advert to "belief," "justification," "validity," and other mentalistic terms of description. The eliminativist may admit (indeed, insist) that such a development amounts to "cognitive suicide" in the sense that our descendants will no longer describe themselves as thinkers and knowers. But the eliminativist may point out that in such a case the species *homo sapiens* will not be dead; rather, human beings will have transmigrated to a new plane of consciousness. Of course, *they* won't describe their new state as a plane of *consciousness* (having abandoned mentalistic talk), and we, from our present vantage point, are unable to foresee how they will describe what we'd call their epistemic lives. The eliminativist, however, accepts as a matter of faith that a new, properly materialistic understanding of the species will be delivered unto our progeny.

The contemplation of "cognitive suicide" confronts only those hard naturalists who regard the vocabulary of a hard naturalism as incompatible with that of traditional epistemology. Those who pursue the second, reductivist, strategy for naturalizing the mind retain the mentalistic and epistemic vocabulary. Their goal is to accommodate and explain epistemic ascriptions within the vocabulary of natural science. This strategy for naturalizing the mind need not reduce cognition to physics and physiology but only to some "naturalistic" vocabulary, however defined.

There have been two variants of this program. The first may seem perilously close to the Wolffian strategy of simply dubbing "thought" natural. It comprises the attempts of some philosophers and cognitive scientists to make a direct empirical assault on mentalistic notions such as judgment and understanding. Such thinkers grant that the cognitive is not reducible to physiology or physics, but they maintain nonetheless that it is a proper object of empirical science. Attempts to empiricalize the mental include efforts toward general theories of problem solving, concept formation, and (eventually) belief-fixation.[36] The fact that these mental activities typically imply success, that they meet a criterion, has led some philosophers to dismiss such efforts as obviously confused, on the grounds that the criterion in question would have to be presupposed by—and so could not be determined by—empirical study. However, the second variant of reductive naturalism provides a *prima facie* response. It comprises various trends in recent epistemology toward causal or informational analyses of knowledge, including Dretske's information-based analysis of knowledge, Nozick's conception of knowledge as a tendency to "track the truth," and Goldman's analysis of justification in terms of reliable belief-producing mechanisms.[37] These philosophical theories characterize knowledge normatively as true belief produced "in the right way," and they try to provide naturalistic description of "the right way." Typically, the vocabulary of (biological) functional analysis is called into play in attempts to naturalize the normative standards of success embodied in cognition. If we assume that the mind has evolved to produce cognitive states that reliably indicate the external world, then, on the dubious assumption that "truth"—the standard of reliability— poses no problem, the normative would have been captured in the domain of the natural without being eliminated.

As used by the nineteenth-century authors examined in chapters 4 and 5, the functional approach provided a means for analyzing the "harmony" between sensory mechanisms and the world without appealing to divine institution and without adopting transcendental idealism. Although such an approach could not satisfy Kant's demand for an explanation of the possibility of transcendent truth, it could provide a framework for explicating the veridical perception of spatial properties within a prag- matic commonsense realism. Describing the "proper functioning" of the mechanisms of spatial perception might provide the basis for distin- guishing accurate from inaccurate representation, or "veridical" from illusory perception of spatial properties. With the advent of natural selection to explain the etiology of mental and physical adaptation, the positing of a "fit" between mind and world was no longer open to the suspicion that it must appeal to a benevolent creator. Natural selection provided a framework for adopting the "normative" vocabulary of function and misfunction, of perception and misperception, while re- maining thoroughly naturalistic.

Even beyond the epistemological endeavors just described, there has been a trend in contemporary philosophy toward using the notion of biological function to develop a naturalistic account of mentalistic notions such as "content" and "belief." The biological approach to content explains the formation of basic concepts or mental categories through the organism's interaction with its environment: a field mouse, for example, might develop concepts such as "food," "mate," and "predator"; human beings might develop a more articulated stock of concepts, including concepts of social structure and interpersonal relations.[38] Some have suggested that the mind has evolved a reliable belief-forming mecha- nism, say, of induction, under the pressure of natural selection.[39] Such a mechanism is a naturalistic counterpart to the early attribution of truth- detecting faculties to the human mind, that is, to one of the central notions of the metaphysically normative account of the mind. It would be foolhardy for evolutionary naturalism to attribute to this mechanism the ability to discern transcendent truth, but it might substitute a more pragmatic conception of truth, equating a "true" belief with a cognitive state that guides action by responding to the survival-enhancing or endangering aspects of a given situation. However, the question would remain of whether any mechanism of belief formation and any set of conceptual structures that need be adequate only for survival are sufficient to determine and explain the cognitive achievements that have been gained by human minds during the five thousand years of historical time. That the discovery of natural scientific truths can be exhaustively explained by such mechanisms or structures, without further appeal to the contingencies of cultural development, is dubious at best.

What is the relation of hard naturalism in contemporary philosophy to the methodological and metaphysical naturalism of previous centuries? Both the eliminativist and the reductive varieties of contemporary hard naturalism are continuous with the methodological naturalism of the nineteenth century; they provide as well a counterpart to metaphysical naturalism. They are continuous with methodological naturalism in endorsing its claim that the techniques of natural science may be exhaustively extended to all subject-matters. They provide a counterpart to the materialism of metaphysical naturalism in claiming that all cognitively significant descriptions of the human organism and its activities can be adequately expressed in the vocabulary of physics and physiology.

In my view, the appropriate response to the methodological claim of hard naturalism was provided by Helmholtz, who flatly denied that the techniques of natural science are adequate to all subject-matters. Over the past decade this point has been argued at length by a number of authors, including Putnam and Taylor,[40] who insist that the interpretive practices central to the study of human lives and human artifacts resist assimilation to the techniques of the natural sciences with their ideal of objective convergence based upon repeatable experiments. They attempt to show that the natural-scientific approach to such studies either avoids the interesting questions, or, in the very act of addressing them, reveals its own inadequacy. They do not claim that there are no "objective facts" about lives and artifacts that are as fixed as the "facts" of natural science; rather, they claim only that in human sciences the objects and the modes of explanation differ significantly from those of the natural sciences. As Helmholtz realized, the object of explanation in the human studies is historically constituted: it is not the product of uniformly acting natural forces but of historically particular paths of development. The explanation of historically constituted practices takes place against a background that is indefinite in extent. Given the constraints of investigation, individual researchers may legitimately disagree on the important objects and the important contexts. The kind of idealization of the object of study and regularization of the standards of observation that favor experimentation and lawful explanation in the natural sciences would eliminate the object of study in the humanities.

Arguments such as these probably can persuade only eliminativist and reductive hard naturalists who have seriously engaged in historical or critical interpretation; or rather, anyone who has seriously engaged in such studies probably would not be tempted to claim that the techniques of natural science can be extended to all domains. Of course, a versatile but committed naturalist could claim that historical, critical, and legal studies are "not of intellectual interest" in the way that scientific studies are. The hard naturalist's claim would thus be revised to assert that the techniques of natural science are exhaustively adequate for all subjects of "genuine intellectual interest." Only in rare cases would anyone claim that the object of study in the human sciences, human lives and their products, is of lesser value than subject-matters that admit of law-based explanations; more likely, the revised claim would reflect the belief that the "universal" findings of the natural sciences are by their very generality of greater intellectual value than the particularistic products of humanistic research (where "greater" is not simply a pun on the scope of universally applicable laws). The hard naturalist might further observe that humanists concede from the start that agreement of the sort ideally expected among unbiased investigators in mature natural sciences will not occur, and that this admission amounts to conceding that humanistic "findings" are less objective and permanent than those of the sciences. Of course, the humanist may think that this lack of convergence is an inherent feature of the subject-matter, one that does not render otiose all prospect of "progress" in the humanities. Previous or conflicting interpretations contribute to the ongoing process of understanding human culture and its artifacts, and a variety of critical perspectives has value for the act of interpretation itself. In any event, if the naturalist were to claim the greater inherent interest of those subject-matters to which natural-scientific techniques are readily applicable, the exchange would seem headed for a standoff, with each side accusing the other of begging the question. The nonreductivist simply points to his experience with the study of cultural products. The hard naturalist beats the drum of science and dismisses the "fluffy" products of such studies. Mutual question-begging seems inevitable and insurmountable if each side simply sticks to its guns. Given the present state of knowledge, each side is simply placing a bet on the future course of knowledge, one not backed by extrapolating specific results from past and present science but merely expressing a judgment of the likelihood that future scientific theories will make reduction or elimination possible.

A more effective strategy for the nonreductivist would be to attempt to meet the hard naturalist on her own ground by finding a naturalistic argument, based on current science rather than science fiction, that leads to a conclusion undercutting hard naturalism. In recent years, just such an argument has been advanced in support of soft naturalism.[41] It claims that, according to our best natural-scientific description of the human animal, there is no objectively correct single naturalistic interpretation of the cognitive life of human beings. Consequently, there is no sound naturalistic argument to the conclusion that the use of mentalistic vocabulary in describing ourselves is incorrect. A consistent and thorough application of hard naturalism leads to the conclusion that there is no one true naturalistic description of human cognition.

The argument proceeds from evolutionary theory. It begins from the fact that from a biological point of view the human organism is unformed at birth and becomes cognitively formed only through the process of acculturation. Stephen Gould discusses this aspect of human development under the title of "neotony," the view that humans are "developmentally young" at birth.[42] Human beings have evolved in such a way that an important part of our development remains in the hands of contingent environmental influences. We are not only relatively immature at birth compared with other primates but comparatively less of our subsequent development is "programmed in" than is that of other animals who are immature at birth (e.g., certain marsupials).[43] Thus, our language, social customs, ontological categories, and many of our concepts are contingently acquired through acculturation. It is likely that we have some innate cognitive structure, but even so, it must mediate the process of acculturation. It does not determine our entire conceptual structure, and so not the full range of concepts we use to describe our cognitive practices. If it did, all human cultures would share the same basic conceptual vocabulary.[44]

The hard naturalist claims that there is one true description of the whole of human cognition, including the concepts with which we describe our epistemic practices. The reductivist and eliminativist versions of this claim differ. The most plausible reductive account that would fulfill the goal of the hard naturalist holds that the content of our beliefs, including those that pertain to ourselves as believers, can be given an evolutionary explanation. What is the likely success of an evolutionary explanation? Opponents could grant from the beginning that there is a correct scientific description of basic human perceptual and motor skills, and further, that we have some inborn cognitive structures that facilitate various reactions to the environment. The conceptual content ascribed to such structures could include various innate tendencies toward avoiding pain, seeking nourishment, and procreating. With these assumptions, an evolutionary account could explain the "content" of the smell of rancid food and the subsequent adaptive response of refusing to ingest it.

But a naturalistic account of these simple acts of perception and cognition does not even approach the goal of a fully naturalistic account of cognition. If a naturalistic description of the human animal is to correctly describe our epistemic practices, the evolutionary account would have to be extended to the full range of human cognition. But at least one group of conceptual categories is a contingent attribute of human cognition: the description of ourselves as knowers. The cognitive practices that characterize mature sciences and the vocabulary of objectivity, evidence, theoretical consistency, and so forth that describes such activity are clearly contingent, for advanced civilizations have developed that did not include the conceptual vocabulary of natural science. Further, the theoretical vocabulary of science itself is clearly a contingent feature of our conceptual structure. Members of the same species, all inhabiting the terrestrial environment, have developed a variety of concepts for describing nature, including Aristotelian physics, Cartesian physics, Newtonian physics, and Einsteinian physics (to keep to a short list). Any common cognitive mechanisms we might possess as a result of evolution must allow for the "natural" acquisition of the variety of extant human beliefs. But then, they will provide no basis for singling out some one set of these beliefs as the "most natural," and therefore as the one true naturalistic description of either the conceptual content or the culturally conditioned epistemic activities of the human mind, including the activity of science.[45]

The eliminativist, on the other hand, suggests that we give up our current descriptions of our epistemic practices in favor of "eliminative materialist" descriptions, which presumably will be in terms of neural events in the brain. Again, opponents could grant the eliminativist an ideally completed neurophysiological description of learning and concept formation. Let us suppose that all learning occurs as the result of a neurophysiological process of "gradient descent," a process in which a pattern of input "settles down" to an equilibrium state. Let us further suppose that this process mediates concept formation (say, by facilitating the formation of prototypes and a similarity metric, the latter of which is "tuned" through experience), and of the detection of regularities in behavior of objects and animals. Think of the process as a natural inductive mechanism.

Imagine that such a process has been convincingly demonstrated to be the mechanism of human learning and concept formation. What constraint would this put on the sort of cognitive life we could have, and therefore upon the descriptions of our cognitive practices that could be true?

Presumably, it could tell us something about, say, the dimensions of the chromatic and spatial properties of things that are likely to enter into early applications of the similarity metric. But what could it tell us about the cognitive structures of a random sample of *homo sapiens* from around the globe and across time? A miniscule amount, compared to what there is to be known about the cognitive structures of these various individuals. If human beings truly have a single mechanism for learning, then the variability in acquired concepts must come from what is learned. And what is learned depends to some extent on geography, but to a larger extent on the cultural context into which a child is born. But cultural contexts are no more constrained by biology than the empirically manifest variety of belief systems would indicate—that is, not very much. A great deal is subject to variation within and across cultural groups, including the refined similarity metrics underlying scientific or other sophisticated judgments, the system of predicates in which such judgments are expressed, and even various conceptions of what counts as a good scientific explanation. These variations must be explained as a result of the joint operation of cultural development and the learning mechanism. In order to understand the concepts learned by an organism, including concepts about how better to learn about nature, the investigator would have to examine the learning context defined by various developed cultures. A complete account of the physiology of the brain would not determine what this context is, and so would not specify the "correct" conceptual vocabulary for the human organism. Indeed, the only case in which the neurophysiological vocabulary would adequately describe all human cognition is that in which all human beings are neurophysiologists who think only about neurophysiology, and who think about their thinking about neurophysiology only in neurophysiological terms. If the eliminativist could bring about such a state of affairs, the eliminativist description of human epistemic practices would be factually correct. That, however, would be an historical fact, not a fact of natural science.

In short, the hard naturalist's "metaphysical" view that we are evolved creatures standing wholly within the natural order leads to the conclusion that there is no one correct naturalistic description of our conceptual structures and cognitive practices, except the description that such structures and practices are variously refined and, in the case of sophisticated practices, are virtually constituted through acculturation. If this result is now supplemented with the claim that cultural products as rich and diverse even as those produced over the course of the history of science are not amenable to study solely through the techniques of natural science but require the interpretive practices of the humanities, then it follows that, to discover what sort of creatures we are, our cognitive lives must be understood through those interpretive practices.

While local disputes over eliminative materialism continue, a normative epistemic vocabulary remains deeply embedded within our epistemic practices. And yet we are without a normative metaphysics upon which to ground it. However, if the preceding argument is correct, such a metaphysics is not required to ground our practices. Our epistemic vocabulary is the product of the cultural process that has also given us both natural science and eliminativist metaphysics. To decide whether to accept eliminativism we need only use the resources for judgment we possess as a result of the tradition that has led to natural science in general. According to the above argument, eliminativism is not acceptable. No natural-scientific imperative can eliminate normative epistemology.

Helmholtz was not in a position to foresee this result, but it is one with which he would agree. Helmholtz was deeply drawn to the forms of explanation of classical physics, and he extended them to human cognition. And yet his experience with investigating historically conditioned subject-matters led him to posit a division in kind between the methods of the natural sciences and of the *Geisteswissenschaften*. It is a small irony of history that the findings of natural science should provide an argument for acknowledging limits to the extent natural science applies to human mentality. It is not, however, an irony Helmholtz would have found bitter, for the arguments just reviewed provide reasons for a conclusion that he acknowledged through sheer good sense. Even if his guiding conceptions in both physics and physiology have been surpassed, the conclusions of his considered judgment—conclusions he was willing to affirm independently of natural scientific support—have remained sound. Helmholtz was a natural scientist who eschewed the parochial claim that natural-scientific patterns of thought exhaust human rationality. His good judgment can serve as a norm for us all.

Appendix A

Nativism-Empirism and Rationalism-Empiricism

Because my treatment of the history of theories of spatial perception and of mind posits a distinction between nativism and empirism on the one hand and rationalism and empiricism on the other, the history of the terms that name these distinctions and of the conceptual geographies associated with them deserves notice. Both pairs of distinctions first achieved widespread use during the nineteenth century. The opposition between nativism and empirism, or *Nativismus* and *Empirismus*, has been used to organize the history of theories of spatial perception—and, more generally, the history of psychology—since Helmholtz thrust it into prominence in the second half of the nineteenth century. Similarly, the opposition between rationalism and empiricism, or *Rationalismus* and *Empirismus*, has been treated as a subtheme in the history of modern philosophy since at least the time of Hegel. Organization of the history of early modern philosophy around the now familiar opposition between the "rationalists" Descartes, Spinoza, and Leibniz and the "empiricists" Locke, Berkeley, and Hume became standard in the nineteenth century (though the two lists often were longer), even if it was not the only way histories were organized.

Psychologists and philosophers have understood the distinctions somewhat differently. Histories of psychology typically treat the terms "rationalism" and "nativism" as virtually synonymous, and they make no distinction between "empiricism" and "empirism" (where the latter term indicates the doctrine that some or all abilities are learned).[1] In contrast, many philosophers consider "rationalism" and "empiricism" to mark a distinction that is importantly different from that marked by the other two terms. These philosophers would like to enforce a sharp separation between questions about the genesis or causal origin of a thought or idea (innate or learned) and questions about the justification of knowledge (through reason alone or by appeal to sensory experience). Accordingly, psychologists and philosophers have sometimes disparaged their counterparts in the other discipline. Thus, psychologists, believing that classical epistemological discussions boil down to a question about whether cognitive abilities are innate or acquired—a question about genesis or causal origin, subject to empirical investigation—may think that philosophy (or at least epistemology) should go out of business. If the classical questions of epistemology are really empirical, then philosophers who think about such questions are attempting to answer from the armchair what should be decided in the laboratory.[2] In opposition, philosophers are likely to accuse psychologists of confusedly attempting to use empirical means to settle normative questions about warrant, confusing a question about evidential basis with a question about causal etiology. Any link between rationalism and nativism—whether by an historical figure such as Descartes, by an historian discussing his philosophy, or by a psychologist engaged in research—would therefore rest on a confusion, because the innateness of an idea or mental ability gives no warrant for supposing that the idea provides a basis for knowledge or that the mental ability yields truth. Innateness by itself does not guarantee truth; postulating an innate tendency toward error is not incoherent.

This mutual name-calling has been going on for over one hundred years. I should like to propose an *Aufhebung*. By my lights, both sides are correct on some points, but neither is correct in characterizing the historical-conceptual relations between nativism and rationalism or empirism and empiricism. Against the psychological and in favor of the philosophical interpretation, nativism and rationalism should not be equated. Although Descartes and Locke respectively argued for and against "innate ideas," and thus were concerned with the "origin" of certain ideas and principles of knowledge, their disagreement about innateness was not the crux of their dispute. Descartes did not believe that certain ideas provided a basis for knowledge simply because they were innate; he contended that they provided a basis for knowledge because they were perceived clearly and distinctly by the intellect. According to his analysis of the relation between the senses and the intellect, the ideas that provide the materials for clear and distinct perception have their origin in the intellect, independently of the senses, and so independently of what Locke would call "experience," by which he meant "sensory experience."[3] It was whether purely rational or intellectual knowledge existed that separated Descartes and Locke, not the question of innateness per se.

On the other hand, to say that an idea or a principle has its origin in the intellect is to say something about its origin. Thus, psychologists have correctly discerned that early modern philosophers were talking about origins. Indeed, philosophers themselves, looking back upon the early modern period, often suspect that their predecessors were discussing origins, but they typically chalk this up to an early manifestation of the conceptual confusion recounted above. However, the distinction between "origins" and "justification" does not make it improper to link "innate" with "sound basis for knowledge"; rather, it reveals the need for argument or justification in such a case. Contemporary arguments appeal to the possibility that at least the "ecological validity" (or soundness?) of an innate disposition might be established by showing its origin through natural selection. Accordingly, etiological considerations could (arguably) be relevant to the epistemic status of an innate mechanism or belief structure. Similarly, seventeenth- and eighteenth-century philosophers appealed to the source of an innate idea or ability to provide it with epistemic credentials, the most notorious being Descartes' appeal to a divine warranty for the intellect. Not innateness per se but a more primary fact about the "origin" of knowledge was alleged to provide the warrant. Locke's concern with the origin of knowledge also need not be seen as a confused foray into mere causal etiology. In arguing that *experience* is the only credible source of justification, he was making a point about the type of warrant to be given for belief, not about the psychology of belief formation.

Placing these recent disputes to one side, let us look more closely into the history of the distinctions involved, and especially into the origin of the paired dichotomies between rationalism and empiricism and between nativism and empirism.

The use of the terms *Rationalismus* and *Empirismus* to indicate a division between epistemological positions dates from the eighteenth century.[4] The positions so denoted have their modern origin in the seventeenth century. Although the precise historical origin of the terms remains shrouded, they were well entrenched by the middle of the nineteenth century. Early in that century, the historian of philosophy Wilhelm Gottlieb Tenneman divided the history of modern philosophy prior to Kant into empiricist and rationalist schools.[5] Near the same time, Hegel treated the interaction between rationalism and empiricism as a subtheme in the development of modern philosophy.[6] Still, neither the terminology nor the historiography was uniform. Nineteenth century historians of philosophy used a variety of divisions and historical stages in describing early modern philosophy, including various combinations of empiricism, idealism, dogmatism, skepticism, and criticism.[7] Some writers preferred the term *Sensualismus* to indicate the philosophies of Bacon and Locke, with their emphasis on the senses as the source of knowledge.[8] Ueberweg preferred a division into empiricism, dogmatism, skepticism, and criticism, but he allowed that with equal right the early modern period might be characterized in terms of the opposition between empiricism and rationalism, which he defined in the usual epistemological terms: empiricism is the position "that all knowledge has its origin in experience," and rationalism, the position "that reason is the source of all knowledge."[9] Toward the end of the century Falckenberg remarked that the division of pre-Kantian philosophers into empiricists and rationalists—depending upon whether the philosopher considered sensory experience or reason to be the "source" of (or warrant for) knowledge—had become standard.[10]

The original use of the terms *nativistisch* and *empiristisch* to indicate opposing theories of the psychogenesis of spatial perception may be dated precisely. Helmholtz first used the terms in this way.[11] However, the dispute about whether the ability for spatial vision—in either two or three dimensions—is innate or acquired was already old when Helmholtz wrote. Seventeenth-century writers on optics had attributed distance perception to an innate "natural geometry";[12] Berkeley opposed this view with the theory that the ability to perceive depth and distance is acquired by associating visual data with information about the third dimension gained through the tactile sense. His position became predominant during the eighteenth century, though it was not unchallenged.[13] Late in the century, Dugald Stewart explicitly maintained that the distinction between the "original" and "acquired" perceptions of sight was funda- mental within visual theory.[14] Further, there were differing views on the character of the psychological operations involved in vision, whether acquired or innate. The variety of positions increased in the nineteenth century as the physiology and psychology of vision came to be studied by practitioners of the new disciplines of physiology and psychology, in addition to its study by metaphysicians, natural and mental philoso- phers, and physicians.

Into this complex and ongoing controversy over the psychology of spatial vision entered Helmholtz. His task was to survey the field of physiological optics for the purpose of writing a work that would appear as volume nine of Karsten's *Encyklopaedie der Physik*. The format of the Karsten series dictated that Helmholtz provide a thorough review of both past and present work, a task he performed diligently. Wishing to provide a systematic theoretical framework for his survey of the field, Helmholtz divided the major theoretical positions with respect to spatial perception into two groups, which he characterized with the adjectives *empiristisch* and *nativistisch*, introducing the distinction as follows:

> It may be hard to say how much of our intuitions [sensory experi- ences] as derived from the sense of sight is due directly to sensation, and how much of them, on the other hand, is due to experience and training. The main point of controversy between various investigators in this territory [the theory of vision] is connected with this difficulty. Some are disposed to grant to the influence of experience as much

scope as possible, and to derive from it especially all intuitions of space. This view may be called the *empiristic theory*. Others, of course, are obliged to admit the influence of experience in the case of certain classes of perceptions; still, with respect to certain elementary intuitions that occur uniformly in the case of all observers, they believe it necessary to presuppose a system of innate intuitions not based on experience, especially with respect to spatial relations. In contradistinction to the former view, this may perhaps be called the *nativistic* theory.[15]

The distinction between nativism and empirism seems clear enough: the classification of a theoretical position into one or the other camp depends upon the degree to which the perception of spatial relations is attributed to innate or acquired factors. However, the term *empiristisch* is ambiguous. Helmholtz clearly intended it to indicate theories according to which the ability to perceive spatial relations is acquired. According to terminological convention the German term is often translated as "empiristic." But did he also mean to draw upon its connection in German with *Empirismus* and *empirische Naturwissenschaft*? He did in fact seek to connect his empiristic theory of spatial perception with an empiricist epistemology and an experimental scientific methodology; however, it is not clear that he wanted his choice of terminology to imply such a connection. This point is of some importance not only for understanding Helmholtz but also for understanding the meanings of *empirisch* and *empiristisch* in German, together with their English cognates.

A very useful myth has grown up around the relation between the use of the word "empiristic" in English and its origin in the German term *empiristisch*. The mythical view is that German *empiristisch* indicates the empiristic theory of the spatial perception, while *empirisch* indicates the epistemological doctrine of empiricism.[16] It is correct that on occasion, *empiristisch* was used to indicate the empiristic theory; Helmholtz introduced the opposition between that term and *nativistisch* with this lexicographic purpose in mind. Moreover, the evidence indicates that Helmholtz considered the empiristic theory to be distinct from Locke's empiricist epistemology. This would be consistent with the mythical view that *empiristisch* and *empirisch* distinguish empiristic theories from epistemological empiricism.

Be that as it may, the contrast between *empirisch* and *empiristisch* described by the mythical view is not available in German. Here are the general relations among *Empirie, Empirismus, Empirist, empirisch*, and *empiristisch*. *Empirie* means experience, or knowledge based upon experience; *empirisch* is its adjective. *Empirismus* typically means empiricism, though it is sometimes used to mean empirism; in either case, *empiristisch* is its adjective. This last is the important point: *empiristisch* always means "pertaining to the doctrine of *Empirismus*" (however that doctrine is understood); *empirisch* always pertains to experience itself, or to knowledge based upon experience. Similarly, German *Empirist* means "one who holds the doctrine of *Empirismus*," whatever that doctrine may be. Thus, although *Empirismus* and *empiristisch* may be ambiguous in German, sometimes denoting a psychological theory about the origin of spatial perception, and sometimes an epistemological theory about the basis of knowledge,[17] they provide no basis for the contrast in German that English provides in the terms "empiricism" and "empirism," along with "empiricist" as opposed to "empirist" and "empiristic." Which is no reason for us not to take advantage of the terminological possibilities English provides, which I propose that we continue to do.

Then, did Helmholtz intend to play upon the ambiguity of the German term *empiristisch*? It would seem not. In the *Handbuch* he characterized Locke's *Erkenntnistheorie* as attempting to ground "alle Erkenntnis auf Empirie," without using the terms *Empirismus* or *empiristisch*; in pigeon-holing Locke, he placed him among the *Sensualisten*, as opposed to the *Empiristen*, following one accepted usage in his time.[18] When Helmholtz used the term *empirisch*, he used it the way that we would use "empirical": to indicate an empirical investigation, or to indicate evidence that is based upon experience. Apparently, he intended the term *empiristisch* to apply only to the theory that spatial perception is the result of learning.

Helmholtz's opposition between the terms *nativistisch* and *empiristisch* was widely adopted in the psychological literature: Ribot (1879) and Klemm (1911) used it in the portions of their respective works devoted to the history of theories of spatial perception, and it has been used to organize histories of such theories ever since.[19] Although the terms have not always been used with as much precision as might be desired, their primary senses as applied to theories of the psychology of spatial perception are reasonably clear. *Nativism* is the view that at birth (or sometime thereafter, depending upon the the ontogenetic timetable) the organism, upon visual stimulation, will experience a spatially organized array of colored areas. The key notion is that the disposition to have spatially organized visual experience is inborn and does not depend upon a process of learning. A nativist need not assert that an innately endowed organism will be visually competent at birth, for the visual system might become functional only after a further period of maturation. *Empirism* is the view that at least some of the spatial organization found in the visual experience of the adult is the result of learning; it asserts that some or all of the ability to perceive a spatially organized visual world is acquired.

Nativism and empirism admit of degree. They are not mutually exclusive except at the extremes. Any degree of native ability short of the full spatial competence of an adult may be consistently paired with an empiristic account of the processes that make up the differences. Nonetheless, nativism and empirism in their most radical forms are strictly incompatible. The most extreme form of nativism (with respect to spatial vision) would hold that percipients are innately endowed in such a way that the experience of a three-dimensional visual world arises from innate disposition alone, without the aid of learning. The most extreme form of empirism would deny that even a bidimensional pattern is innately perceived. According to extreme empirism, all spatial organization found in visual experience must be acquired; the visually naive percipient does not localize color sensations, even in two dimensions. At either extreme, *all* spatial organization in visual experience is explained either nativistically or empiristically.[20]

A great many writers took positions between the extremes, and so it is difficult to bifurcate theorists into sharply opposing camps. In fact, theorists who held nearly identical positions with regard to what is learned and what is innate have been cross-categorized. Helmholtz and later writers classified Johannes Müller as a nativist, indeed, as one of the chief founders of nativism. Müller held that the innately given spatial organization in vision is a mental representation of the retinal image, and that the perception of depth is acquired through learning. Müller's position might therefore with equal cause be labeled a form of empirism. In fact, Berkeley held the same position with respect to what is learned and what is innate, and yet he is known as an empirist, perhaps indeed as the chief founder of empirism.[21] Müller and Berkeley have been assigned to opposite camps for understandable historical reasons. When Berkeley wrote, the idea that the perception of depth and distance result from learning challenged current dogma, whereas when Müller wrote a century later, the nativistic component of his position would have been accentuated in comparison with the more radical forms of empirism then extant, especially since he opposed extreme empirism. Nonetheless, that individual thinkers who largely agreed upon what is learned and what is innate have been assigned to opposite camps is a mark of the latitude with which the terms "nativism" and "empirism" (or "empiricism," in some cases) have been used. This suggests both the need to specify degrees of nativism and empirism, and to understand prior classifications of theorists in relation to then-contemporary standards of what counts as an extreme version of the two positions.

Because of the importance of the paired oppositions between nativism and empirism as well as between rationalism and empiricism in the history of philosophy and in the history and philosophy of psychology, the content and implications of assigning a belief or ability an origin as innate or learned, especially in relation to assigning a rational or an empirical basis to a knowledge claim, requires some further sorting. Here as throughout, let us denote each member of the two pairs of positions by the four "isms," using "nativism" and "empirism" for the doctrines that our perceptual abilities are innate or learned, and "rationalism" and "empiricism" for the doctrines that the evidence for knowledge claims derives from reason or from sense experience.

What is it for an ability or a belief to be learned, or acquired on the basis of experience? Clearly it is not for the said ability or belief to have "experience" as the sole factor in its causal etiology. All learning presupposes some innate ability or predisposition, if only the bare ability to learn.[22] We may learn a particular fact on the basis of experience, but the concepts by which we recognize that fact must already have been in place. If the concepts were previously learned, then we must at least have had the ability to learn concepts. This last ability might be viewed as specific to concepts, or as part of a more broadly characterized learning ability. Historically, the most extreme empirist positions have posited a small number of general "laws of learning," such as laws of association, or, more recently, laws of conditioning. This form of empirism is extreme in keeping the structure of innate ability to a minimum.

What is it for a particular ability or belief to be innate? Although it may be sufficient that the individual have the belief or ability in question prior to birth (and independent of intrauterine experience), it clearly is not necessary. An innate ability might manifest itself only after a period of preprogrammed maturation that proceeds independently of experience (except where experiential deprivation is severe enough to cause death or clinical impairment).[23] However, something further than the mere *capacity* to manifest the belief or ability is required for a belief or ability to be innate; if mere possession of a capacity were sufficient, then everything would be innate, since, as Locke observed, one has the capacity to manifest whatever beliefs or abilities that one in fact manifests.[24] A bare capacity is counted as an innate ability only if it is more particular than a general learning capacity, as in the alleged "domain specific" innate ability of children to learn the grammatical structure of their first language.[25] This capacity does not fall prey to Locke's rejoinder, because it is language-specific: it is less general than the minimum general learning abilities that must be attributed to any learner.

In contrast with nativism and empirism, rationalism and empiricism are not specifically directed to the question of whether particular abilities or beliefs are innate or learned but whether some knowledge claims have rational, nonempirical warrant. Rationalism admits such knowledge, empiricism denies it. Of course, empiricists typically did not deny innate rational ability that is independent of sense experience, just as empirism did not deny all innate contributions to learning. Empiricists and rationalists could also agree that some propositions, such as the principle of contradiction or the proposition that equals from equals yield equals, are known by reason alone. The issue between them concerned reason's ability to grasp a mind-independent reality without appealing to sensory experience. Rationalists maintained that some claims to knowledge, such as knowledge of the essences of God, of matter, and of mind, can be justified by appeal to pure rational or intellectual intuition, independently of sensory evidence. Typically, this rational knowledge did not extend to individual existences; rationalists recognized that experience is necessary for coming to know particular facts. God was the only individual existent considered knowable by reason alone, through the connection between divine essence and existence. Empiricism, by contrast, denied that the essences of God, matter, and mind can be known by reason alone, holding that these and other general concepts must be acquired by abstracting from particular instances.[26] More generally, empiricists maintained that reason requires material from the senses in order to exercise its power; it cannot provide its own intellectual objects.

How, then, do the various "isms" fit together? Although the four positions can be held independently of one another, they are not conceptually independent in every context. Thus, rationalism shows a strong conceptual relation with some forms of nativism. The rationalist Descartes includes among the contents of reason the idea of God, and makes recognition of this idea a condition on the justification of all other truths of reason. His rationalism is committed to an innate idea as a component. But a rationalist need not be committed to particular innate ideas in order to exhibit a form of nativism; one might hold that only the faculty of reason itself is innate, but that reason is by itself sufficient for establishing a variety of principles pertaining to the essences of mind and matter. Such innate principles could either extend to or exclude the "natural geometry" of spatial perception. A rationalist need not be a nativist with respect to spatial perception; he or she could hold that reason comes equipped to grasp (say) the essence of matter, but that vision must learn to infer the third dimension on the basis of the retinal representation.

Similarly, empirism and empiricism might be related to the extent that the latter provides the motivation for the former. Thus, a perceptual theorist who holds that we perceive a three-dimensional layout by drawing an unconscious inference from a two-dimensional image in accordance with certain rules might argue that the rules of vision must be acquired through experience, since all knowledge must be so acquired. Conversely, a theorist who believes we move from two to three dimensions through associative habit-formation, as opposed to inference, could be an empirist without empiricist motivation, since the associative acquisition of a habit need not be counted as a case of knowledge. Berkeley saw no problem in divorcing the psychological processes of vision (involving association or "suggestion") from intellectual activities such as judgment and inference, thereby effectively sundering empirism and empiricism.

Appendix B

Sensation and Perception: Epistemological and Psychological

When he drew his distinction between nativism and empirism, Helmholtz seemingly equated nativism with the view that much of visual perception is attributable "directly to sensation."[1] The resulting implication that sensations are innate seems uncontroversial. But should one also infer that perceptions are acquired? Helmholtz seems to have used the terms in this manner, but not everyone did. Although the terms "sensation" and "perception"—or *Empfindung* and *Wahrnehmung*—sometimes have been used to indicate that the first aspect of sensory experience resulted from innate mechanisms and the second, from intellectual operations, usage was not uniform throughout the early modern period. These various uses can, however, be divided into two groups, which may, without excessive anachronism, be characterized as epistemological and psychological.

In its epistemological use the distinction between perception and sensation parallels the division between ideas of primary qualities and ideas of secondary qualities: *sensations* are ideas of secondary qualities, *perceptions* are ideas of primary qualities. Sensations are mere subjective states that do not constitute belief or knowledge, whereas perceptions are epistemic acts involving a belief about an external object. This version of the distinction motivated the initial terminological distinction, widely credited to Thomas Reid. A similar distinction had been proposed earlier by Malebranche, who distinguished between merely subjective sensory states or *perceptions sensible* (which included sensations of the Lockean secondary qualities) and perceptions that convey ideas of properties bodies can actually have, or *perceptions pures* (which included ideas of size, shape, and motion).[2] Despite Malebranche's priority, we shall consider Reid's version, for he provided a more sustained discussion.

Reid maintained that a sensation is something whose character is entirely dependent upon the mind (although it can have as its cause a state of the nervous system that has itself been caused by an external object). A sensation is "something which can have no existence but in a sentient mind, no distinction from the act of mind by which it is felt."[3] He assimilated sensations to pains: he found no distinction between sensations and the act of sensing, just as a pain is equivalent to the act of feeling a pain. Reid included under "sensation" the traditional list of secondary-quality ideas (sensations of smell, taste, sound, color, and touch), but excluded all primary-quality ideas, including extension, figure, and motion. In the case of perception, he allowed that the act of mind by which we perceive a thing can be distinguished from the object of perception; we can distinguish our perception of a tree from the tree itself, and we generally suppose the latter to exist independently of our mental states. He found two elements present in any perception of an object: a conception of the object's form (a primary quality) and a belief in the object's present existence.[4] According to Reid, sensations all have an innate basis, but perceptions may be either innate or acquired. Reid believed that all the spatial perceptions of touch are innate, but that by vision "we perceive originally the visible figure and colour of bodies only, and their visible place: but we learn to perceive by the eye almost everything we can perceive by touch," including three-dimensional spatial properties.[5] Thus, spatial perceptions could have an innate basis and yet not be termed sensations, for these innate perceptions are nonetheless perceptions of external objects, and not, as are sensations, mere subjective states caused by external objects.

The second or psychological version of the sensation-perception distinction was common among many physiologists and psychologists in the eighteenth and nineteenth centuries; it is akin to the signification that the terms are sometimes given today. According to this version, a sensation is the immediate result upon consciousness of the stimulation of the sensory nerves; a perception is mediated by psychological operations performed upon sensations. It is difficult to distinguish the origin of this use of the term "sensation" from that of the epistemological use, because early authors such as Descartes used the term both with the implication of subjective state, and with the current implication of an immediate result of physiological stimulation.[6] Only later did "perception" come to indicate the product of psychological processes. As theorists began to emphasize the distinction between immediate and mediate sensory and perceptual states, the terminology became entrenched; Helmholtz's junior colleague Wilhelm Wundt provides a characteristic statement:

> We have defined *sensation* as the first mental act that occurs through the immediate transformation of physical nervous processes in a still unknown manner, and which, as the most elementary process of a psychological kind, may not be further analyzed. We have shown in

great detail how perception is generated from sensation by way of logical processes.[7]

Wundt here distinguishes between psychologically primitive elements and further mental events derived from them. The "logical" processes mediating between sensation and perception arise through learning. However, not everyone who agreed that perceptions were mediated considered the mediating processes to be acquired; Tourtual and Lotze posited innate psychological mechanisms mediating between sensation and perception. Moreover, not all authors who considered the mediating processes to be acquired characterized them as "logical" (intellectual or judgmental); among authors who provided associational accounts of the acquisition process, Berkeley and Steinbuch distinguished associative processes from the intellectual operations of the understanding.

Because classifying an aspect of visual experience as a sensation in the psychological sense implies only that it is the immediate product of sensory stimulation, the psychological distinction between sensation and perception is potentially, though not necessarily, orthogonal to the epistemological one. On the one hand, if one believes that the immediate effect of visual stimulation is a mental representation of the retinal image, then one is committed to there being *spatial* sensations (a contradiction in terms in Reid's usage). On the other hand, if one does not believe that there are spatial sensations, then the psychological distinction between sensation and perception parallels the epistemological division between ideas of secondary qualities and ideas of primary qualities.

Further consideration of the psychological distinction between sensation and perception can help clarify the possible relationships between nativism and rationalism, and empirism and empiricism. As observed in Appendix A, it is possible to be a nativist without being committed to rationalism. A nativist about spatial perception could conceivably assert that the ability to perceive space is innate without thereby being committed to the postulation of innate knowledge of any type. The nativist might interpret the innate mechanisms responsible for spatially ordered visual experience by analogy with the innate mechanisms responsible for color sensations. Color sensations have generally been thought to be generated as a result of physiological, noncognitive mechanisms; a given stimulus affects the nervous system in one way to produce a sensation of red, while a different stimulus produces a sensation of green. The innate mechanism for generating these sensations conforms to specifiable rules describing the relation between the stimulus and the phenomenal character of the sensation, and yet no one believes that these rules are innately understood by the percipient or are used by the percipient's visual system in producing color sensations. Innate mechanisms for spatial vision might be viewed in the same manner, as in Descartes' "psychophysical" account of distance perception. Since it is possible to construe the nativist's hypothesized innate mechanisms for generating spatial perception as noncognitive, conceptual considerations do not drive the nativist to the doctrine of innate knowledge. Innate mechanisms for spatial perception need not constitute innate knowledge, and hence knowledge independent of sensory experience need not be attributed to percipients. The same point can be extended to the division between empirism and empiricism. The learning mechanisms underlying the acquisition of spatial perception can be construed noncognitively, as Berkeley and Steinbuch did.

The connection between spatial sensations and nativism implied by Helmholtz now emerges more clearly, as does the connection between empirism and the denial of spatial sensations. In its psychological sense, "sensation" refers to the mental elements of the sensory process, those elements out of which, by one process or another, everyday perceptual experience arises. In accordance with this usage, an empiristic position is one that minimizes the degree to which our perceptual experience is supposed either to be given directly in sensation or to be derived from sensation by innate processes. The empirist program is to show how we acquire the ability to perceive a three-dimensional visual world, given that we begin with a set of visual sensations that are nonspatial (or at the most, organized in only two dimensions). By contrast, a nativist might seek to show that there are innately given spatial sensations. The division into nativists and empirists would thus reduce to a disagreement over the spatiality of sensations. However, the division need not reduce to this disagreement. Some nativists have agreed that sensations are nonspatial; they have postulated innate processes for producing the perception of a three-dimensional visual world starting from nonspatial sensations. Thus, within either an empiristic or nativistic framework, one may adopt various positions, depending on the attributes one assigns to sensations and the character of the processes one postulates as mediating between sensation and perception.

Notes

Chapter 1, Introduction

1. Examples of the epistemic sense of the term "normative" may be found in Harman, "Epistemology," 47, and Putnam, *Realism and Reason*, 245–7.

2. The characterization of verbs such as "to perceive," "to understand," or "to know" as "success" verbs, the application of which implies an achievement (or the capacity for one), became widespread following Ryle, *The Concept of Mind*, chs. 2, 5. Ryle maintained that ordinary cases of, say, perceiving require no explanation and are not subject to empirical study, although cases of misperception might be (ch. 10).

3. On the attempt to naturalize the mental without eliminating it, see Fred Dretske, *Knowledge and the Flow of Information*, Preface; Dennett's Foreword to Millikan, *Language, Thought, and Other Biological Categories*, as well as Millikan's Introduction and ch. 5; and Fodor, *Psychosemantics*, 98–9. For an enthusiastic prediction of the demise of the mental, see Patricia Churchland, *Neurophilosophy*, chs. 7–9.

4. Descartes needed to refute the Aristotelian metaphysics and its attendant theory of knowledge in order to establish the metaphysical foundations of his own physics. He attempted to refute the Aristotelian theory of mind and establish his own theory of mind—together with a justification of the claim that the mind can intuit the essence of matter—through the cognitive exercises contained in his *Meditations on First Philosophy* (published in Latin, 1641; English trans., Cottingham, Stoothoff, and Murdoch, *Philosophical Writings of Descartes*, vol. 2). See my "Senses and the Fleshless Eye."

5. The tendency to give an account of the mind of the knower in connection with a discussion of scientific knowledge was especially pronounced among persons who are regarded as philosophers; hence, one finds little of this type of discussion in Galileo or Newton (the Queries to the *Optics* notwithstanding), but a great deal of it in Descartes, Locke, Berkeley, Hume, and Kant. Among the major early modern philosophers, Spinoza and Leibniz stand out because they did not make the theory of mind, or of human understanding, fundamental. In this regard, Helmholtz would count as both physicist and philosopher.

6. I leave aside the cosmological posturings of ancient atomists such as Epicurus and Lucretius. Although they certainly may be seen as urging the "naturalization" of mind, they did not have the idea of developing a general natural science of the mind, perhaps because they had no examples of a comprehensive natural science such as those provided by Descartes and Newton.

7. The best known call for a naturalistic epistemology is Quine's "Epistemology Naturalized," in *Ontological Relativity and Other Essays* (New York: Columbia, 1969),

69–90; Quine is avowedly nonfoundationalist (74–6). Goldman, *Epistemology and Cognition*, has mounted a quite different case for the relevance of psychology to epistemology, based upon the claims of cognitive science, as has Giere, *Explaining Science.*

8. Churchland (*Neurophilosophy*, 46) has suggested that study of the nervous system might lead to the discovery of constraints on human knowledge of the type proposed by Kant, inasmuch as the sensory receptors constitute a filter between the mind and the world—one example being the limited sensitivity of the eyes to a small spectrum of electromagnetic radiation. Of course, physical theory has long posited regions of the electromagnetic spectrum for which the eyes are not sensitive. For the claims of scientists about the implications of naturalistic theories of mind, see Reidel, *Biology of Knowledge,* and the literature he cites.

9. Boring, *History of Experimental Psychology*, 2d ed., devoted equal space to "prepsychological science" (the origin of psychology in sensory and motor physiology) and to "prescientific psychology" (the origin of psychology in early modern philosophy). Subsequent histories have tended to devote somewhat more attention to the the latter than the former; this holds true both of histories largely inspired by Boring, such as Schultz and Schultz, *History of Modern Psychology*, 4th ed., and of those that are more independent, such as Robinson, *Intellectual History of Psychology,* rev. ed., and Lowry, *Evolution of Psychological Theory,* 2d ed., the latter nonetheless including an interesting discussion of the relation between psychology and the rise of modern science. By the way, Ebbinghaus' remark is the first sentence of the introduction to his *Abriss der Psychologie* (Leipzig, 1908).

10. Some histories of psychology allow that early modern philosophical discussions of the mind differed from later psychological treatments not only in method but in aim, inasmuch as the focus of attention was the theory of knowledge and the "question of the validity of knowledge" rather than the "empirical search for laws of mental functioning"—Watson, *Great Psychologists,* 219. But all the histories I have examined, including Watson's, agree in treating the problems, solutions, and conceptual vocabulary of philosophy and psychology as continuous. One of the the most explicit—and most sophisticated—is Gardner's *The Mind's New Science,* ch. 4.

11. In speaking of "borrowing," I purposely leave open (for reasons that will soon be apparent) the question of whether such borrowings were a direct translation of problems and concepts from philosophy into psychology. Psychology also borrowed conceptually and adapted laboratory methods from physics and physiology; see Marshall, "Physics, Metaphysics, and Fechner's Psychophysics," and Turner, "Helmholtz, Sensory Physiology, and the Disciplinary Development of German Psychology."

12. See Boring, *Sensation and Perception,* 28. This treatment is very widespread—see Lowry, *Evolution of Psychological Theory,* 18–21, 65–6, 80; Wertheimer, *Brief History of Psychology,* rev. ed., 37–8, 41; and Leary, "Immanuel Kant and the Development of Modern Psychology." Even historians of psychology who show some appreciation of the ways in which Kant's work was not psychological emphasize his "nativism" or doctrine of innate ideas and suggest that his attitude toward the a priori, insofar as it was not psychological, was a hindrance to progress—see Robinson, *Intellectual History,* 284. Pastore, *Selective History,* and Morgan, *Molyneux's Question,* are exceptions. Pastore scrupulously avoids discussion of what he acknowledges to be the philosophical side of authors such as Descartes and Locke, and he avoids discussion of Hume and Kant altogether. Morgan emphasizes that the aims of Locke and Kant were distinct from those of present-day psychology (2, 128–30), and he largely eschews discussion of their philosophical sides.

13. Hochberg, "Nativism and Empiricism in Perception," and McCleary, "Ancient History and Recent Past," in McCleary, ed., *Genetic and Experiential Factors in Perception*, 1–7; see also Hothersall, *History of Psychology*, xv-xvi. An exception is O'Neil, *Beginnings of Modern Psychology*, 2d ed., ch. 2.

14. A textbook rendition of the above picture of the role of skepticism in modern epistemology may be found in Dancy's *Introduction to Contemporary Epistemology*, 1–2. This picture remains influential, and provides one guiding thread through Rorty's popular *Philosophy and the Mirror of Nature*.

15. Green, *Works*, ed. Nettleship, 1:7–8, 159–166, charges Locke and Hume; Ward, "Psychological Principles," levels the charge against Locke, Berkeley, and Hume, as does Sellars, "Empiricism and the Philosophy of Mind," 131, 162–9; Rorty, *Mirror*, 3–4, 126, and ch. 3, names Locke, Hume, and Kant. Hamlyn, *Sensation and Perception*, having made clear his view that philosophical problems are conceptual, not empirical (60, 70, 185), accuses Descartes and Hume of confusing psychological with philosophical concerns, but not Locke (60, 136, and 94, respectively); he also charges Helmholtz with confusing a question about the experience of space with a question about the concept of space (153–5). The authors named here do not apply the term "psychologism" to the fallacy they describe.

16. For example, Hamlyn, *Sensation and Perception*, 60.

17. Goldman, *Epistemology and Cognition*, 6; the attitude described is implicit in Fodor, *Representations*, 1, 273–83.

18. Hothersall provides a clear statement of the use of the history of psychology to impart disciplinary identity: "Knowledge of psychology's past allows students to see, sometimes for the first time, psychology as a whole. The history shared by all psychologists distinguishes them from other social and behavioral scientists and influences current research. In addition, the origin of most contemporary psychological issues, concerns, and questions can be found in psychology's past, which provides the student with a context and framework for understanding recent developments" (*History of Psychology*, vll). See also the discussion and quotations in Wertheimer, "Historical Research— Why?"

19. This tactic is apparent in "systems and theories" introductions to psychology—see Chaplin and Krawiec, *Systems and Theories of Psychology*, 3d ed., chs. 1, 4. It has been adopted in the deservedly successful introductory textbook of *Basic Psychology* by Henry Gleitman (2d ed.), which cites Locke more frequently than anyone except Freud, Pavlov, and Piaget (and Descartes, Hobbes, and Berkeley more frequently than most), and which uses Locke and Kant to introduce the distinction between nativism and empiricism. The technique of motivating a problem in psychology by stating its origin in an early modern philosopher is too widespread to require documentation—readers interested in a recent example might examine Katz, *Cogitations*. Consider also Chomsky, *Cartesian Linguistics*, and *Knowledge of Language*, Preface, on "Plato's problem." Finally, Gardner, in his *Mind's New Science*, lists "rootedness in classical philosophical problems" as one of five defining criteria of cognitive science (7, 42–3). This statement in based primarily upon his wide reading in the current literature, and he acknowledges that a significant number of cognitive scientists would not include philosophical rootedness among the defining criteria of their field; nonetheless, Gardner is certainly correct that the names of such philosophers as Locke and Kant are frequently cited in the literature of cognitive science by authors who claim to have refuted or confirmed a statement or a general position from their writings.

20. A full discussion of the paired terms nativism-empirism and rationalism-empiricism is provided in Appendix A, and an elaboration of epistemological and psychological senses of sensation and perception is presented in Appendix B.

21. "Psychologism" is a nineteenth-century term and is misleading when applied to seventeenth and eighteenth-century authors. Dewey, writing in Baldwin, ed., *Dictionary of Philosophy and Psychology*, (New York, 1901-05) provided the following definition of the term: "The doctrine of Fries and Beneke . . . which translates the critical examination of reason (of Kant) into terms of empirical psychology" (2:382). The German historian of philosophy Johann Eduard Erdmann is widely credited with originating this use of the term; he applied it to Beneke, though it also fit his discussion of Fries: the appellation appeared in *Grundriss der Geschichte der Philosophie*, 2d ed. (Berlin, 1870), 2:636; on Fries, see 2:378–9. Fries's work will be discussed in chapter 4. The conceptual framework in which it would make sense to accuse a philosopher of psychologism is older than the term; the framework was provided in the works of Tetens, Kant, and Herbart, among others (see chapters 3 and 4).

22. Some philosophers may associate the charge of "psychologism" with the reaction of Frege against Mill and others, thus limiting its application to the attempt to found mathematics and logic on psychology; however, in its original use, and in its scattered use throughout the twentieth century, the meaning of the term extended to the "naturalistic fallacy" of attempting to found epistemology on psychology. See the entry in vol. 6 of Edwards, ed., *Encyclopedia of Philosophy*, and the Introduction to Kornblith, ed., *Naturalizing Epistemology*, 8–9.

23. For a recent discussion, which contains references to several others, see Brown, "Normative Epistemology and Naturalized Epistemology."

24. For an interesting discussion of the origin of the notion of the statistical norm, see Hilts, "Statistics and Social Science."

25. Dewey, "Naturalism," in Baldwin, ed., *Dictionary of Philosophy and Psychology*, 2:137–8. Dewey also contributed the entries for "Nature" and "Natural"; for the latter term, he emphasized the sense of "regular" or "usual" (or "normal"!).

Chapter 2, Descartes to Hume

1. See Park and Kessler, "Psychology."

2. Although Aristotle set the perception of spatial properties apart from the perception of color, sound, odor, etc., denoting the former as "common sensibles," he provided no theory of size or distance perception. The chief discussions of such psychological topics occurred in the optical tradition; among the most extensive prior to the seventeenth century was that of the eleventh-century Islamic philosopher Alhazen, on which see Sabra, "Sensation and Inference in Alhazen's Theory of Visual Perception."

3. Standard Aristotelian doctrine held that the human intellect requires a bodily phantasm to provide materials for thought but that the operations of the intellect itself were immaterial; see Kessler, "The Intellective Soul."

4. There were seventeenth-century authors who foresaw extending a mechanistic account to include the intellectual powers of the mind, including, most notably, Hobbes; but the anathema with which his program was met suggests that he is one of those exceptions that proves a rule. On the reaction to Hobbes, see Mintz, *Hunting of Leviathan*.

5. According to one widely received opinion, the human soul is the form of the body, which makes it the form of a corporeal thing, and in that sense part of nature (even if the intellect, which is a power of the soul, is not the "act" of any bodily organ)—

Aquinas, *Summa Theologica*, Pt. I, qu. 76, art. 1. Others, however, following Avicenna and Averroes, argued that either the agent intellect alone, or both the agent and patient intellects, are one for all humans, perhaps standing outside the created world (according to those who equated these intellects with God himself); see Kessler, "The Intellective Soul." In the late seventeenth and early eighteenth centuries the Cambridge Platonists and other British authors refused to count the soul (or the intellect) as "natural" by contrast with "supernatural," insisting that the soul is divine. Ephraim Chambers, *Cyclopaedia: or, An Universal Dictionary of Arts and Sciences*, 2d ed. (London: 1738), defined "soul" as follows: "the philosophers, many of them, allow of two, others of three kinds of *souls*, viz., a *rational* soul, which they hold to be divine, and infused by the breath of God. See Reason. . . ." (vol. 2); see also Zachary Mayne, *Two Dissertations Concerning Sense and the Imagination* (London, 1728), 226–7.

6. Descartes appeals to the "light of nature," or the "natural light," at several points in the third, fourth, and sixth of his *Meditations on First Philosophy*, in *Philosophical Writings of Descartes*, trans. Cottingham, Stoothoff, and Murdoch, 2:27–8, 32, 41–2, 57 (hereafter cited as CSM plus volume and page numbers). He contrasts the light of nature with the light of grace in the second set of *Objections and Replies* to the *Meditations*, in reply to the fifth objection (CSM 2:105).

7. Many Aristotelian authors identified the soul with the *form* of humans, whose nature was taken to include both form and matter; but this qualification does not affect the spirit of my point.

8. *Principles of Philosophy*, Part III, in CSM 1. References to this work will be by part and article number.

9. Descartes may be said to have engaged in "physiological psychology" in the *Treatise on Man*, inasmuch as that work presents a model of the bodily basis of sensation and imagination (as well as circulation, digestion, sleep and waking, and other organic functions); but the soul and the "mental" side of sensation and perception are excluded from the treatise (36, 86–7, 95, 112–113).

10. Intellectual historians rightly warn that attributions of "materialism" to seventeenth and eighteenth authors must be made with care—for instance, one must guard against being taken in by charges of "materialism" leveled against an author by his opponents, for such charges may reflect no more than a perception of heterodoxy. I have restricted the label to those who held that the universe contains only matter, whether they did so on the basis of metaphysical argumentation or merely by way of indicating what they considered to be the most likely hypothesis. For Hobbes' materialist pronouncements, see *De Corpore* (London, 1655), ch. 25, art. 2; his *Objections* to Descartes' *Meditations*, second and fourth objections (CSM 2); and *Leviathan* (London, 1651), Pt. I, ch. 1. See, further, La Mettrie, *Man a Machine* (published in French in 1748), 85, 140–1, 148–9; d'Holbach, *System of Nature*, trans. Robinson, vol. 1, chs. 6–7; and Priestley, *Disquisitions Relating to Matter and Spirit* (London, 1777), Preface, Introduction, sec. 3.

11. Although Hobbes, d'Holbach, and Priestley espoused associationism, they did not produce the detailed accounts of mental life characteristic of methodological naturalism. Diderot was an exception among the materialists; see Anderson, "Encyclopedic Topologies."

12. Hartley, *Various Conjectures on the Perception, Motion, and Generation of Ideas* (Bath, 1746) and *Observations on Man* (London, 1749). On Hume, see the following paragraphs.

13. Mayne's *Two Dissertations* exemplifies this common assumption about sense and imagination and argues against the reduction: "if it can be proved to satisfaction

... that the faculties of perceiving, which brutes have, as well as men, namely sense, and imagination, are not intellectual, or the mind's powers of understanding things; it undeniably follows that the human intellect must differ essentially, or in its very nature, from the perception and capacity of brutes" (unnumbered page from the preface). Mayne argues against the reduction on the basis of the characteristic activities of the understanding, including that of "consideration," or judgment, which he claims are absent from what he calls the "perceptive powers," sense and imagination (63–5, 92–8). In the attached essay on consciousness, Mayne argues for the immateriality of the soul from the unity and indivisibility of consciousness and from the fact of selfconsciousness, which he claims are inconsistent with the equation of the soul with body (214–18).

14. *Treatise of Human Nature,* (London, 1739–40), xx (page references are to L. A. Selbe-Bigge's 1888 reprint). Among commentators, emphasis on Hume's naturalism has been in and out of favor, and has come into favor again; until recently, the predominant treatment of Hume's work among philosophers emphasized his skepticism. See sec. 3 for further discussion.

15. Ibid., xxi. See also Hume's *Inquiry Concerning Human Understanding* (London, 1748), sec. 1.

16. On mental geography and dynamic laws, see *Treatise,* Bk. I, Pt. i, secs. 1–4, and *Inquiry,* secs. 1–3; on punctiform sensations in vision, *Treatise,* Bk. I, Pt. ii, sec. 1–3 (hereafter cited by book, part, and section, as in I.ii.1–3, plus page numbers where needed).

17. *Treatise,* I.iii.7 (97, 628–9), and also I.iii.13 (153), I.iv.7 (265); *Inquiry,* sec. 5.

18. *Treatise,* III.i.1 (458).

19. *Treatise,* I.iii.9 (108); *Inquiry,* sec. 4, pt. 1, para. 3.

20. There are two points about the truth-neutrality of belief formation in Hume that should be kept distinct. First, as had many philosophers, he contended that the untutored or "natural" (as opposed to artificially trained) habits of mind can easily lead into error; see the section on "unphilosophical probability" in the *Treatise* (I.iii.13) and the discussion, in *Inquiry,* sec. 10, pt. 2, of the passions that naturally incline some to the belief in miracles against, on Hume's account, the weight of evidence. But, second, Hume further contended that the most judicious attempts to determine non-occurrent sensory impressions was also liable to failure; or, as he also put it, the success of such attempts was a wholly contingent matter, there being no necessary connection between the "vivacity" of our ideas and the truth. As an example of this contingency, consider the associative law of cause and effect (the law that the constant conjunction of two impressions or experiences will lead to an associative link, such that upon presentation of the one in sensory experience, the other is drawn forth in the imagination and so is expected in sensory experience). Clearly, this associative law is truth neutral, even when applied in accordance with Hume's "rules" for judging causes and effects (*Treatise,* I.iii.15). Although Hume spoke of a "preestablished harmony" arising between expectations based upon past constant conjunction and the course of future experience, he recognized that such harmony was, from the standpoint of reason, fortuitous (*Inquiry,* sec. 5, pt. 2, end; see also *Treatise,* Bk. I, Pt. iii, sec. 6 [91–3]).

21. In the *Treatise* Hume argued that intuitive certainty extends to arithmetic and algebra but not to geometry, although he maintained that the latter "excels the imperfect judgments of our senses and imaginations" (I.iii.1, 69–71); in the *Inquiry,* he allowed that arithmetic, algebra, and geometry all admit of intuitive and demonstrative certainty (sec. 4, pt. 1). As Michael Ayers has pointed out to me, Hume's associative account of abstract ideas—developed in the *Treatise,* I.i.7—was

an inroad into the the traditional domain of the intellect, and Hume used it to argue against special nonsensory *objects* of mathematics; he seems not, however, to have wholly avoided positing special powers of intuitive perception as directed toward ideas originating in the senses (*Treatise*, I.iii.1; *Inquiry*, sec. 4, pt. 1). Indeed, the perception of truth—through perception of the "resemblance" or "contrareity" of ideas—would seem to be the most fundamental of such powers.

22. *Inquiry*, sec. 4, pt. 1, first two paragraphs.
23. Hartley, *Observations on Man*, 1:325 (Pt. I, ch. 3, sec. 2, prop. 86).
24. An example of the liaison from experiment to experience is found in Chambers, *Cyclopaedia*, entry under "experiment" (vol. 1), which gives the following as one definition (attributed to the Schools): "a comparison of several things observed by the senses, and retained in the memory; in some one similar convenient instance. See Experience."
25. *Essay Concerning Human Understanding*, ed. Nidditch, Bk. I, ch. i, art. 2 (citations hereafter contract book, chapter, and article numbers into one, as in I.i.2).
26. The phrase is suggested by Rorty in his *Philosophy and the Mirror*, 137, 140.
27. Locke, *Essay*, II.xii.
28. Ibid., II.xxiii.6–12; III.vi.8–20.
29. See Stich's introduction to Stich, ed., *Innate Ideas*.
30. As an example of an argument for the relevance of evolutionary etiology, see Goldman, "Innate Knowledge."
31. On the optical tradition, see Lindberg, *Theories of Vision*, and Smith, "Getting the Big Picture in Perspectivist Optics."
32. On Alhazen's optics and its connection with a theory of knowledge, see Sabra, "Sensation and Inference," and Smith, "Big Picture." The seventeenth century Jesuit Frances Aguilon included summary discussion of the role of vision in cognition, following Aristotelian lines, in his textbook *Opticorum libri sex* (Antwerp, 1613), 94-100.
33. For an example of this sort of account of psychophysical interaction, see Descartes' *Optics*, in his *Discourse on Method, Optics, Geometry, and Meteorology*, trans. Olscamp, 101, and Fourth Discourse.
34. This three-part division of the theory of vision is implicit in Descartes, *Objections and Replies*, in CSM 2:294–296 (the "three grades of sensory response" distinguished by Descartes do not directly map onto the tripartite schema given above; my physical stage is prior to Descartes' exclusively physiological first grade of sense—but it is a stage he could accept—and his second and third grades are a division within what I label the mental stage), and in Newton, *Optics*, 4th edition (London, 1730), Queries 12, 23, 28. This division became the standard organization for treatises on the theory of vision; e.g., Cornelius, *Theorie des Sehens und räumlichen vorstellen* (Halle, 1861), and Helmholtz, *Handbuch der physiologischen Optik* (Leipzig, 1867). For a characteristic description of the causal chain in perception as it was conceived in the eighteenth century, see Reid, *Inquiry into the Human Mind* (London and Edinburgh, 1764), ch. 6, sec. 21, in his *Works*, ed. Hamilton (Edinburgh, 1895), vol. 1; Reid describes this chain under the heading "Of the process of nature in perception."
35. The term "sensory given" as used here implies psychological as opposed to epistemological givenness. The sensory given comprises the first properly mental stage of perception, which often was described as sensation; on epistemological as opposed to psychological senses of the terms "sensation" and "perception," see Appendix B.
36. See Reid, *Inquiry*, ch. 6, sec. 21.

37. See Hatfield and Epstein, "The Sensory Core and the Medieval Foundations of Early Modern Perceptual Theory."

38. Hume, *Treatise*, I.ii.5 (56); Reid, *Inquiry*, ch. 6, sec. 2; Lambert, *Neues Organon* (Leipzig, 1764), 2:219–21 (Phänomenologie, secs. 3–5); Tetens, *Philosophische Versuche* (Leipzig, 1777), IV.vi.3, VI.i.3 (1:343, 439). On the place of Lambert and Tetens in German philosophy, see Beck, *Early German Philosophy*, ch. 16.

39. Although Berkeley was precluded from providing a causal account of the relation between the material retinal image (considered as an object of touch) and the "proper object" of vision (light and color with visual form and magnitude), in secs. 50–7 of his *Theory of Vision Vindicated and Explained* (London, 1733) he engaged in an heroic effort to explain the "proportion" that must hold between the two, given the obvious facts of empirical correspondence between the character of physical (tactile) events at the retina and temporally subsequent visual experience.

40. See Hatfield and Epstein, "Sensory Core," 380–2.

41. Gehler, *Physikalisches Wörterbuch* (Leipzig, 1787–1796), 6 vols. Gehler's primary interest lay in chemistry, but he discussed visual theory with clarity and comprehension. I have chosen to use his work in characterizing eighteenth century visual theory not because it was in any way original but because it provides a representative account of accepted doctrine near the end of the century. In composing the articles on vision in his dictionary, Gehler drew from a variety of seventeenth and eighteenth century authors writing in English, French, and German. This breadth of citation—both temporal and linguistic—was characteristic of writings on vision throughout the eighteenth century.

42. Gehler, *Physikalisches Wörterbuch*, 4:11–12.

43. Müller's views are discussed in chapter 4; Descartes' most explicit discussion of the process of physiological transmission may be found in the *Treatise on Man*, trans. Hall, 83–86, but see also the *Optics*, trans. Olscamp, 100; his most explicit discussion of the character of the immediate sensory given occurs in the sixth set of *Objections and Replies* (CSM 2:295). For additional discussion of these and other passages from Descartes, see Hatfield and Epstein, "Sensory Core," 374–7.

44. Gehler, *Physikalisches Wörterbuch*, entries under "Entfernung" and "Gröss, Scheinbare." On Alhazen, see Sabra, "Sensation and Inference"; other authors who provided accounts of what have come to be known as the "constancies" of size and shape include John Pecham (born ca. 1230, died 1292), *Perspectiva communis*, ed. and trans. Lindberg, 145–7, and Descartes, *Optics*, trans. Olscamp, 107.

45. Berkeley, *New Theory of Vision*, sec. 2. Berkeley's statement is virtually a paraphrase of a sentence found in prop. 31 of Molyneux's *Treatise of Dioptricks* (London, 1692), 113. For further discussion, see Pastore, *Selective History*, chs. 4–5, and Donagan, "Berkeley's Theory of the Immediate Objects of Vision."

46. See, for example, Reid, *Inquiry*, ch. 6, sec. 21.

47. For characteristic instances of lists of "signs" for distance perception, see Reid, *Inquiry*, ch. 6, sec. 22, and Porterfield, *Treatise on the Eye*, Bk. 5, ch. 5.

48. It was common to distinguish "inflow" and "outflow" theories of proprioception, even in the eighteenth century; see Reid, *Inquiry*, ch. 6, sec. 22.

49. Although it is clear that Descartes posited a psychophysical relation between brain events and sensory perceptions, it is less clear that this relation should be conceived as one of direct causal interaction. Several ways of conceiving this relation are canvassed by Smith, *New Studies in the Philosophy of Descartes*, ch. 6. Moreover, it may be observed that Descartes' metaphysics of mind and body had deeply entrenched occasionalist tendencies; see Specht, *Commercium mentis et corporis*, ch. 2, and my "Force (God) in Descartes' Physics," especially 135–7.

50. For Descartes' account of distance as directly elicited by brain events, see *Optics*, trans. Olscamp, 104–6, and *Treatise of Man*, trans. Hall, 94–100; for his invocation of judgment in the perception of size, shape, and distance, see *Optics*, 107, and *Objections and Replies*, CSM 2:295. Pitcher, *Berkeley*, discusses the general characteristics of a psychophysical account (19–20), without attributing such an account to Descartes.

51. There are two sources for this mode of explanation by eighteenth century optical writers. One is the appeal to faculties in explaining cognition that was characteristic of writing on the soul in general; seventeenth century writers could appeal to the authority of Aristotle and Augustine in positing such faculties. This type of faculty psychology reached its pinnacle in the work of Wolff; see Dessoir, *Geschichte der neueren deutschen Psychologie*, 2d ed., 377–91. In the optical tradition, the appeal to faculties took on a life of its own (even if it originated in the same ancient authorities that dominated writing on the theory of the soul in general). A tacit appeal to faculties in the optical tradition derived from Ptolemy's and Alhazen's use in their theories of adjudication as a primitive psychological operation; later optical writers simply accepted the notion that adjudication ("estimation," "judgment," etc.) is a primitive mental process involved in the perception of distance and other visual properties. Summers, *Judgment of Sense*, recounts the history of such theories in relation to aesthetic judgment.

52. Gehler, *Physikalisches Wörterbuch*, 4:13–14. Dugald Stewart gave a similar account in the *Outlines of Moral Philosophy* (Edinburgh, 1793), Pt. 1, sec. 2, art. 1 (*Collected Works*, ed. Hamilton, 2:16): "Prior to experience, all that we perceive by this sense [vision] is superficial extension and figure, with varieties of colour and of illumination. In consequence, however, of a comparison between the perceptions of sight and of touch, the visible appearances of objects, together with the correspondent affectations of the eye, become signs of their tangible qualities, and of the distances at which they are placed from the organ. In some cases our judgment proceeds on a variety of these circumstances combined together; and yet, so rapidly is the intellectual process performed, that the perception seems to be perfectly instantaneous." See also Hume, *Treatise*, I.ii.5, I.iii.9 (56, 112); Lambert, *Neues Organon*, 2:242–3 (Phänomenologie, secs. 43–5); Tetens, *Versuche*, VI.i.2–3 (1:43140).

53. *Physikalisches Wörterbuch*, 2:467.

54. Berkeley, *New Theory of Vision*, sec. 16. A philosophically penetrating account of Berkeley's theory of vision may be found in Pitcher, *Berkeley*, chs. 2–3.

55. Berkeley, *Theory of Vision Vindicated*, sec. 42. The distinction made so explicit here is less clearly drawn in the *New Theory of Vision*, in which Berkeley sometimes speaks unreservedly of visual judgments (secs. 20, 24, 27, 38, 41, etc.), but in which he explains the mental operations underlying these "rapid judgments" in terms of suggestion (secs. 17, 25, 53, etc.). A contrast between judgment and suggestion is more clearly implied in the *Treatise Concerning the Principles of Human Knowledge* (Dublin, 1710), Pt. I, sec. 43 (hereafter cited by Part and section).

56. *Principles*, I, 43; *New Theory of Vision*, secs. 17 and 51; *Theory of Vision Vindicated*, secs. 39 and 40.

57. Berkeley attributes the operation of suggestion to the imagination in the *Three Dialogues between Hylas and Philonous* (London, 1713), in his *Works*, eds. Luce and Jessop, 2:204, and again in the *Theory of Vision Vindicated*, secs. 9–10. This attribution is consonant both with his frequent denial that lines and angles are used to calculate distance and magnitude in vision and with his rejection of any intelligible connection (appropriate for the understanding) between the objects of sight and the objects of touch. Note that although Locke explains the transition from two to three dimen-

sions by contending that "the judgment presently, by an habitual custom, alters the appearances into their causes" (*Essay*, II.ix.8), it is *judgment* that acts habitually in such cases; Berkeley distinguished habitual suggestion from judgment.

58. Hartley, *Observations on Man*, 1:56–84 (Pt. 1, ch. 1, sec. 2).

59. Ibid., 1:75.

60. This chemical analogy for the association of ideas is well known in its nineteenth-century instantiations (e.g., J. S. Mill's "mental chemistry"); it is found among earlier writers as well, for example, Reid, *Inquiry*, ch. 1, secs. 2 and 4, and Kant, *Anthropology from a Pragmatic Point of View*, trans. Gregor, 52–3 (marginal page number to German edition: 177).

61. Reid listed three ways in which signs may be linked psychologically to their referents: "by *original principles of our constitution,* by *custom,* and by *reasoning*" (*Inquiry*, sec. 21, in his *Works*, ed. Hamilton, 1:188a). Of these three, the first accounts for our original perceptions; the principles of our constitution determine that a retinal image will give rise to a "visible appearance" that corresponds to the perspective projection contained in that image (sec. 2, 135–6). These appearances, together with information from the ocular muscles, constitute the signs for distance. Our experience of the connection between signs for distance and tactual perception of distance "produces that habit by which the mind, without any reasoning or reflection, passes from signs to the conception and belief of the thing signified" (sec. 21, 188b). Reid followed Berkeley in distinguishing between those mental connections between ideas that arise from custom and habit, and those that result from ratiocination (sec. 22, 191 and sec. 23, 193b). (However, although Reid credited Berkeley with the origin of the doctrine that visible appearances are signs—sec. 2, 135a—he did not attribute the above tripartite division to Berkeley.) Reid did not delve further into the nature of the process by which ideas become conjoined through custom, other than to remark that it is simply a property of our nature, "acknowledged by all, that when we have found two things to have been constantly conjoined in the course of nature, the appearance of one of them is immediately followed by the conception and belief of the other" (sec. 24, 197b; see also 198–201).

62. Some form of the doctrine of association was assumed as a matter of course by most nineteenth century writers who discussed psychology, at least in Germany, France, and Britain. For a survey see Warren, *History of the Association Psychology*. On association in Germany, see Dessoir, *Neueren deutschen Psychologie*.

63. CSM 2:295.

64. CSM 2:296; see also 57. For discussion of the distinction between the third grade of sense and the mature judgments of the intellect, see sec. 2 of my "Senses and the Fleshless Eye."

65. *Meditations*, III, IV (CSM 2:26–7, 56–8). Of course, the "teachings of nature" also must be attributed to God's creative power; nonetheless, Descartes regularly distinguished the natural light from the teachings of nature and consistently warned that truth must not be expected from the latter, explaining that God instituted the teachings of nature for the preservation of the body, not for the perception of the foundations of metaphysics (CSM 2:56–7).

66. The *products* are normatively conceived, because *qua* judgments they are either true or false; the *processes* are, by hypothesis, indifferent to truth or falsity: their normal functioning may lead to true or false judgments (indifferently). Using this distinction between process and product, Hume gave a naturalistic account of belief formation, but retained a methodologically normative conception of truth itself (founded on the perception of relations between ideas), whereas Hartley was a naturalist even about truth.

67. Descartes attributes the error in such cases not to the body, but to the mind-body union (CSM 2:59); it is attributed to what he sometimes called the "institution of nature," by which a brain motion of a certain sort yields a sensation of a certain sort (*Optics*, trans. Olscamp, 101; see also Wilson, *Descartes*, 207–20). The "teachings of nature" depend upon this "institution of nature," for it is by this institution that sensations of color, which allow for discrimination among objects, and pains, which warn of bodily harm, are produced.

68. Rorty presents a similar diagnosis of the origin of modern philosophy in *Philosophy and the Mirror*, esp. chs. 1 and 3; he explicitly endorses Reid's criticism (60, n. 32, and 142–6). Of course, neither Rorty nor Reid thought that the term "idea" was new in the seventeenth century—just its use to denote an existent with the properties described above.

69. Green, *Works*, ed. Nettleship, 7–8, 165–6; Sellars, "Empiricism and the Philosophy of Mind," 131, 160–9; Rorty, *Philosophy and the Mirror*, 140 (Rorty cites both Green and Sellars). None of these authors uses the term "psychologism," but they level the appropriate charge.

70. Although the term "epistemology" is a nineteenth-century coinage, and the related phrase "theory of knowledge" also came into common use only in that century, nonetheless the central works of the major early modern philosophers may aptly be characterized as contributions to the theory of knowledge. In the case of Descartes, this interest was manifested as a concern for the proper method of acquiring knowledge; it involved an investigation of what might be termed the "cognitive powers" of the mind or soul, and especially of the intellect or understanding. Locke, Berkeley, Hume, and Kant each undertook projects at least nominally connected with that of Descartes, in that they wrote essays, treatises, inquiries, and critiques of human understanding, human knowledge, or human reason.

71. Various contemporary analyses of the conditions for knowledge are discussed in Pappas and Swain, eds., *Essays on Knowledge and Justification*.

72. The claim for intersubjectivity typically supposes a *ceteris paribus* clause, which may include assumptions about background knowledge or other cognitive features of the agents.

73. Yolton, in *Perceptual Acquaintance from Descartes to Reid*, has emphasized the contribution of theories of the senses to the doctrine of the veil of perception, while at the same time questioning the standard reading of such authors as Descartes and Locke, claiming that it has been overly influenced by Reid.

74. Newton, *Opticks*, 4th ed. (London, 1730). Johannes Müller, writing over a century later, remarked that "the sensory nerves are immediately sensible only of their own states, or the sensorium is sensible of the states of the sensory nerves"; *Handbuch der Physiologie des Menschen* (Coblenz, 1840), 2:262.

75. Descartes, *Optics*, trans. Olscamp, 101. Descartes did not always speak of the brain as causing sensations in the mind; on occasion, he spoke as if the mind directly perceives the states of the brain, e.g., *Objections and Replies*, CSM 2:295.

76. Porterfield states the latter point as follows: "Having shewed, that the Mind does not see any Pictures in the *Retina*, or judge of Objects from what it observes in these pictures, it will be asked, what Connection there then is betwixt Vision and these Pictures, and how it comes to pass, that Objects are seen perfectly or imperfectly, accordingly as these Pictures are perfect or imperfect. To this I answer, That tho' all our Perceptions are Modifications of the Mind itself arising from the Motions or Vibrations excited in the *Sensorium*, to which the Mind is present; yet the Mind never considers them as such, but always ascribes them, either to the object, the Organs, or both" (*Treatise on the Eye*, 1:371–2).

77. The notion of "resemblance" as applied to the relation between image and object has been called into question in recent years, most notably by Goodman in *Languages of Art*, and it remains to be seen whether any notion of natural resemblance is defensible. Partially because of the difficulty of cashing out the notion of resemblance, several Locke commentators have suggested reading the *Essay* in such a way that the distinction between primary and secondary qualities, and indeed other discussions of the representational relation between ideas and objects, is cast into language that does not invoke the notion of resemblance—see, e.g., Curley, "Locke, Boyle, and the Distinction between Primary and Secondary Qualities." But Locke did in fact repeatedly invoke the notion of resemblance in his discussion of these topics (see, e.g., *Essay*, II.viii.7, 15, 22, 25; II.x.7; II.xii.17), as had Descartes in his discussion of corresponding topics. In Descartes' case, his primary use of "resemblance" is to deny that ideas of color, sound, etc. resemble anything in their external causes—see, e.g., *Meditations*, III and *Principles*, I, 66, 69–71—although the comparison of color with, say, shape in such passages suggests that the notion of resemblance at least makes sense in the case of shape, even if the resemblance generally is not perfect—e.g., an oval projected into the eye, transmitted to the brain, and sensed by the mind may lead to the judgment that a circle is present—as Descartes explains in the *Optics*, Sixth Discourse (Olscamp, 107). For further discussion of Locke on resemblance, see Alexander, *Ideas, Qualities, and Corpuscles*, ch. 10. Even if we ultimately reject the substantive claims of Descartes and Locke that certain ideas accurately represent through resemblance, that will not change the fact that this claim played a central role in their attempts to justify the new science.

78. Even if no precise criterion of "likeness" can be given— e.g., in a given case, are we speaking more of capturing a person's spirit, or of accurately depicting the bone structure of her face?—the notion still has its use; judgments of likeness can be made and will allow of discussion, disagreement, and the presentation of evidence, in the sense of drawing a viewer's attention to various aspects of the image and of the visage of the person in question.

79. I am taking the notion of the *tertium quid* to imply a real existent, separate from a state of mind, and separate from external objects. This understanding accords with classical discussions of "sense data" (and related terms), and with expositions and criticisms of the notion of an idea in Descartes, Locke, and others. It agrees with what Broad, *Scientific Thought*, called the "object theory" of sensible appearance (in opposition to the "multiple relation theory," 237), which he defended at some length (chs. 7–8, 13). Several authors have proposed versions of this conception of sense data (notwithstanding disagreements among themselves and subsequent changes in position): Bertrand Russell, in "The Relation of Sense-Data to Physics"; Moore, in "The Status of Sensa," who also attributed this conception to Locke (195); and Price, *Perception*, ch. 5. Although these theorists introduced sense data (or sensa, or sensibilia—the terminology can matter, but does not for the present point) in order to avoid skepticism (in some cases, by embracing phenomenalism), early modern philosophers have regularly been criticized, with varying degrees of severity, for abetting skepticism by positing "third things" as here conceived: see Reid, *Essays on the Intellectual Powers of Man* (1785), Essay 2, ch. 14, in his *Works*, ed. Hamilton, 1:298–306; Bennett, *Locke, Berkeley, and Hume*, 31–2, 124–5; Kenny, *Descartes*, ch. 5 (116, 122–5); and Rorty, *Philosophy and the Mirror*, 45–51.

80. Thus, after reading Descartes' formal definition of "idea" as "the form of any given thought, immediate perception of which makes me aware of the thought" (CSM 2:113), and Locke's similar remark that he uses the term "to stand for whatsoever is the object of the understanding when a man thinks" (*Essay*, I.i.8; see also II.viii.8), it

remains uncertain whether or not we should attribute to these authors the doctrine that ideas are separate from the mind that has them, constituting objects of perception analogous to the way in which external objects are objects of perception to the sense of sight. A decision about these matters must depend upon an appeal to a broader context.

81. Descartes, *Principles*, I, 53, 56, 65. Actually, Descartes considered God to be the one true substance; minds and body are substances in a derivative sense.

82. For a penetrating discussion of this point, see Sellars, "Berkeley and Descartes."

83. Second and Third *Meditations* (CSM 2:18–19, 28–30), and *Principles*, I, 66–8.

84. In the Sixth Discourse of the *Optics*, Descartes explains that the images in the eye (and hence in the brain) "usually contain only ovals and diamonds, yet they cause us to see circles and squares" (Olscamp, 107); the shape in the brain is different from the phenomenal shape, and so phenomenal shape cannot be accounted for by supposing that the shape in the brain is "inspected" by the nonextended mind. However, other passages would seem to militate against this conclusion, passages in which Descartes says that the mind, when imagining a shape, "turns towards the body and looks at something in the body which conforms to an idea understood by the mind or perceived by the senses" (CSM 2:51; see also *Treatise on Man*, 86–7). On the basis of such passages some commentators have proposed that Descartes considered shape in the brain to provide the figural content of shape perceptions; see, for example, Smith, *New Studies*, ch. 6. It must be admitted that Descartes' language is confusing, even contradictory, on this point. Thus, in a famous passage from the Sixth Discourse of the *Optics*, he explicitly warns against the view that we see by inspecting corporeal pictures in the brain "as if there were yet other eyes in our brain" (Olscamp, 101). Moreover, he re-affirms the point about the difference between retinal-image/brain shape and phenomenal shape, in the passage from the sixth set of *Objections and Replies* (CSM 2:295; quoted above in sec. 2), where he equates phenomenal shape with the shape that results from a rapid judgment based upon the "sensory given" (the latter being figurally congruent with the shape in the brain). No matter how one wishes to read Descartes, other philosophers attributed phenomenal shape (and color) to the content of ideas, without holding that the ideas are themselves actually extended (or colored); see Berkeley, *Principles*, I, 49; also secs. 10 and 99.

85. See Cronin, *Objective Reality in Descartes and Suarez*. The most extensive and the best discussion of objective reality and representational content in Descartes that I know is Sellars, "Berkeley and Descartes."

86. O'Neill, *Epistemological Direct Realism in Descartes' Philosophy*, has adopted this reading, as has Yolton, *Perceptual Acquaintance*, ch. 1. Rorty, who adopts the standard interpretation (or, the standard interpretation from the middle decades of the twentieth century), acknowledges Yolton's alternative reading with the remark that if it is correct, "we shall have to look further along in history for the emergence of what is now thought of as the epistemological problematic created by Descartes" (*Philosophy and the Mirror*, 49, n. 19). Reid notwithstanding, the real emergence of the problematic as Rorty describes it dates from the introduction of sense data by Moore, Russell, Broad, and others.

87. This reading was defended by Mackie, *Problems from Locke*, 37–51; see also Douglas Greenlee, "Locke's Ideas of 'Idea'." Instances of Locke's reifying talk include *Essay*, II.iii.1, where Locke speaks of ideas as things that "make their approaches to our minds," are conducted by the nerves "to their audience in the brain, the mind's presence room," and are perceived inasmuch as they "bring themselves into view"; also II.vii.10, II.x.14, etc.

88. *Essay*, II.ii.2. See also II.viii.7, in which Locke distinguishes between "ideas" as properties of objects (by which he means the properties in objects that regularly cause ideas of a certain kind in perceivers) and "ideas" as "perceptions in our minds." In these passages he does not treat ideas as things perceived in the mind's inner chamber but as perceptions the mind has.

89. Reid, *Intellectual Powers*, Essay II, ch. 11, in *Works*, ed. Hamilton, 1:292b. In ch. 6, sec. 3 of the *Inquiry*, Reid distinguished the "visible appearances" of objects, which are two-dimensional, from the objects themselves (*Works*, 1:135–7); since these appearances are perceptions (not sensations), by Reid's own account they have an object distinct from the mental act of perception itself. (For elaboration of the sensation-perception distinction as drawn by Reid, see Appendix B.)

90. See Aquila, *Intentionality: A Study of Mental Acts.*

91. In this connection, Berkeley's skepticism regarding matter, while invoking "veil of ideas" arguments, is in the end based on the comparative intelligibility of an alternative causal chain for the production of sensory ideas; he argues that *matter* cannot coherently be ascribed the causal role, whereas an infinite active spirit (God) can (*Principles*, I, 25–30).

92. The tendency to blame the possibility of skepticism on the "veil of ideas" doctrine may be partly explained by the fact that earlier in this century the positing of sense-data as third things was accepted as sufficient grounds for posing a skeptical problem, and that the writers who did so turned to Descartes, Locke, or Berkeley for brief stretches of text to illustrate one point or another about sense data; see, e.g., Ayer, *Foundations of Empirical Knowledge*, chs. 1–2.

93. Rorty, *Philosophy and the Mirror*, chs. 1 and 3. Bennett, *Locke, Berkeley, and Hume*, 67–9, disparages Locke's willingness to entertain general questions about representational accuracy.

94. The account of color perception by the reception of sensible species was not the central feature of the Aristotelian account of knowledge, which focused on the role of the intellect in abstracting the intelligible essence of objects, as opposed to merely accidental qualities, such as color. But the doctrine of color as a real quality, perceived through the intromission of species, was an abiding feature of Aristotelian theories of the senses. See Suarez, *De anima* (Madrid, 1978–81), disp. 7, qu. 1, art. 5–10.

95. Confusion can arise regarding exactly what the ideas of primary qualities are alleged to resemble. The "resemblance" is between sensory ideas and the "macro" properties of objects—my sensory idea of the table imagistically represents it as rectangular and as brown; the table actually is rectangular, but it is only derivatively brown (by virtue of other properties, such as the microscopic shape, or texture, of its surface). Confusion can arise because the justification for denying that the table is brown (primitively, not derivatively) involves appeal to the microscopic particles that mediate perception (e.g., that are reflected from the surface of the table into the eye). Ordinary sensory ideas of "macro" shape do not "resemble" these corpuscles, unless by happenstance (e.g., a sensory idea of a sphere resembles the shape of all spherical particles); yet the positing of corpuscles as causally efficacious in the case of color leads us to *deny* resemblance in that case, on the grounds that the corpuscular account tells us how to understand the physical property of being colored. I shall return below to the question of the warrant for holding that the real macro- and microproperties of objects are the primary qualities.

96. *Meditations*, CSM 2:44, 49.

97. *Principles*, I, 48, 53, 69. I have argued for this reading of Descartes' justification of the distinction between primary and secondary qualities in sec. 4 of "Senses and the Fleshless Eye."

98. In particular, Locke speaks of the primary qualities as properties that "the mind finds inseparable from every particle of matter" (*Essay*, II.viii.9); it is locally tempting to treat this as an a priori deliverance of the intellect.

99. The "materials of reason and knowledge," as Locke puts it (*Essay*, II.i.2), derive from sensation and reflection; but since ideas of reflection have as their source the operation of the mind in connection with sensory ideas, the latter are necessary if the mind is to have any "materials" whatsoever. That reason is unable to provide its own content independently of the senses is apparent from Locke's denial of innate ideas, as well as from his discussion of the operation of the rational faculties (II.xi).

100. II.viii.11–14 spell out the mechanical account of the operation of bodies upon the senses; sec. 1 asserts the resemblance thesis as a conclusion from this discussion. See also IV.iii.16, which Alexander discusses in "Boyle and Locke on Primary and Secondary Qualities."

101. *Essay*, II.xxxii.14.

102. Descartes scholars have been emphasizing the methodological character of Descartes' skepticism for some time; see Wilson, *Descartes*, ch. 1. For Locke's curt dismissal of skepticism, see *Essay*, IV.ii.14, IV.x.2, and IV.xi.8.

103. *Principles*, I, 18.

104. A close relation to this argument would have been accepted by Berkeley's more metaphysical opponents, such as Descartes; the reason for its acceptance would have arisen not from "veil of ideas" considerations, but from insistence that reason alone can establish knowledge of causal relations, such as that involved in sense perception. Locke would have denied Berkeley's argument; he countenanced a degree of knowledge known as "sensitive," which was not up to the standards of metaphysical knowledge, but which he considered adequate to justify a belief in external objects.

105. *Principles*, I, 19.

106. The argument against the conceivability of material substance comes in two forms. At *Principles*, I, 17, Berkeley attacks the abstract idea of substance in general. Elsewhere, he counters the notion that *extension* or *motion*, two fundamental concepts of mechanistic science, can be conceived abstractly, that is, as distinct from particular ideas of touch or vision (Intro., 6–8; I, 97–99). At I, 43–44, the theory of distance perception from the *New Theory of Vision* is used to deny that the objects of touch and sight exist in a common, external space. Hume outlines this argument briefly, and attributes it to Berkeley, near the end of sec. 12, pt. 1 of his first *Inquiry*.

107. Earlier in this century, arguments from perceptual relativity provided a standard means for introducing sense data and the veil of ideas; see, for example, Warnock, *Berkeley*, rev. ed., 147–8, 230–1.

108. Locke also doubted whether the essence of matter can be known, and he rejected the "real use" of the intellect (see n. 99, above); at the same time, he was willing to countenance the assertion that mind-independent objects have spatial properties, which neither Hume nor Kant would do.

109. Hume, *Inquiry*, sec. 4, pt. 1.

110. Hume's and Kant's conclusions about this restriction are, of course, not to be equated—e.g., Kant argued that we are in possession of synthetic a priori principles pertaining to all possible experience, whereas Hume contended that we at best form good habits that we hope will serve us well on future occasions—but they may be grouped together in contrast to the claims of realist metaphysicians, such as Descartes, Spinoza, Berkeley, or Leibniz.

111. This story of the hopes and failures of the intellect is historically separable from the problematic of veil-of-ideas skepticism, in that traditional skepticism with regard to

the senses can be posed independently of intuitionism, as is evident from the fact that Spinoza and Leibniz could adopt intuitionism with little or no connection to skepticism. Moreover, even in Descartes' case, it is possible to analyze his talk of "clear and distinct (intellectual) ideas" on analogy with sensory ideas, as clear and distinct *perceivings*, thereby avoiding the postulation of an intellectual *tertium quid*.

112. Descartes, CSM 2:21; see also 41 and 50–1 on intellectual light and the apprehension of mathematical objects through pure understanding. Locke, *Essay*, II.xxi.5, IV.ii.1. Hume, *Treatise*, I.iii.1; *Inquiry*, sec. 4, pt. 1.

113. One might be tempted to suppose that Locke and Hume denied the existence of the intellect or understanding by subsuming all its operations under the imagination. Locke discusses the various powers of the understanding in the *Essay*, II.ix-xi, where he considers the powers of comparing, compounding, and abstracting, by which human beings are distinguished from brutes. These discussions, I think, show that Locke would not reduce understanding to imagination. Hume, in contrast, explicitly suggests such an equation in the *Treatise*, I.iv.7 (265, 267). However, it is unlikely that he denied the existence of the faculty of understanding; he seems rather to have put very narrow limits on its powers, restricting its scope to the perception of intuitive relations between ideas, as in mathematical knowledge, while reducing causal reasoning to imagination.

114. In fact, Locke's and Hume's discussion of "intuitive and demonstrative certainty" may be understood as starting from reflection upon mathematical demonstration. Both authors agree in defining such certainty in terms of the perception of agreement or disagreement among ideas: Locke, *Essay*, IV.ii.1–2 (of course, Locke argued that intuitive and demonstrative certainty could be extended beyond mathematics, IV.ii.9); in the *Treatise*, Hume argued that intuitive certainty extends to arithmetic and algebra but not geometry (I.iii.1), but in the *Inquiry*, he allowed that arithmetic, algebra, and geometry all admit of intuitive and demonstrative certainty (sec. 4, pt. 1).

115. "Abstraction" need not entail the formation of abstract ideas such as the mythic idea of a triangle that is neither scalene, nor equilateral, etc.; it implies the (not inconsiderable) ability to attend to the mathematically relevant features of a sensory idea (or an idea in the imagination), and to make the particular stand for the general. It is clear that both Locke and Hume were committed to the power of abstraction in this latter sense: *Essay*, II.xi.9; *Inquiry*, sec. 12.

116. On the crucial role of spatial constructions in geometrical demonstrations during the period up to and including Kant, see Friedman, "Kant's Theory of Geometry."

117. For a comparison of Descartes' conception of mathematical abstraction with that of the Aristotelians, together with their respective conceptions of metaphysics, see my "Metaphysics and the New Science."

118. *Essay*, IV.ii.1–5; III.vi.9.

119. *Principles*, I, 11, 26–9, 139–40.

120. The Cartesian doctrine that animals are machines precluded attributing qualia to animals; but he provided the functions of the senses in detecting danger, and of the imagination in forming associative connections, a mechanistic account without appealing to inner experience: *Treatise on Man*, trans. Hall, 1–2, 87–105, 113–15, including notes.

121. Recall the point from section 1 about truth and the standard for knowledge. According to Descartes, the intellect provides a criterion or standard for truth in this sense: it is not the fact that a judgment has been properly made by the intellect that makes the judgment true; rather, it is a fact about the will and the intellect that

properly made judgments are true, where truth is understood in the usual way, as the correspondence of thought with its object.

122. *Inquiry*, sec. 9, esp. n. 1.

123. On the proviso regarding properties of objects and ideas of secondary qualities, see Locke, *Essay*, II.viii.13.

124. On the Cartesian theory of a "harmony" between created minds and the created world, see Bréhier, "The Creation of the Eternal Truths in Descartes' System." See also my "Reason, Nature, and God in Descartes."

Chapter 3, Kant

1. The following abbreviations are used in the text and notes of this chapter to refer to Kant's works:

    Ak   Kant, *Gesammelte Schriften*, Akademie Ausgabe (Berlin, 1902-); cited by volume and page number.

    Critique   Kant, *Critique of Pure Reason*, trans. Norman Kemp Smith (New York: St. Martin's, 1964); Page citations are preceded by the letter "A" or "B," referring to the first and second editions, 1781 and 1787. Citation of a single edition means that the passage appeared only there; a "/" separates page references to passages unchanged between the editions.

    Prolegomena   Kant, *Prolegomena to Any Future Metaphysics*, trans. James W. Ellington (Indianapolis: Hackett, 1977); pagination from Ak 4.

    Where context removes ambiguity, page numbers alone are given, in both text and notes. The translations have been modified on occasion.

2. Quotations from Kant, *Metaphysical Foundations of Natural Science*, trans. Ellington, 471 (pagination from Ak 4, as in the margin of the translation); on self-observation, see also Kant, *Anthropology from a Pragmatic Point of View*, trans. Gregor, 120–1, 133–4 (page numbers are to Ak 7). On the interaction between Kant's pronouncements and the development of early nineteenth-century German psychology, see Leary, "The Philosophical Development of the Concept of Psychology in Germany" and "Immanuel Kant and the Development of Modern Psychology."

3. In the conventional mid-eighteenth-century course on metaphysics at a German university it was customary to lecture on both rational and empirical psychology, and Kant followed the convention. The available sources of information regarding Kant's lectures are discussed in Satura, *Kants Erkenntnispsychologie*, 3–24. Various student notes and other pertinent materials have been collected in Ak 28:221–305. On rational and empirical psychology in eighteenth-century Germany, see Dessoir, *Geschichte der neueren deutschen Psychologie*, 2d ed., 165–209.

4. In his Inaugural Dissertation of 1770, *On the Form and Principles of the Sensible and Intelligible World*, trans. Kerferd and Walford, Kant allowed a "real use" for the intellect in application to the soul as an immaterial being: secs. 5, 6, 8, 23, 27, and 30, note (Ak 2:393–5, 411, 414, 419). For a review of Kant's relation to rational psychology, see Ameriks, *Kant's Theory of Mind*, 11–17.

5. As noted, page citations to the *Critique* and other works will appear in the text. The principal criticisms of rational psychology occur in the Paralogisms (A and B versions). It is possible that Kant continued to present the doctrines of rational psychology in his lectures into his so-called "critical period" (Ak 28:226, 265); Satura, *Kants Erkenntnispsychologie*, 14, reports that these lectures have been variously dated from 1773 to 1785, while Ameriks places them earlier, perhaps in the 1760s, on the basis of their content (*Kant's Theory of Mind*, 31, 236). Fortunately, we

need not enter further into questions of dating at this point, nor need we consider the question of whether Kant presented his newest thoughts in his lectures or was conservative in what he presented.

6. *Critique*, A379, 383; B417–18, 420; see also A778–9/B806–7.

7. Against the claim that Kant brought the psychological within the domain of the natural, it might be objected that these passages equating the laws of association with laws of nature occur in the context of Kant's contention that something beyond mere empirical regularity is required to explain how laws of nature can be established, namely, Kant's notion of the transcendental unity of apperception. Hence, it might be argued, Kant shifts the natural laws of the mind from psychological to transcendental status. But by Kant's lights an appeal to the transcendental is also required in order to explain the possibility of universal laws of outer sense, that is, of corporeal nature (not as a ground for the practice of physics, which Kant thought was healthy as matters stood, but in order the better to understand the limits of reason—A87–9/B119–21). Thus, laws of physics and laws of psychology are on a par in this regard: both require a transcendental explanation of their possibility. But since this consideration does not remove physics from the domain of the natural, it does not remove psychology, either.

8. *Critique*, A387, B427–8. For an extended discussion of mind-body interaction in Kant, see Ameriks, *Kant's Theory of Mind*, ch. 3.

9. Fries is discussed at some length in chapter 4, sec. 1, where references to Beneke are also given.

10. Wolff, *Kant's Theory of Mental Activity*, 176–7; Bennett, *Kant's Analytic*, 7.

11. Strawson, *Bounds of Sense*, 31–2, 88–9; see also 15.

12. Christian Wolff, *Psychologia empirica*, new ed. (Frankfurt and Leipzig, 1738), and *Psychologia rationalis*, new ed. (Frankfurt and Leipzig, 1740). Alexander Baumgarten did not write a separate treatise on psychology, but he devoted over one third of his textbook on *Metaphysica*, 7th ed. (Halle, 1779), to it. Johann Nicolaus Tetens described his *Philosophische Versuche über die menschliche Natur und ihre Entwicklung* (Leipzig, 1777) as a "psychological analysis of the soul based on experience" (Preface, 1:iv). Christian August Crusius devoted one of four parts of his *Entwurf der nothwendigen Vernunft-Wahrheiten* (Leipzig, 1745) to "pneumatology," or the doctrine of spirits, otherwise known as rational psychology.

13. David Hartley, *Betrachtungen über den Menschen* (Rostock, 1772–3); Charles Bonnet, *Betrachtung über die Natur*, trans. Titius (Leipzig, 1766), and *Analytischer Versuch über die Seelenkräfte*, trans. Schütz (Bremen, 1770–1); David Hume, *Philosophische Versuche über die menschliche Erkenntniss* (1755), vol. 2 of Hume's *Vermischter Schriften*, trans. and ed. Sulzer (Hamburg and Leipzig, 1754–5). Kuehn has documented the fact that German intellectuals had extensive knowledge of Scottish philosophy (and especially of "common sense" philosophy) in the time of Kant: *Scottish Common Sense in Germany, 1768–1800*.

14. Lossius' work was discussed by Tetens, *Versuche*, 1:532–46. For a description of Lossius' work, with citations of other commentators, see Kuehn, *Common Sense*, ch. 5. In denying that Lossius was a materialist I depart from Kuehn and others.

15. *Physische Ursachen des Wahren* (Gotha, 1775), 41–2, 195.

16. Ibid., 210, 220–1, on the reduction of reasoning.

17. Ibid., 58, 248.

18. On the analogy with aesthetics, ibid., 249ff.; on intersubjective agreement, 263–4.

19. Wolff, *Philosophia rationalis sive logica*, 3d ed. (Frankfurt and Leipzig, 1740), Discursus praeliminaris, ch. 3, secs. 55–62. Various divisions of philosophy can be found in this and other of Wolff's works. The division given in the text accords with much of what

Wolff says, as well as with his practice in producing the individual volumes of his Latin works; it also accords with the division provided in other textbooks in the Wolffian tradition, such as Baumgarten's *Metaphysica*, which was long used by Kant in his lectures on metaphysics.

20. *Psychologia rationalis*, secs. 643, 645.

21. Baumgarten extended cosmology to include all things in the world (as opposed to the being outside the world, which was treated under theology), among which were included spirits, as well as the *commercium* or interaction of the parts of the world, including that between spirit and body: *Metaphysica*, secs. 402–405, 448–465.

22. The quotation is from Wolff, *Psychologia empirica*, Prolegomena, trans. Richards, sec. 9. On the greater trustworthiness of empirical psychology, see secs. 4–5.

23. *Critique*, A305–6/B362–3, A341–8/B399–406.

24. Wolff, *Psychologia empirica*, Prolegomena, trans. Richards, sec. 5.

25. On the use of empirically based, or "historical," propositions in philosophy, see *Philosophia prima, sive, ontologia* (Frankfurt, 1736), Discursus praeliminaris, secs. 7–12; on the grounding of logic in empirical psychology, see secs. 89–91.

26. Tetens speaks of seeking to discern the *Grundkraft* and the original operation of the *Naturkraft* of the soul (e.g., *Versuche*, 1:737). Although *Kraft* might simply be understood as a power or active principle of a substance in the best tradition of German scholasticism, among Wolffian thinkers it became connected with the Newtonian conception of understanding a substance through knowledge of the laws of its basic forces.

27. "Die äussern Empfindungen schwächen die innern, und umgekehrt," Johann August Eberhard, *Allgemeine Theorie des Denkens und Empfindens* (Berlin, 1776), 158.

28. Ibid., 183–4.

29. Tetens, *Versuche*, 1:3, iv.

30. Materialism is rejected in the preface to volume 1 of the *Versuche*, where Tetens set his project in empirical psychology apart from what he characterized as a particular "metaphysical" approach, which "insists on a reduction [*Reduktion*] of that which is observed in the soul to modifications of the brain" (1:v).

31. In introducing the laws of association, Tetens explicitly remarks that they can be violated by "das selbstthätige Dichtungsvermögen" and by "die Denkkraft," thereby rejecting the claim that those laws universally determine the succession of ideas in the soul; later he makes the point that accidental associations of ideas can cause errors of the senses, common sense (*gemeine Verstand*), and reason, all three (Tetens, *Versuche*, 1:576–7). Virtually all German writers on psychology recognized associative laws as the "laws of imagination": Wolff, *Psychologia empirica*, sec. 117; Baumgarten, *Metaphysica*, sec. 561; Tetens, *Versuche*, 1:106–14; Eberhard, *Allgemeine Theorie*, 111–14.

32. *Versuche*, Bk. XI (esp. 1:754–66). On the positing of "*Denkkraft*" as "the third simple ingredient of the human faculty of knowledge," (1:298), see Bk. IV, pts. 1–2; on spontaneity, or *Selbstthätigkeit*, see Bk. II, esp. pt. 3 (1:752–66). In accord with Wolffian doctrine, these distinct powers are all regarded as expressions of the one basic "power of representation" (1:594–9).

33. Crusius, *Entwurf*, secs. 4–5, 424, 459. Kant, too, contended that empirical psychology did not properly belong to metaphysics. He conjectured that it retained its place there in the curriculum only for pedagogical convenience: *Critique*, A848/B876–7; see also his lectures on metaphysics, Ak 28:223, 541, where the inclusion of empirical psychology in metaphysics is attributed to previous failures to appreciate that metaphysics is the "science of pure reason" and excludes empirical findings.

34. Crusius, *Entwurf*, secs. 425–426, 471–474.

35. On the Wolffian position according to which, in effect, empirical knowledge of the soul as a thing in itself is attainable, see *Psychologia empirica*, sec. 10, and *Psychologia rationalis*, sec. 6.
36. *Critique*, A92–3/B124–6; also *Prolegomena*, sec. 19: "objective validity and necessary universal validity (for everybody) are equivalent concepts, and although we do not know the object in itself, yet when we consider a judgment as universally valid, and hence necessary, we understand it thereby to have objective validity" (298).
37. Readers of Kant will recall that the wording here is crucial: Kant states that every event or alteration has a cause, not that every effect has a cause, because he is interested in establishing a substantive claim about the course of events, not in trading upon the definition of "cause" and "effect" in order to establish a trivial tautology (B4–5).
38. Kant wrote: "It is a merely empirical law, that representations which have often followed or accompanied one another finally become associated, and so are set in a relation whereby, even in the absence of the object, one of these representations can, in accordance with a fixed rule, bring about a transition of the mind to the other" (A100).
39. See also A766/B794; *Prolegomena*, preface, 257–8; and *Critique of Practical Reason*, trans. Beck, Preface, 12–3 (page references to Ak 5).
40. *Prolegomena*, sec. 36 (319, n. 24); see also *Critique*, B167–8.
41. For a discussion of the role of necessity in Renaissance philosophy and science, including the ascription to the intellect of the power of achieving necessity through induction, see Jardine, "Epistemology of the Sciences."
42. See also, among others, A1, B3–4, and *Prolegomena*, sec. 14.
43. The question of whether transcendental argumentation is salvageable, or is irrevocably question-begging, has been much discussed of late, and especially in connection with Kant's transcendental idealism. Recently Allison, *Kant's Transcendental Idealism*, has mounted an impressive defense of Kant's transcendental idealism and his transcendental arguments against a long line of previous commentators. Guyer, *Kant and the Claims of Knowledge*, has sought to extricate Kant from the allegedly question-begging pursuit of transcendental arguments that take geometry or Newtonian physics as given bodies of knowledge by showing that the arguments of the transcendental deduction, and the attendant doctrine of transcendental idealism, were perversions of Kant's original, realist-inclined, project. See also the various articles in Bieri, Horstmann, and Krüger, eds., *Transcendental Arguments and Science*, and Förster, ed., *Kant's Transcendental Deductions*.
44. See *Critique*, B4–5, 20, 128, and *Prolegomena*, secs. 6, 15. Allison, *Kant's Transcendental Idealism*, ch. 5, contends that Kant's argument from the assumed validity of Euclid's geometry was not his main argument; see sec. 4, below, for discussion of the relation between the argument from geometry and Kant's other arguments.
45. See *Critique*, B5, A93/B126, and *Prolegomena*, secs. 20, 22. Even if Kant took merely "the possibility of experience" as his given element, it is necessary to examine carefully what he meant by "experience." The conception of "experience" I adopt stresses that experience is of public objects about which intersubjectively valid judgments can be made—a conception that is suggested by Kant's discussion of "consciousness in general" in the *Prolegomena*, secs. 20, 22, 29 (300, 305, 312), see also secs. 17–19. For a discussion of various conceptions of experience in Kant, see Guyer, *Kant and Knowledge*, ch. 3.
46. Kant ranges the objects of metaphysics under these three heads at A334–5/B391–2. But see also B395, note, where he explains that since the science of nature can take care of itself the proper interest of the speculative employment of reason pertains to

God, freedom, and immortality; see further A798/B826, and second *Critique*, preface (Ak 5: 4–6). This qualification does not remove questions pertaining to "the world in itself" from the proper domain of metaphysics, for the question of freedom arises under the heading of "cosmological" questions of causality (A410/B447).

47. *Prolegomena*, sec. 40 (327); *Critique*, A711/B739; on geometry, see also A87/B120 (notice that in the continuation of this latter passage, it is claimed that a deduction is required for the pure concepts of the understanding, including the concept of cause, inasmuch as they purport to relate to objects apart from possible sensory experience, something that the practitioners of physics need not claim for the law of cause).

48. See the articles in Pt. 1 of Bieri et al., *Transcendental Arguments*, and Henrich, "Kant's Notion of a Deduction."

49. See the methodological remarks at Axi-xiv, Bxv-xxiv, A105/B25–9, and A782–8/ B810–16.

50. Transcendental knowledge is not analytic in the usual way—it is not knowledge arising from the analysis of concepts—nor is it synthetic (and a priori) in the usual way—it is not knowledge arising from conditions manifest in pure intuition, or by consideration of the possibility of experience of objects. Transcendental knowledge results from "transcendental analysis," which does not analyze concepts such as cause or substance themselves, but rather analyzes the "faculty of understanding itself" as the "birthplace" of concepts such as cause, or substance (A64–6/B89–91). Transcendental philosophy is not to be classified with either the trivially analytic propositions that come from the analysis of given concepts, nor with the synthetic a priori knowledge that transcendental analysis itself explains. One could say that it is "analytic," in that it analyzes the possibility of knowledge; however, it does not analyze the mere concept of knowledge, but investigates instead the possibility of knowledge for a finite intellect with certain forms of sensibility. The role played by appeal to the notion of a finite intellect with certain forms of sensibility in transcendental argumentation might seem to parallel the appeal to the notion of possible experience in synthetic a priori knowledge of the ordinary variety; that might make such argumentation "transcendentally synthetic." For present purposes there is little need to settle upon a description, so long as it is recognized that such knowledge is distinct from ordinary analytic and synthetic a priori knowledge (see A783/B811 and A787–8/B815–16).

51. Kant wittingly conceived sensibility as a passive faculty and accepted the implication that it is "acted upon"; see A30/B45, B72, A92/B125, B145, etc., and *Prolegomena*, secs. 9, 11, Remarks II-III (282, 284, 289–90). Such passages have, of course, raised a good bit of discussion, because they raise the problem of how Kant could justify application of the concept of cause to the thing that "affects" the faculty of sensibility (which is, presumably, the "thing in itself"); for an overview and a proposed solution, see Allison, *Kant's Transcendental Idealism*, ch. 11, sec. 3.

52. "Synthesis" covers a multitude of cognitive acts in both transcendental and empirical psychology, as conceived by Kant (A77–8/B103). The integration of sensations into perceptual images is a synthesis (A99, 120; B152, 160), as is the combination of concepts in a judgment (A78–9/B104–5). There is also a synthesis of all the representations of "inner sense" to yield a unified consciousness. As will be discussed more fully below, each of these syntheses has an empirical and a transcendental counterpart.

53. Patricia Kitcher, "Discovering the Forms of Intuition," sec. 3.

54. Kant draws the line between empirical psychology and logic in terms of lower and higher faculties in the *Anthropology*, 140–1. While this might suggest that he would

also draw the same line between empirical and transcendental psychology, on p. 142 he equates the object of psychology with all the phenomena of inner sense, including the "I" of empirical apperception, and restricts transcendental apperception to "pure" consciousness, which accords with the reading I shall shortly give to the *Critique*.

55. A94, 115. Even the synthesis effected by the understanding has its empirical, associative counterpart (A119–23). There are no direct correlates in B to Kant's statement at A115 that all the faculties "can be viewed as empirical." Moreover, at B152 he equates association with imagination, in contrast with understanding. But it remains the case that imagination and sensibility both have their "transcendental" sides, and that the synthesis of the understanding is conceived as a transcendental counterpart to the empirical synthesis of imagination (B162, and note).

56. Ward, "Psychological Principles," 162–6; see also C. I. Lewis, *Mind and the World Order*, 98–100.

57. Thus, even though Kant, in dismissing empirical psychology, talks about its application to the particular conditions under which judgments take place (A53–4/B77–8), the force of these remarks is not aptly captured by treating empirical psychology as limited by its particularity. Because it cannot speak to questions of right, the only service that it can provide to logic is the explanation of cases in which, owing to circumstances, the epistemic conditions of objective judgment are not met (cases for which there is no normative account, because the norms of logic are violated). Of course, it can also provide causal explanations corresponding to correct judgments; but it is unable to distinguish the two cases.

58. See note 7. Although all the representations subject to the unity of apperception are also subject to the laws of association, the converse does not hold. Animal consciousness, at least, is subject to the laws of association, but not to the unity of apperception (Kant to Marcus Herz, 26 May 1789, in Kant's *Philosophical Correspondence*, ed. and trans. Zweig, 153–4; Ak 11:152).

59. *Critique*, A532–58/B560–86; see also Bxxvii-viii. Even though he speaks of "noumenal freedom," free acts are manifest in the chain of appearances: A537/B565 and A544/B572.

60. Kant asserts this point at B161, note: "In this manner it is proved that the synthesis of apprehension, which is empirical, must necessarily be in conformity with the synthesis of apperception, which is intellectual and is contained in the category completely a priori. It is one and the same spontaneity, which in the one case, under the title of imagination, and in the other, under the title of understanding, bring combination into the manifold of intuition."

61. The discussion in secs. 24–5 of the B Deduction (and the related discussion at A106–8) reveals a deep tension in the argument of the *Critique*. When Kant says that the self, as agent of transcendental apperception, is an object that can be "thought" but not known in itself, he seems to contradict his own practice; for the Deduction posits transcendental apperception as an object of transcendental knowledge. However, one can read his denial of knowledge of this self as a denial of knowledge of the traditional metaphysical variety, as well as a denial that the self as it is in itself can be known through introspection; there remains transcendental knowledge, not of the self as a thing in itself but as the ground of the possibility of knowledge. (For an extensive discussion of this problem, see Allison, *Kant's Transcendental Idealism*, chs. 12–13.)

62. Inaugural Dissertation, sec. 15. On the development of Kant's views on space, see Buroker, *Space and Incongruence*, ch. 2, sec. 3, ch. 3, sec. 1, and ch. 4. See also Garnett, *Kantian Philosophy of Space*, ch. 5, sec. 4, and ch. 6.

63. It is widely recognized that Kant thought his theory of space and time received confirmation through its role in addressing the first and second Antinomies. Beyond this, there is some disagreement over the precise importance of the Antinomies in motivating the critical philosophy itself. At one point, Kant attributed the Antinomies a motivating role similar to that he earlier had attributed to Hume (Kant to Christian Garve, 21 September 1798, in Kant's *Philosophical Correspondence*, 252; Ak 12:255). But it is unlikely that the Antinomies provided the initial motivation for Kant's doctrine of the ideality of space, for that doctrine appears in the Inaugural Dissertation, where it is motivated in part through the need to explain the possibility of geometrical knowledge (sec. 15C-E). For a discussion of the geometrical and nongeometrical motivations to Kant's arguments in the work of 1770, see Garnett, *Kantian Philosophy of Space*, ch. 6; see also Buroker, *Space and Incongruence*, ch. 4, sec. 1.

64. Kant characterized the arguments pertaining to both space and time as a "transcendental deduction" at A87/B119, where he acknowledged the central role played by geometry in the deduction of space (A87/B120).

65. Kant added the subheading just quoted in B and rearranged his arguments. The effect was to isolate and emphasize the argument that depends upon geometry, which in A was placed in the midst of the other arguments. My discussion follows the order in B.

66. This argument was also presented in the Inaugural Dissertation, sec. 15A.

67. Allison also interprets the passage in this manner: *Kant's Transcendental Idealism*, 86–9.

68. Although Kant's arguments in the Aesthetic rely on the distinction between intuitions and concepts, the distinction can best be understood by considering passages from subsequent portions of the *Critique*, including A50–2/B74–6, A68/B93, and A712–17/B740–5.

69. B39–40, revised from A25. In the A edition, this argument and the preceding one came after the treatment of geometry, and therefore could have relied upon the fact that the space in question was space as described geometrically.

70. In both earlier and later works, Kant used his argument from incongruent counterparts to support the point that space is an intuitive, not a conceptual representation; see Buroker, *Space and Incongruence*, chs. 3–5.

71. The question quoted in the text was stated in this explicit form only in B, which incorporated material from the *Prolegomena* into the Introduction in order to explain more fully the project of the *Critique*. But the description of this project (as portrayed in section 2 of the present chapter) was present, if briefer, in the A version of the Introduction: A2–4, 10.

72. *Critique*, B20 and A711/B739. See also A87/B120, where the status of geometry is contrasted with that of metaphysics: "Geometry, however, proceeds with security that is completely a priori, and has no need to beseech philosophy for any certificate of the pure and legitimate descent of its fundamental concept of space." The same point is mentioned again at A149/B189; it is asserted or assumed at various points in the *Prolegomena* (see especially the First Part). As mentioned earlier in this section, in both A and B Kant described the argument of the Aesthetic as a "deduction" (A87/B119), and that is how I shall read the passage quoted in the text. It might seem odd that Kant should describe the Aesthetic as providing geometry with a deduction, while he nonetheless maintains that geometry need not appeal to philosophy in order to secure its foundations. But his strategy was to chart the boundaries of metaphysics by explicating the conditions that made possible sciences that were uncontroversially actual, and then to see whether the same conditions extended to the proper objects of metaphysics (God, the soul, and freedom).

73. Such dicta may tempt one to speculate that Kant was privy to the possibility of non-Euclidean geometries (especially given the extensive work on the parallel postulate by Kant's contemporaries). The careful and thorough study of the origins of non-Euclidean geometries by Torretti, *Philosophy of Geometry*, ch. 2 (including discussion of Kant's contemporaries), should remove the temptation.

74. *Critique*, A713/B741; A838/B866. Kant contended that his transcendental philosophy was propaedeutic to knowledge of this sort: A845–6/B873–4, A66/B91.

75. See, e.g., Kant's assertion that geometrical axioms can be known with certainty because they are intuitively "evident" (*evident*), A753/B761, and his discussion of appeals to intuition in establishing geometrical propositions at A47–8/B65–6. Philip Kitcher, *The Nature of Mathematical Knowledge*, ch. 3, describes Kant's theory of mathematical knowledge as appealing essentially to figures constructed in thought and inspected "with the mind's eye" (49). As Kitcher observes, Kant's theory is (and has been, especially by nineteenth century authors) subject to various objections, not the least of which is the "exactness problem": why should we suppose that mentally constructed figures can give us precise knowledge of angular equalities, etc., when measuring physically constructed diagrams won't do so? Kant might have answered by contending that we attend only to the "relevant" properties in our mental constructions (see A713–4/B741–2). (Of course, such an answer would not satisfy the nineteenth century critic, who requires that the "inspection" be sufficiently accurate to measure the curvature of space.) Partly in order to avoid the exactness problem, Friedman, "Kant's Theory of Geometry," has interpreted Kant's theory of geometrical proof in such a way that it relies entirely on procedures of construction, with intuition playing an exemplifying rather than a demonstrative role. Friedman's account is compelling as a reconstruction of Kant's conception of geometrical proof; the problem comes in accounting for the "evident" status of the axioms and of closely related propositions, such as that a straight line traverses the shortest distance between two points, or that two lines cannot enclose a figure. At the least, Kant certainly seems here to be following the "visual inspection" model (A47–8/B65–6), notwithstanding Friedman's proposed reconstruction of the straight-line case, 495, n. 59.

76. A47–8/B65–6; B16. See also Inaugural Dissertation, 15C: "Geometry uses principles which are not only indubitable and discursive, but come before the gaze of the mind, and the *evidence* in demonstrations (which is the clarity of certain cognition in so far as it is assimilated to sensual cognition) is not only greatest in geometry but is also the only evidence which is given in pure sciences and is the *exemplar* and means of all *evidence* in other sciences." Again, Kant stressed that, in "inspecting" the products of construction, we attend only to the "relevant" properties (see previous note). This should not be taken to imply that concepts are adequate fully to determine these relevant properties; some basic geometrical truths can only be determined through an appeal to intuition.

77. It may be helpful for those inclined to think of "the knower" as a concrete object of empirical investigation to recall that Kant's transcendental investigation takes as its object not the knower considered as a part of the causal order of nature but an impersonal "transcendental subject," the characteristics of which cannot be established empirically.

78. As has been frequently remarked, Kant apparently overstates his conclusion; that is, he should be entitled only to the conclusion that the space of which we have necessary knowledge is ideal, not to the conclusion that things in themselves lack spatial properties. Buroker provides an excellent discussion of Kant's motives for denying spatial properties to things in themselves, according to which these

motives originate in his paired beliefs that things in themselves could possess spatial properties only in accordance with a Leibnizian theory of space and that the existence of incongruent counterparts refutes the Leibnizian theory (*Space and Incongruence*, ch. 5, sec. 2).

79. At B160 Kant says "space and time are represented a priori not as forms of sensible intuition, but as themselves intuitions," which might seem to conflict with the idea that the space of pure intuition is not by itself an object of intuition. However, this idea is confirmed by the note to B160, in which Kant explains that he is here speaking of space regarded as an object of intuition that requires the activity of the understanding in combining or synthesizing the manifold (one might then read B160 as positing pure a priori objects of intuition, constructed in conformity with the forms of intuition). Kant explicitly notes that he here adds to the doctrine of the Aesthetic, though without altering the fundamental point, that the form of intuition "precedes" (i.e., is independent of) the concept of space and the synthetic activity of the understanding.

80. In the Inaugural Dissertation, Kant spoke of space, considered as a "formal principle of the sensory world," as providing "a law which is stable and is implanted in [the mind's] own nature," in accordance with which the mind coordinates all sensations (sec. 15E). In the first *Critique*, Kant spoke of space, considered as a form of intuition, as providing, "prior to all experience, principles which determine the relations" of all objects of outer sense (A26/B42), and also as providing a "constant form" of receptivity (A27/B43), and a "condition" on all outer appearances (A89/B121). Whether considered as providing laws or principles, the point is that space, considered as a form of intuition, is conceived not as a particular a priori space but as an a priori rule for constructing spatial intuitions (see previous note).

81. See also B147, A224/B271, A239/B299, and *Prolegomena*, sec. 11 and Remark I (283–4, 287–8). Hume proposed that the objects of the senses might not conform to geometrical rules of construction, such as the infinite divisibility of lines or angles: first *Inquiry*, sec. 12, pt. 2.

82. See Kant's letter to Herz, 26 May 1789, in his *Philosophical Correspondence*, 152–3 (Ak 11:50–1).

83. *Metaphysical Foundations*, ch. 3.

84. Kant emphasizes the first point in the A and the second in the B version of the Deduction. In A, he contends that association as a law of nature rests on the "affinity" of the manifold; but the necessity of this affinity is a consequence of the original unity of apperception as the condition for experience of a coherent outer world: he concludes that "all appearances stand in thoroughgoing connection according to necessary laws, and therefore in a transcendental affinity, of which the empirical is a mere consequence" (A113–4; see also A100–1, 121–2). At B140, he asserts that the "empirical unity of consciousness, through association" (where the example of association given pertains merely to inner sense) is "derived from" the original unity of apperception; see also B161, note. Incidentally, in order to avoid confusion it is important to distinguish the claim in section 1 above that the representations of inner sense are subject to the causal law, which Kant makes in discussing the organon of the critical philosophy, after his work is done, from his repeated assertions in the Deductions that the the law of association is subjective and contingent (e.g., A100, B140). In the latter case, he is speaking of the law as a mere empirical generalization, based upon experience (as in Hume), independent of the deliverances of transcendental philosophy. As such, the empirical law of association cannot guarantee the unity of apperception.

85. Ak 11:79. Kant's dismissal of concern with the causal origin of the faculty of sensibility should not be read to imply more than it says; for we have seen that Kant argued in the Aesthetic that the ability to represent space could not in principle be acquired through experience. Thus, transcendental psychology presumably requires some measure of nativism regarding spatial representation within the domain of empirical psychology.

86. *Critique*, A88/B120: the pure concepts of the understanding "speak of objects through predicates not of intuition and sensibility but of pure a priori thought, they relate to objects universally, that is, apart from all conditions of sensibility" (though not, for finite intellects, apart from the condition that thoughts without content provided by intuition are empty).

87. B148 and 150–1. *Qua* finite intellect, a being must have a faculty of sensibility to provide it with intuitions; but Kant saw no reason why the forms of this faculty must be space and time.

88. The classification of Kant as a nativist (or as one who believed that human visual spatial abilities are inborn) was common among nineteenth-century sensory physiologists (see the discussions of Steinbuch and Tourtual in chapter 4, section 2, and of Helmholtz in chapter 5). It was made commonplace by Boring, *Sensation and Perception*, 28 and 233, and has been widely repeated in recent literature (for citations, see Chapter 1, section 2, and Appendix A). Pastore has questioned Kant's supposed influence on the later course of nativism without questioning whether Kant was himself a nativist: "Reevaluation of Boring on Kantian Influence."

89. Kant's lectures on anthropology from 1791: *Die philosophischen Hauptvorlesungen Immanuel Kants*, from the student notebook of Heinrich zu Dohna-Wundlacken, ed. Kowalewski, 94. Kant goes on to cite experiments with the newly-sighted blind, saying that they "saw, e.g., a globe as only a circle," but that touch allowed them to make up the deficit. He explicitly cites the case of an Englishman, the details of which match the Cheselden case.

90. *Anthropology*, trans. Gregor, 155.

91. A303/B360; see also A293/B350.

92. A297/B354. The explanation in parentheses is suggested by the discussion in the *Anthropology*, trans. Gregor, 146.

Chapter 4, Spatial Realism

Abbreviations used in the text and notes are explained as they are introduced throughout this chapter.

1. *Philosophisches Magazin*, ed. J. Eberhard (Halle, 1788–95), sought to downgrade Kant's philosophy; *Neues philosophisches Magazin, Erläuterungen und Anwendungen des Kantischen Systems bestimmt*, eds. J. H. Abicht and F. G. Born (Leipzig, 1789–91), sought to promote it.

2. Beiser, *Fate of Reason*, examines the historical background to and the intellectual development of the major currents of post-Kantian philosophy in the period from 1781 to 1793. Translation of selected post-Kantian texts may be found in Giovanni and Harris, trans., *Between Kant and Hegel*.

3. Mandelbaum chose these two currents of thought as organizing themes in his *History, Man, and Reason*; see e.g. 6–20. Mach did hold a chair in the history and theory of inductive science for six years, after having held a chair in physics for 28 years.

4. In 1884, Helmholtz wrote "I was a faithful Kantian at the beginning of my career, as I am now; or rather, that which I wanted to see changed in Kant were unimportant

details": *Vorträge und Reden*, 4th ed., 1:viii. Helmholtz and his contemporaries did not originate the call to return to Kant. In a work written in 1831 to celebrate the publication of the first *Critique* five decades earlier, F. E. Beneke had called for a return to the true Kant, or to Kant understood as promulgating an antimetaphysical stance that placed the theory of knowledge (or of cognition) at the center of philosophy: *Kant und die philosophische Ausgabe unserer Zeit* (Berlin, 1832), 12, 73–89. On Beneke's role in the "back-to-Kant" movement, see Köhnke, *Entstehung und Aufstieg des Neukantianismus*, esp. ch. 2, sec. 2.

5. For a discussion of such conflicts, complete with additional references, see Ash, "Reflections on Psychology in History."

6. See H. W. Cassirer, *Kant's First Critique, 63–4*; Bennett, *Kant's Analytic, 7*; and Strawson, *Bounds of Sense*, 15–16.

7. The movement against reading Kant psychologically became prominent during the 1870s, when J. E. Erdmann apparently coined the term "psychologism" to describe Beneke's psychological interpretation of Kant's project. Hermann Cohen and Alois Riehl (members of the so-called "Marburg neo-Kantians") produced a sustained interpretation of Kant's first *Critique* that argued explicitly and at some length against equating Kant's concepts and methods with those of empirical psychology: Cohen, *Kants Theorie der Erfahrung* (Berlin, 1871), esp. ch. 7, and Riehl, *Philosophische Kriticismus und seine Bedeutung für die positive Wissenschaft*, (Leipzig, 1876–9), preface (1:v), introduction, sec. 3 (1:5–8), and part 2, ch. 2, sec. 3 (1:294–315). Heinrich Rickert (a principal "Southwestern neo-Kantian") also adopted an antipsychologistic reading: *Gegenstand der Erkenntnis* (Freiburg, 1892).

8. Fichte, as far as I know, did not use the term "psychological" to characterize Kant's work, but he contended that Kant's philosophy was in fact based on "intellectual intuition," which he distinguished from metaphysical intuition of an intelligible world and equated with the mind's awareness of its own activities (while acknowledging that Kant himself refused to admit that his philosophy was grounded in intellectual intuition): J. G. Fichte, "Zweite Einleitung in die Wissenschaftslehre," in his *Sämtliche Werke*, ed. I. H. Fichte, 1:471–9; trans. Heath and Lachs in Fichte, *Science of Knowledge*, which gives the pagination to the above edition, as is standard practice. Inasmuch as Fichte's own position may be characterized as "psychological" (see the subsequent note), so may his reading of Kant. Hegel explicitly characterized Kant's method as psychological: *Vorlesungen über die Geschichte der Philosophie*, in his *Sämtliche Werke*, 19:561, 574, trans. Haldane and Simson, *Lectures on the History of Philosophy*, 3:432, 443.

9. Fichte's "science of knowledge," with its starting point in the analysis of the pure ego and its claim to have discovered the true "laws of thought," has been characterized as a "constructive psychology" (*konstructive psychologie*): Falckenberg, *Geschichte der neueren Philosophie*, 2d ed. (Leipzig, 1892), ch. 11, sec. 1, subsec. 3 (352), trans. Armstrong, *History of Modern Philosophy*, 3d ed. (433). It was not a natural-scientific psychology, to be sure, but a particular way of continuing Kant's own transcendental psychological investigation of the knowing self; as such, it was not, in its intention, "psychologistic" in the late nineteenth-century sense of that term. Similar remarks apply to Hegel's phenomenological investigation of the path of natural consciousness in the development of "scientific cognition." Although both these authors placed the structure of the mind at the center of philosophical investigation, neither approached the mind as a natural scientist; and although they each emphasized a philosophy based upon inner or phenomenal "experience," it would be misleading to label their investigations "empirical."

10. Jakob Friedrich Fries, *Neue oder anthropologische Kritik der Vernunft*, 1st ed., 3 vols. (Heidelberg, 1807); 2d ed., 3 vols. (Heidelberg, 1828–31). Page citations will appear in the text of the present section; they are to the second edition, which has been reprinted in vols. 4–6 of Fries's *Sämtliche Schriften*.

11. For a characteristic statement on transcendental idealism, see Fries, *Neue Kritik*, 1:xxiii; Fries rejected Fichte's metaphysics of the ego (e.g. 1:91).

12. Fries is not here repeating the charge, common during the 1790s, that Kant's claim to have provided an explanation of objective knowledge depended upon his assumption of the thing in itself. Rather, his criticism is akin to the charge of K. L. Reinhold (and others) that Kant begged the question against Hume by positing that objectively valid experience is actual, and then inquiring into the manner in which it is possible. On Reinhold's charge, see Beiser, *Fate of Reason*, ch. 8 (esp. 241–2).

13. Fries explicitly contrasts his inductive procedure with Kant's attempted proof, *Neue Kritik*, 1:25–6, 320–1; see also 1:381–3 and 2:15, 72–3.

14. Fries recognized that the formal unity of our cognitive processes is not proven by individual bits of intuitive knowledge, such as might be found in particular instances of mathematical intuition or single observations of nature (ibid., 2:43–6, 61). But, he maintained, the transcendental unity of apperception is demanded by the claims to knowledge that we in fact do make: not only by claims to know universal laws of mathematics or universal principles of physics, but also by the idea of truth itself, "in the demand that there should in general be truth, and only *one* truth, for our mind" (2:51). The positing of original formal apperception explains how mathematical demonstrations can proceed from merely particular intuitions but achieve universal validity: such demonstrations can achieve universality only given that one and the same cognitive activity acts always in accordance with the same general rules in the formal determination of the matter of intuition, thereby providing intuitions conforming in each instance to a universal rule (2:65, 113–20). The explication of our ability to find causal sequences in perception is more complicated (2:47–70), as is the explanation of the demands of the "idea of truth" (2:52–3, 65).

15. Kant, *Critique of Pure Reason*, Bxvi.

16. Fries's discussion of mathematical knowledge in the *Neue Kritik* was minimal (see 1:344, 353 and 2:113–120); he provided a much fuller treatment in his *Mathematische Naturphilosophie nach philosophischer Methode bearbeitet* (Heidelberg, 1822), on which see Gregory, "Neo-Kantian Foundations of Geometry."

17. For a concise discussion of Herbart's philosophy in its context, see Falckenberg, *Neueren Philosophie*, ch. 14, sec. 2. Träger, *Herbarts realistisches Denken*, provides a summary of Herbart's metaphysics with special emphasis on his response to Fichte.

18. Johann Friedrich Herbart put forth his views on metaphysics, psychology, and epistemology in a number of works. His first major philosophical work was the *Hauptpuncte der Metaphysik* (Göttingen, 1808). His *Lehrbuch zur Einleitung in die Philosophie* (Königsberg, 1813; 4th ed., 1837) was a much fuller statement of his metaphysics, fleshing out the foundations of his psychology as adumbrated in the *Hauptpuncte*. His metaphysics received its fullest statement in the *Allgemeine Metaphysik nebst den Anfängen der philosophischen Naturlehre* (Königsberg, 1828–29); his psychology was more fully developed in the *Lehrbuch der Psychologie* (Königsberg and Leipzig, 1816; 2d ed., 1834) and received its most detailed treatment in *Psychologie als Wissenschaft neu gegründet auf Erfahrung, Metaphysik und Mathematik* (Königsberg, 1824–1825). References are given to section numbers as these appear in Herbart's *Sämtliche Werke*, ed. Hartenstein, which follows the later editions of the above named works. Page numbers to the Hartenstein edition are given in parentheses.

19. *Einleitung*, sec. 1 (1:27–30); 1st ed., sec. 4 (1:43).

20. *Hauptpuncte*, introduction (3:5–14); *Einleitung*, secs. 4, 11–16, 34–8 (1:43–7, 53–9, 77–80); *Allgemeine Metaphysik*, secs. 165–72 (4:17–30).
21. *Hauptpuncte*, sec. 1 (3:15–17); *Einleitung*, secs. 1, 117 (1:28, 175–7); *Allgemeine Metaphysik*, sec. 166 (4:18–19).
22. Herbart allied causality with substance: "No substantiality without causality" (*Allgemeine Metaphysik*, sec. 220; 4:110). By this he meant that we must posit substance as the ground of appearances as "given." Appearances are, in his view, simultaneous with their causes; he therefore argued that another word should be used for connections between appearances through time, and chose the word "alteration" (*Veränderung*): ibid., secs. 223–4 (4:115–18). See also *Einleitung*, secs. 123, 125, 130 (1:187–9, 194–5, 210–4).
23. *Einleitung*, sec. 34 (1:77–8); see also *Lehrbuch zur Psychologie*, sec. 180 (5:126–7): "General notions, which are thought merely through their content, without slipping the act of representing into their compass, are, as remarked above, *logical ideals*; just as logic as a whole is *an ethics of thought, but not a natural history of the understanding.*"
24. The content of Herbart's logic was largely conventional, including Kant's table of judgments; but he gave particular prominence to *modus ponens* and *modus tolens*, and sought to unify the syllogistic forms (see *Einleitung*, second part; 1:77–113).
25. Herbart did not regard the "*Begriffene*" as objects in the world; he distinguished among (a) actual acts of thought (occurring within the natural history of a particular mind), (b) real objects, external to the mind, and (c) the objects of concepts regarded as subject-matter for philosophical reworking (*Einleitung*, secs. 34–5; 1:77–9). The aim of philosophy is to achieve a correspondence between the world as we conceive it and the world as it is in itself; we may be limited in our ability to achieve this aim, but, he contended, we should regard our efforts as attempts in this direction.
26. Kant, first *Critique*, Axvi–xvii. See Herbart's own comparison of his position with the starting points of Spinoza and Kant, the leading metaphysical opponents in German philosophy at the end of the eighteenth and beginning of the nineteenth centuries (*Allgemeine Metaphysik*, end of preface to first part; 3:64–5). Herbart sided with Kant in taking concepts of objects, rather than attributes of substance, as his starting point; but he made it clear that he disagreed with Kant's overall position in many respects. On the importance of Spinoza in German philosophy of this period, see Beiser, *Fate of Reason*.
27. *Einleitung*, secs. 11–13, 30–33 (1:53–6, 71–6); see also *Allgemeine Metaphysik*, introduction and sec. 170 (3:66–71 and 4:25–6).
28. In the *Hauptpuncte*, he used the term *Wesen* to name the simple substances he posited as the real (secs. 2–3; 3:17–20), and he spoke often of *einfache Wesen* or *reale Wesen*. In the *Allgemeine Metaphysik*, he came to speak of the multitude of *Realen* (secs. 214, 217; 4:101, 107). It is usual in English to characterize the simple beings or simple substances of Herbart's ontology as "Reals."
29. *Einleitung*, secs. 132–5 (1:216–221); *Allgemeine Metaphysik*, secs. 213–18 (4:98–108).
30. *Allgemeine Metaphysik*, sec. 142 (3:421–3); *Einleitung*, secs. 131, 157 (1:215, 289).
31. *Allgemeine Metaphysik*, sec. 142 (3:421–5). Herbart considered Kant's argument from the necessity of geometry to the postulation of "pure intuition" to be circular; for, he pointed out, in Kant's view geometry itself is based in pure intuition: sec. 241 (4:149–50).
32. *Einleitung*, sec. 137 (1:223–4); *Lehrbuch zur Psychologie*, sec. 156 (5:110).
33. *Einleitung*, sec. 152 (1:263–5); *Allgemeine Metaphysik*, sec. 143 (3:425).
34. *Allgemeine Metaphysik*, secs. 244, 253–63, 278, 297 (4:158–9, 181–202, 223–5, 255–6).
35. *Allgemeine Metaphysik*, sec. 243 (4:154–6).

36. Herbart's discussion of the need for continuity in mathematics reaches its peak in the *Allgemeine Metaphysik*, secs. 258–60 (4:191–7). He was not satisfied with what he called (in line with mathematical practice at the time) the "continuity" of the manifold of irrationals, constructed from the system of roots (sec. 242; 4:153–4); these did not, as he put it, possess a sufficient degree of density (*Grad der Dichtigkeit*). According to his argument, the most actual (*eigentlichste*) continua arise from functions that change (grow) in accordance with a rule, as the arc of the secant. He maintained that there can be no thought of constructing such a function out of points. He also rejected any attempt to construct such functions by "summing" differentials (as if these could be considered as parts of the arc) and reminded the reader that differentials themselves are not to be equated with points but indicate the growth of a function in accordance with a rule (secs. 258–9; 4:192, 195). He here reflected the standard interpretation of differentials in terms of "flowing" entities. Even as Herbart was publishing his *Metaphysik*, A. L. Cauchy was formulating the notion of a limit upon which H. E. Heine and K. Weierstrauss later built in replacing the notion of a "flowing" differential with a purely arithmetic notion (the latter two authors are credited with introducing the now familiar "epsilontic" definition of the limit); Weierstrauss distinguished continuity from differentiability.

37. Herbart discusses the distinction between intelligible and psychological space in *Einleitung*, sec. 157 (1:289–92). His account of the origin of psychological space will be described in subsequent paragraphs.

38. *Einleitung*, sec. 160 (1:309–13); *Psychologie als Wissenschaft*, sec. 143 (6:306).

39. *Einleitung*, sec. 157 (1:289); *Allgemeine Metaphysik*, sec. 142 (3:423).

40. *Hauptpuncte*, sec. 7 (3:25–30); *Einleitung*, sec. 157 (1:289).

41. *Psychologie als Wissenschaft*, sec. 144 (6:307); *Allgemeine Metaphysik*, sec. 141 (3:416–21).

42. *Einleitung*, sec. 157 (1:288); *Psychologie als Wissenschaft*, secs. 109–10 (6:114–19). This argument needs to be filled out with the premise that the relations among representations are not themselves attributable to mental structure but arise through experientially induced interactions among states of soul-Reals.

43. *Lehrbuch zur Psychologie*, preface (5:5). G. F. Stout provides an exposition of "The Herbartian Psychology," Mind (1888). For an historical treatment, see Leary, "Herbart's Mathematization of Psychology."

44. *Lehrbuch zur Psychologie*, secs. 179–180 (5:125–27).

45. Herbart, *Lehrbuch zur Psychologie*, secs. 10–32 (5:15–29); a more extensive treatment is provided in *Psychologie als Wissenschaft*, Pt. I, ch. 1, sec. 4, and chs. 2–3. For a brief account of Herbartian mental mechanics, see Ribot, *German Psychology of Today*, trans. Baldwin, ch. 1.

46. *Lehrbuch zur Psychologie*, sec. 29 (5:26–7).

47. Herbart, *Psychologie als Wissenschaft*, secs. 110–13 (6:117–29); *Lehrbuch zur Psychologie*, secs. 168–9, 174–5 (5:117–8, 119–21).

48. *Lehrbuch zur Psychologie*, sec. 173 (5:119). See also *Psychologie als Wissenschaft*, sec. 109 (6:116).

49. *Psychologie als Wissenschaft*, secs. 102, 110 (5:506, 6:118).

50. *Lehrbuch zur Psychologie*, sec. 173 (5:119). This brief passage includes virtually Herbart's entire account of the actual process by which the ability to represent a two-dimensional spatial manifold is acquired; in the more extensive *Psychologie als Wissenschaft*, sec. 111 (6:120–3), he gave an only slightly more detailed account of the process just described, and added an account of the acquisition of the concept of space containing movable objects, which he described as taking place by motion of an object relative to the background (sec. 114, 6:129–35), and an account of the acquisition of three-dimensional representation (ibid., 6:135–41).

51. Ibid., sec. 111 (6:122), first emphasis mine.
52. *Lehrbuch zur Psychologie*, sec. 174 (5:120); see also *Psychologie als Wissenschaft*, sec. 110, where Herbart attributes actual spatial relations only to "the represented" (the object of representation), not to the representation (6:117–8).
53. *Psychologie als Wissenschaft*, sec. 113 (6:127–8); on the conceptual character of space, see also *Lehrbuch zur Psychologie*, sec. 175 (5:120–1).
54. Beneke contested the charge that he had not properly acknowledged Herbart's influence: *Neue Psychologie* (Berlin, 1845). Although denying that Herbart influenced the original development of his position (he mentioned Schleiermacher, Jacobi, Tetens, Garve, Rousseau, Kant, and Fries, among others), he granted that he had found Herbart's psychology to be of interest (78–87).
55. Beneke, *Kant*, 89–98; *Neue Psychologie*, 91–4.
56. Beneke, *Lehrbuch der Psychologie*, 2d ed. (Berlin, 1845), chs. 3–6; *Neue Psychologie*, 64–7, 114–17. A useful study of Beneke's philosophy and psychology has been provided by Brandt, "Friedrich Eduard Beneke: The Man and His Philosophy."
57. I here use the term "psychology" advisedly, but in opposition to received opinion. While the existence of physiology is commonly recognized as dating from the early nineteenth century, the founding of present-day scientific psychology is dated from the third quarter of the nineteenth century when experimental techniques first began to be applied to psychological questions with some regularity. But a case can be made that psychology as a natural science had its conceptual origins in the eighteenth century. Kant acknowledged the presence of an empirical psychology that had pretensions to being a natural science, Fries, Herbart, and Beneke claimed either to be founding or to be furthering the aims of the nascent science, and Steinbuch and Tourtual distinguished "psychological" portions of their work from physiology and philosophy. Some histories set two starting dates for scientific psychology: the birth of the conception of psychology as a science in the work of Herbart and his contemporaries in the second and third decades of the nineteenth century, and the birth of experimental psychology after 1860—e.g., Flugel, *A Hundred Years of Psychology* (New York, 1933), Part I, ch. 1 and Part II, ch. 1. I certainly agree that by the early decades of the nineteenth century a number of thinkers conceived psychology as an autonomous scientific discipline; see the following subsections and chapter 7, sec. 1.
58. See Löw, *Philosophie des Lebendigen*, ch. 2; and also Hall, *History of General Physiology*, secs. 25–35.
59. See Roe, *Matter, Life, and Generation*, and Lenoir, *Strategy of Life*, ch. 1.
60. Kant, *Critique of Judgment*, trans. Meredith, Second Part. Original German edition published in Berlin, 1790; 2d ed., 1793.
61. Ibid., secs. 75, 78; see also first *Critique*, A687–8/B715–16.
62. Third *Critique*, secs. 65, 80–1.
63. On the relationship between Kant and Blumenbach, whom Kant mentions explicitly in the third *Critique* (sec. 81) and to whom he sent a copy of that work along with a laudatory letter, see Lenoir, *Strategy of Life*, 17–35.
64. See Johannes Müller, *Zur vergleichenden Physiologie des Gesichtssinnes* (Leipzig, 1826), for a not-unbiased discussion of the various theoretical positions in German physiology in the early decades of the nineteenth century. On the question of whether any forces or powers outside of those known to physics and chemistry were required to explain organic structures, opinion was divided especially over the question of morphological development. One group, which has been called "teleomechanist" (Lenoir, *Strategy of Life*, 24), denied that the development of the living organism could be explained through physical and chemical processes; they, as Kant, ac-

cepted a special formative force that directs the development of the organism. This force does not cause matter to violate the laws of physics and chemistry (indeed, it acts by directing physical and chemical processes in a manner appropriate to development), but it is not itself a merely physical or chemical force. A second group, which may be called "organic materialist" (I introduce this term to distinguish between Blumenbach and Reil, both of whom Lenoir includes under his term "teleomechanist"), accepted no forces or powers in living things except those known to physics and chemistry. Organization was understood to involve no special soul stuff or vital power, but it itself was thought to go beyond physics and chemistry in the sense that its origin in nature could not (at present) be explained as arising through ordinary physical and chemical processes: see Johann Christian Reil, "Von der Lebenskraft," *Archiv für die Physiologie*, 1 (1796), 8–162, (the issue actually appeared in July of 1795), reprinted as *Von der Lebenskraft*, ed. Sudhoff, esp. sec. 11. Both groups accepted functional organization as an irreducible part of nature; they differed over whether the laws of physics and chemistry are sufficient to explain the development of this organization.

65. Among the authors treated below, Tourtual explicitly maintained that the important aspect of comparison among the various sensory organs was not such incidental features as their shape but rather their functional organization: the "mode of operation" or "function" of the parts of a thing. For this reason, he proposed to compare the parts of the sensory organs according to their "role in the sensory process" (*Die Sinne des Menschen* [Münster, 1827], xiv-xv).

66. Steinbuch described his work as a contribution to "physiology," and specifically, to the "theory of the senses" (*Beytrag* [Nurnberg, 1811], v); he stated that he wished especially to rework the "psychological side" of the theory (vi). The character of his theory itself gives the meaning of "psychological" here, which is contrasted with the rest of physiology: Steinbuch focuses especially on the processes and laws involving mental representations, as opposed to the accompanying activity of the nervous system.

67. Steinbuch was cited prominently by both Tourtual and Müller. A more thorough review of the German literature on vision in the 1820s and 1830s would show that his work was widely discussed.

68. *Philosophisches Magazin*, 1 (1789–90), 360–71. In his presentation of the Kantian theory of the senses, Abicht characterized space as a *Grundvorstellung* of human sensibility, by which he seems to have meant that it was an innate component of sensibility.

69. Even Gehler had adopted the analysis of the retinal image into single sensations, and Tourtual accepted Steinbuch's nonspatial sensations without adopting his empirism (see sec. 2.2 below).

70. *Beytrag*, 1–5. He announces his agreement with "better physiologists and philosophers" that mental activities are always accompanied by brain processes (p. 4), and later (174) he names Reil as foremost among the adherents of this view. His "epigenetic" conception of embryonic development also agrees with that of Reil. On Reil, see note 64 above.

71. Ibid., 4, 23. Hereafter, page citations to Steinbuch's *Beytrag* are given in the text.

72. Steinbuch credited Gehler with proposing this hypothesis (ibid., 146), though Gehler was merely reporting a commonly held view.

73. Steinbuch considered the limbs to be moved by continuously varying degrees of contraction, in such a way that, for each position of the unencumbered limb, there corresponds a specific constellation of simple acts of will and states of muscular contraction, the ideational elements of which are qualitatively most similar to the

ideas that coincide with positions of the limb in the immediate neighborhood of this
first position (ibid., 30–1).

74. Steinbuch stated explicitly that these simple volitional ideas give "the *matter of ideas*,
obtained by the sense organ from without, its *form*, in order that it may be made into
a concrete sensory idea" (ibid., ix-x). He continued by making great claims for the
role of these ideas in perception: "Here, therefore, in these previously unknown
simple ideas of our simple spontaneity in the movement of our sense organs lies the
key that unlocks the door to the inner sanctum of the senses and the mind in general.
The simple fact that parallel to the outer movements of the sense organs there run
inner ideas of will is the principle upon which our sensory intuition is based, and out
of which our *subjective* space (the space in our intuition) arises through successive
development."

75. Steinbuch first introduced the laws of association on the basis of their common
acceptance among physiologists and psychologists. Only later did refer to the
possibility of explaining them by their underlying brain processes, in the manner of
Reil (ibid., 174). Reil's account of association ("Lebenskraft," 142–51) itself referred
to that of Erasmus Darwin, which had been influenced by Hartley. Although
emphasizing the possibility of physiological associations among muscle motions,
Reil admitted that the basis of association was not yet understood.

76. Steinbuch hastened to add that this mental representation is not the idea of an outer
line (external to the body), intuited by the outer sense organs. The intuition of
various *Bewegideen* in a series, as they have been arranged according to the law of the
association of ideas, is no more than the foundation of such a line-idea (*Beytrag*, 36–
37).

77. "If two, three, or more simple ideas come to consciousness simultaneously, they
become associated with one another, by virtue of this law, in such a way that all of
them appear together as but a single idea" (ibid., 39). As examples Steinbuch offered
the idea of green, a combination of blue and yellow, and the simple smell of fine
perfume, which is actually composed of many component smells. He continued:
"The same holds true of my *Bewegideen*, as simple ideas. If two, three, or more of
them simultaneously come to consciousness as sensations, i.e., with fresh liveliness
of appearance, they are raised together to a unity of consciousness. The properties
of their intuitional quality are thereby so thoroughly fused with one another,
amalgamated so to speak, that the combined product appears with a quality that
differs from the qualities of the parts (*Bewegideen*), and the phenomenal qualitative
nature of this product consists of a mixing of the elements or parts in the idea" (ibid.).
Steinbuch remarked that this process, by which elements are combined to yield a
product with properties differing from those of any of the elements, is similar to the
process of chemical combination (40).

78. Steinbuch's own account is quite explicit: "Hence the intuitional quality of the
individual *Bewegideen* within must therefore through similarity of appearance
grade into one another just as, externally, their corresponding spatial points grade
into one another spatially. These individual ocular *Bewegideen*, compounded from
the voluntary activity of all four right eye muscles, must therefore necessarily be
intuited as outside one another, and indeed be intuited outside one another in the
same order as that in which externally their corresponding spatial points lie with
respect to one another; i.e., they must appear as an extended surface, which is similar
to the outer [imaginary] surface [formed by the sweep of the optic axis], and in which
must be contained all of the compounded *Bewigideen* that are possible for this eye"
(ibid., 161–2).

79. One wonders why the muscles involved in accommodation of the eye for various distances, a process of which Steinbuch was aware (ibid., 208–9), could not serve the function of filling in the third dimension; perhaps he believed that the accommodative muscles are not exercised in utero.

80. Steinbuch argued that there could be no innate (he says "a priori") basis for the correspondence between retina and inner representation, either physiological or psychological: "That the connection of single points of the retina with corresponding points of the inner visual space could not be given a priori clearly follows from the fact that this inner visual space is itself acquired. Therefore this connection necessarily must also be acquired. Since however this connection relates the [nervous] *activity* of single retinal points with points of inner visual space, it can itself take place only by way of this activity. The formative period of this connection of the retinal surface with the inner visual space begins therefore *after the birth of the individual*, whereby the exterior light comes to excite the retinal points to their own specific activity" (ibid., 168–9). The difference between an adult and a child "who has not yet learned to see" resides in "inner psychical procedures" developed as a result of visual stimulation after birth (166).

81. In Steinbuch's words, "since the mind intuits the inner visual space as empty extension, there appears to it in the middle of this space a luminous point, which copies [*copirt*] the external luminous point in the world, or represents it to the mind, which must cognize it as the only luminous point at hand" (ibid., 173).

82. Steinbuch explained (ibid., 173–4) that the process of association also extends to—nay, is mediated by—an association between the organic activity underlying muscular contraction (and therefore underlying *Bewegideen*) and the organic activity produced by stimulation of the retina (which underlies a sensation of light and color).

83. Steinbuch makes the point quite explicit: "Once the association of the inner visual space with the retina of its eye has come about, then henceforth any external extended object of vision can, through its mediation, be transported to the mind as such, and intuited in inner visual space as extended, since the retinal points that are affected by the object's retinal image necessarily must transmit their respective activity through the optic nerve to the points of inner visual space that have been associated with them and which manifest themselves in the same relative intuitional situation as obtains among the points contained in the retinal image" (ibid., 180).

84. Ibid., 219: he says that the apparent distance experienced in the case of illusions should be ascribed "not to a false judgment or inference of the soul, but to a deficient intuition" (which is nonetheless the product of normal associative processes).

85. Ibid., 106–35 covers the case of the touch and 263–9 the case of vision.

86. Steinbuch was quite explicit about the faithfulness of this correspondence (268–9):

> Should this individual for example close his eyes, or hold his hand before them, the inner and the outer will occur together: the inner man will (in representation) by means of his hand impede sight of the visual objects of his external world [i.e., *the world intuited as external to his intuited self*] in the same moment of time in which his external hand (the hand or the thing in itself) prevents the external light rays from entering his external eyes; the inner physical world of this individual is a faithful replica [getreue Nachbildung] on a reduced scale [of the outer], and the voluntary movements of the eye and tactual organs, etc. are the instruments through which this picture [Abbildung] of the external (world in itself) came into being. This individual will not be able, however, to detect by means of the senses that he intuits himself and his world only internally, on a reduced scale, since he has no means to discover this; for he can through his sense organs never become acquainted with the external world in itself, and therefore he is never in the position to undertake a comparison

between his inner world (in representation) and his outer world (in itself), between his inner body and his outer body.

The world in itself and our representations of it are in correspondence, inasmuch as the spatial relations true of our hand in itself, physical light rays, and our eyes in themselves are represented faithfully in the intuited spatial relations among our intuited hand, intuited light rays, and the intuited location of our eyes.

87. Steinbuch explicitly relied upon authority in invoking the principle of association (ibid., 35), and it is easy to suppose that he could accept the operation of these laws uncritically (or leave their operation unexplained) because of the extent to which they found acceptance among his contemporaries (see chapter 3, sec. 2 above, and Dessoir, *Geschichte der neueren deutschen Psychologie*, 232–48 and 391–7). His appeal to a physiological basis for associative connections (*Beytrag*, 174) provided no help in explaining the fact that in some instances the product of associating nonspatial sensations is a spatial representation.

88. *Prolegomena to Any Future Metaphysics*, 258.

89. Page citations to the *Sinne des Menschen* (Münster, 1827) will appear in the text. The word "Aesthetik" in the full title does not refer to the theory of beauty or of taste, but to the theory of the senses (as in Kant's use of the same term). Tourtual later published *Die Dimension der Tiefe* (Münster, 1842), on binocular vision.

90. *Sinne des Menschen*, xii, lx, 229–30, 303–4. Tourtual did not use the term "nativism," but spoke of space and time as *angeboren, inwohnend* or *immanent* forms of perception. He contrasted space and time with sensory qualities such as color, denominating the latter as "immanent energies" (lv, 22–3, 139–40). Tourtual considered the doctrine that space is an innate form of sensibility to be what he termed a "physiological and psychological" hypothesis. On the history of term "nativism," see Appendix A.

91. Tourtual's "pragmatic attitude" is best revealed in his discussion of our limitation to our own subjective world. Having admitted that we have no direct access to things in themselves via intellectual intuition (ibid., 12), he maintained that the best we can obtain is a scientific conception of the world through a comparison of the represented objects of the various senses among themselves. He maintained that the understanding "instinctually" refers the objects of the various senses to the spatiotemporal order apprehended through sight. Natural scientific investigation completes and refines this natural tendency, so that a unified account of the physical basis of the objects of the five senses is obtained. He labeled the result of this investigation the "comparative sensory object," contending that "despite our lack of acquaintance with actual things, we can with rigorous certainty make the assertion *that actual things are related to one another as are their comparative objects*" (20). This certainty is not the certainty demanded in traditional metaphysics (nor did Tourtual so consider it); it constitutes the surest conclusion available through the reasoned use of the senses (and so, in Tourtual's view, meets the standards of rigor).

92. Ibid., v. If the general division into physiological and transcendental and the notion of a preformed content of sensibility are not a sufficient sign of Tourtual's Kantian inclinations, he tells us outright that he is familiar with and has been influenced by no other philosophical theory of the senses than Kant's (xii, lx).

93. As he put it, "no science can endure as such without metaphysical foundations" (ibid., vi). He attributed the conception of a purely factual, nonphilosophical science to his "southwestern neighbors," a phrase most likely intended to indicate the French positivists.

94. "I am not subject to the charge of having transplanted psychological facts into the ground of physiology. I have become convinced that the human mind and body are

connected in the unity of a living thing, and that consequently physiology, as the science of vital phenomena, has mental activity as its object, no less than bodily activity" (ibid., vii). He goes on to talk about a connection of "psychic" with "somatic" physiology, which may lead to a better understanding of mind-body interaction. In all of these discussions, it is clear that he considers the mental to be distinct from the bodily sensory processes, though he leaves open whether the two can ultimately be understood under universal laws of sensibility (see x, 3).

95. In comparing the sensory systems within a single type of organism and across species too, Tourtual emphasized similarities in function as opposed to homomorphism; his position was in agreement with the physiology of K. E. von Baer as opposed to that of K. F. Burdach (see Lenoir, *Strategy of Life*, 76).

96. Tourtual considered the point that the perceiver is always active in constructing a sensory representation to be among the fundamental teachings of his theory; *Sinne des Menschen*, 139–40, 319.

97. Tourtual contended that Euclid's geometry applies with necessity to subjective space, but he acknowledged that we have no guarantee that nature always conforms to this space: "Each spatial sensory representation is therefore essentially geometrical, and if there were to be spatial relations in nature which were geometrically incorrect—which we are, from the subjective standpoint, no more able to allow than the existence of a logical absurdity, since geometry is the logic of the faculty of intuition—they could never be found in intuition" (ibid., 229–30; Tourtual considered the possibility that our senses might filter out relations that do not conform to Euclid's geometry to be a limitation on our knowledge of the thing in itself). His account of geometry is discussed below.

98. Tourtual answered "yes" to Molyneux's question (ibid., 264–65). He denied the familiar doctrine that "touch educates vision": each sense has its own spatial abilities, but in the actual case, he observed, vision usually guides touch (ch. 13, especially 305).

99. "The form of the impression in the sense organ gives the higher senses [vision and touch] the inducing impetus; they accordingly accommodate the sensations [resulting from the impression] to their indwelling geometry, and so intuit the object of sensation in geometrical form. The faculty of spatial creation in sensibility [as it acts during perception] is only a closer determining of spatial relations in general, just as corporeal form is itself only a delimitation in the dimensions of real space" (ibid., 142).

100. Ibid., 229. Tourtual differed from Kant as well as from Fries (whose work he frequently cited) on the proper interpretation of this general doctrine. He denied that the laws of sensibility can be revealed independently of actual material sensations—there is no "pure space of sensibility," if that means a space without quality (23–26)—and he saw no reason to distinguish the manner in which the matter and the form of intuition have their foundation in the innate structure of the percipient (22–3):

> Both the matter and the form of outer representation are given neither in a purely a priori nor a purely a posteriori manner. Sensation consists in nothing else than the awakening to consciousness of the energy immanent in the senses; the faculty of sensation accordingly presupposes the existence [Vorhandenseyn] in the mind of the material of representation before all impingements from the external world in the same way as the form of representation must already be given in the faculty by which this material of representation is spatially and temporally intuited.

It is likely that Tourtual misunderstood Kant's notion of the pure space of sensibility: Kant probably would not have allowed for the intuiting of form without some type of matter. However, Kant's position on the "matter" of pure sensibility was by no means clear.

101. Kant gave a similar argument against the position that the ability for spatial representation is acquired (first *Critique*, A23/B38). Tourtual also argued against Steinbuch's claim that spatial representation is always mediated by *Bewegideen* by appealing to cases of insects with fixed eyes that nevertheless discriminate space (226) and persons without arms who can localize on areas of their skin that they have never touched (227), contrary to the results predicted on Steinbuch's account.

102. Tourtual refused to attribute the laws of nature to the understanding innately; rather, he attributed to the understanding a penchant for seeking order in the materials provided by the senses, as a result of which it develops a physics based on experience (*Sinne des Menschen*, ch. 14). By contrast, ascribing laws of spatial organization to the senses does not attribute to them an implicit "theory" of space but a functional endowment suited to a spatially ordered environment.

103. Tourtual's introduction of the notion of a locational sign is worth quoting in full:

Each point of the retina and of the papillary body of the tactual organs has something distinct from all other points, which depends upon its location, and which as such passes into the quality of sensation [produced by stimulation of this point]. This specific sign [Merkmal], then, is that which gives the sense the impulse to determine the [spatial] relation of the sensation according to the appointed laws. This sign cannot, however, already in itself be spatial, but can be only a modification of the perception of the impression, since through this sign the representation of space is first made possible. We have no need, therefore, to postulate a precise relationship of spatiality between the origin of the optic nerve fibers in the brain and the elements of the retinal image. [ibid., 185–6]

104. Compare Tourtual's discussion of continuity with that of Herbart as discussed in sec. 1.2 of this chapter. The sensory apprehension of the spatial relations of an object does not take place instantaneously but in a temporally extended process that usually goes unnoticed. Spatial perception does not proceed like "the impressing of a seal" but like the activity of drawing, which he compared to attention in its active nature (ibid., 174–5). Tourtual reasoned that since, in general, attention is limited to one part of the visual field at a time, the process underlying the production of visual intuition operates by grasping various elements of visual sensation successively and referring them to a spatial location.

105. Here is Tourtual's explication of the process (ibid., 189):

The sensational signs of the points that are given in the nervous structure stand to one another in the same relations of immediate contiguity and continuous succession that actually occur in the lineal abreastness of their object-points [the points of the external object, presumably the retinal image for vision], and the attentional activity of the mind proceeds like the copying hand of the draftsman from one sign to another in continuous temporal succession; by remaining in possession of the previous representation during the representing of each successive point, the mind puts the former together with the latter into a total representation, and genetically develops in this manner its subjective lines, just as objective lines are developed through motion.

Tourtual remarks that since the process of spatial representation occurs in time, any intuition of an extended line must depend on memory. Only simple forms are drawn all at once; with more complex forms, the parts are first "drawn" and then put together.

106. Here is Tourtual's account (ibid., 215):

Since in the case of stationary eyes the representational activity of the visual sense moves here and there over the image [of intuited surface relations], in each moment of motion that degree of contraction of the four right muscles [recti] which corresponds to the point intuited just then arises simultaneously [with the intuition of the point], so that during the [act of determining the intuited] surface-relations, the acts of accommodation [Refraktionsacte] form a continuous series similar to the projection of the lines [from the retinal point to the object] in the third dimension. The [spatial] relation of each point is accordingly determined simultaneously by the sensational sign of the retinal point and the accommodative act accompanying this sign, and thereby lines rendered in three-dimensional space are genetically developed according to length, breadth, and depth.

107. On the charge that nativism is a lazy-person's theory, see Boring, *History*, 249.
108. For our purposes Müller's most important works are *Zur vergleichenden Physiologie des Gesichtssinnes des Menschen und der Thiere* (Leipzig, 1826), hereafter, "PG," and *Handbuch der Physiologie des Menschen*, 1st ed., 2 vols. (Coblenz, 1834–7), hereafter "HP."
109. Müller's methodological discussions occur in PG, ch. 1, and HP, "Prolegomena on General Physiology" (vol. 1).
110. It treated binocular single vision most thoroughly of topics concerning spatial vision (on which, see chapter 5, sec. 2).
111. He further maintained that "before the education of the senses, before the cultivation of other intellectual faculties, at the primordial level, when an animal can in general be regarded only as sensible, a division of the inner alterations of the self into those for which the animal is itself the foundation and those for which the cause of excitation is distinct from the animal is unthinkable" (PG, 40).
112. See also PG, 51, and 54–5: "Was aber den erfüllten Raum betrifft, so *empfinden* wir überall nichts, als nur uns selbst räumlich, wenn lediglich von Empfindung, von Sinn die Rede ist; und so viel unterscheiden wir von einem objectiven erfüllten Raum durch das Urtheil, als Raumtheile Unserer selbst im Zustande der Affection sind, mit dem begleitenden Bewusstsein der äussern Ursache der Sinneserregungen."
113. Writing in 1827, Tourtual described Müller's position as a "physiological idealism" and classified him together with Kant (*Sinne des Menschen*, 144–45); forty years later, Helmholtz adopted the same grouping (*Handbuch der physiologischen Optik*, Pt. 3, historical appendix to sec. 26).
114. Valentin, *Lehrbuch der Physiologie des Menschen* (Braunschweig, 1844), 2:458 (Valentin explicitly rejected Steinbuch's position, 450, 490); Volkmann, "Sehen," in *Handwörterbuch der Physiologie*, ed. Wagner (Braunschweig, 1842–53), 3:316–17, and *Physiologische Untersuchungen im Gebiete der Optik* (Leipzig, 1863–4), 166.
115. Steinbuch is cited by Müller, PG, xv and HP 2:380; Berthold, *Das Aufrechterscheinungen der Gesichtsobjecte* (Göttingen, 1830), 57–68; Erpenbeck, *Das Recht-Sehen trotz des umgekehrten Sehbildes auf der Retina* (Würzburg, 1833), 19; and Hasenclever, *Die Raumvorstellung aus dem Gesichtssinne* (Berlin, 1842), v-vi. Both Steinbuch and Tourtual are cited by Hueck, *Das Sehen seinem äussern Process nach entwickelt* (Riga and Göttingen, 1830), 5, 14, 17–18, 52, 65, etc., and Bartels, *Beiträge zur Physiologie des Gesichtssinnes* (Berlin, 1834), v-vii, 14. Müller was, of course, cited with similar frequency, but several authors took up the side of Steinbuch or Tourtual against Müller on the question of whether spatial sensations are innately given.
116. Weber, "Der Tastsinn und das Gemeingefühl," in Wagner, *Handwörterbuch*, vol. 3, pt. 2, 481–543, and "Ueber den Raumsinn." See also Boring, *Sensation and Perception*, 465–6, 475–7.
117. Weber, "Tastsinn," 489–93, 524–9.
118. Ibid., 486:

wir durch die reine Empfindungen ursprünglich gar nichts über den Ort wissen, wo auf den die Empfindung vermittelnden Nerven eingewirkt wird, und dass alle Empfindungen ursprünglich nur unser Bewusstsein anregende Zustände sind, welche dem Grade und der Qualität nach verschieden sein können, aber unmittelbar keine räumlich Verhältnisse zu unserem Bewusstsein bringen, sondern nur mittelbar, durch die Anregung einer Thätigkeit unserer Seele, mittelst deren wir uns die Empfindungen vorstellen und in Zusammenhang bringen, und zu welcher wir durch eine angeborne Seelenanlage oder Seelenkraft angetrieben werden.

119. Ibid., 482–3, 527–9. Weber's emphasis on the role of experience in localization is more explicit in "Ueber den Raumsinn," 118–25, 131–3, in which his commitment to an innate ability for spatial representation is not made explicit; indeed, in the later article he may have been flirting with the position that the bare ability for spatial representation is itself acquired (see p. 109 on the construction of spatial representations from "Raumelemente" acquired through experience), although he repeatedly spoke of spatial representations being "awakened" in us through experience (e.g., 86, 109).

120. E.g., Ludwig, *Lehrbuch der Physiologie der Menschen*, 2d ed. (Leipzig and Heidelberg, 1858–61), 1:319–21.

121. The role of physiological hypotheses and their interaction with postulated psychological processes was quite complex. Weber himself emphasized the geometrical correspondence between the arrangement of fibers in the periphery and that in the brain, without supposing that this fact by itself explained the spatial ordering of sensations ("Ueber den Raumsinn," 85–6, 129). As we shall see in the subsequent section, Lotze believed that the spatial ordering of the nerve fibers was of no consequence. Johann Czermak reviewed the disagreement between Weber and Lotze and adopted a compromise position that posited a point-for-point anatomical projection into the brain, but ascribed the formation of spatial representations to the physiologically-mediated influence of the (physical) proximity of the nerve fibers on the production of "local signs:" "Physiologische Studien," 15: 507–13, 17:577–81. Virtually every German physiologist who adopted the hypothesis of punctiform sensations around mid-century subscribed to the view that these sensations contain qualitative signs for their location on the basis of which they are localized through a mental process. Some, such as Weber, emphasized the role of learning in this process, whereas others, such as Lotze, postulated an innate mechanism to account for some portion of this localization.

122. For a discussion of the importance of local signs for later theory, see Boring, *Sensation and Perception*, 28–32 and 232–5. A fuller account is available in Ribot, *German Psychology*, ch. 3. Boring and Ribot both treat Lotze as an early radical empirist and their classification has stuck. As we shall see, it is incorrect.

123. Lotze received considerable recognition for his article "Leben, Lebenskraft" in Wagner's *Handwörterbuch der Physiologie* (1842), 1:ix–lviii; for discussion, see Lenoir, *Strategy of Life*, 168–72.

124. See Schnädelbach, *Philosophy in Germany, 1831–1933*, trans. Matthews, ch. 6, sec. 2.i. On Lotze's life and philosophical work, see also Falckenberg, *Geschichte der neuren Philosophie*, ch. 16, sec. 2.

125. Lotze, *Medizinische Psychologie oder Physiologie der Seele*, hereafter "MP," (Leipzig, 1852), Bk. 1, chs. 1–3.

126. See also *Mikrokosmus* (Leipzig, 1856–64), trans. Hamilton and Jones, hereafter "HJ," as *Microcosmus* (Edinburgh, 1885), Bk. 3, ch. 2, sec. 4 (1:306–8). The chapters of the German edition are broken into sections, but these sections are numbered only in the translation.

127. Lotze presented the theory of local signs in three major publications between 1846 and 1856: first in "Seele und Seelenleben," in Wagner's *Handwörterbuch* (1846), then

in MP, and again in the first volume of *Mikrokosmus* (1856), Bk. 3, ch. 2, sec. 4 (HJ 1:309). MP presented the theory in the greatest detail (ch. 4); the term "local sign" was introduced on p. 331.

128. On the reception of the theory of local signs, see Boring, *Sensation and Perception*, 233–235 and Woodward, "From Association to Gestalt." I discuss the views of Wundt, Hering, and Helmholtz in chapter 5.

129. See also *Mikrokosmus*, Bk. 3, ch. 3, sec. 2 (HJ 1:319–20).

130. MP, 359–62; Lotze rejected the Herbartian contention that only one point at a time actually comes to intuition, without mentioning any names (381–3).

131. Lotze maintained that the description of a process by which spatial representations develop from wholly nonspatial beginnings must inevitably presuppose that the mind in question has the ability to form spatial representations; the ability is presupposed even if the representations develop only in the course of experience, just as we must be ascribed the ability to have qualitatively distinct sensations even though such sensations arise only through stimulation: MP, 334–6, 377–82. See also *Mikrokosmus*, Bk. 3, ch. 2, sec. 4 (HJ 1:308).

132. On Berkeley and other early writers, see chapter 2, sec. 2.

133. Ribot, *German Psychology*, 70; Boring, *Sensation and Perception*, 28–32, 232–5; Woodward, "From Association to Gestalt," 574.

134. This aspect of Lotze's position was adopted by Georg Meissner, *Physiologie des Sehorgans* (Leipzig, 1854), sec. 54 (102–7), who accepted a punctiform analysis of visual sensation and attributed the origin of a two-dimensional representation to an innate psychological mechanism, citing his concurrence with Lotze in this regard (105).

135. On the varying degrees of nativism and empirism in nineteenth-century German sensory physiology and psychology, see Appendix A; see also chapter 5, sec. 2.

136. *Mikrokosmus*, Bk. 2, ch. 4, sec. 3 (translation modified from HJ 1:232).

137. Ibid., Bk. 9, ch. 2, secs. 1–2 (HJ 2:605, 611).

138. Ibid., Bk. 9, ch. 2, sec. 2 (HJ 2:611). Lotze characterized Kant as treating "den Raum als subjective Form nur menschlicher Anschauung" (as opposed to the possible forms of intuition of other knowing beings), which he equated with a position that assigns "den Raum zu dem angebornen apriorischen Besitz unsres Geistes."

139. Ibid., Bk. 9, ch. 2, sec. 1 (HJ 2:602–10).

140. Ibid., Bk. 9, ch. 2, sec. 1 (HJ 2:607–8).

141. *Logik* (Berlin, 1874), secs. 350–8; trans. Bosanquet, *Logic*, 2d ed., 2:294–313.

142. Quotations from Lotze's *Logik*, sec. 357 (*Logic*, 2:311).

Chapter Five, Helmholtz

The following abbreviations are used in the text and notes of this chapter to refer to Helmholtz's works:

EW *Epistemological Writings*, ed. Cohen and Elkana (Boston, 1977). This volume translates a collection of Helmholtz's papers on geometry, arithmetic, and perception published in Berlin in 1921 with extensive commentary by Paul Hertz and Moritz Schlick.

PO *Handbuch der physiologischen Optik*, 3d ed., 3 vols. (Hamburg and Leipzig, 1909–11). The third edition is a reprint of the first (Berlin, 1856–66; reissued, 1867). Page citations will be to the third edition (volume and page number), the pagination of which is indicated in James Southall, trans., *Treatise on Physiological Optics*, 3 vols. (Milwaukee, 1925). My translations begin from Southall's but may differ significantly. The third edition contains extensive notes by its editors, A. Gullstrand, J. von Kries, and W. Nagel.

PO2 Second edition of the *Physiologischen Optik* (Hamburg and Leipzig, 1886–96).

SW *Selected Writings of Hermann von Helmholtz*, ed. Kahl (Middletown, Conn., 1971). I have emended the translations on occasion. Kahl provides a chronological bibliography of Helmholtz's publications.

VR *Vorträge und Reden*, 4th ed., 2 vols. (Braunschweig, 1896); cited by volume and page number.

WA *Wissenschaftliche Abhandlungen*, 3 vols. (Leipzig, 1882–95); cited by volume and page number.

1. *Ueber die Erhaltung der Kraft: Eine physikalische Abhandlung* (Berlin, 1847). Reprinted in WA 1:12–75; Kahl's revision of John Tyndall's translation, "On the Conservation of Force: A Physical Memoir," appears in SW, 3–55.

2. Helmholtz, *Die Lehre von den Tonempfindungen als physiologische Grundlage für die Theorie der Musik* (Braunschweig, 1863).

3. The first part of the first edition appeared in 1856, the second—along with several pages of sec. 26 of the third—in 1860, and the rest of the third in 1866; the work was issued as a whole in 1867.

4. The most complete source of biographical information on Helmholtz is Leo Königsberger, *Hermann von Helmholtz*, 3 vols. (Braunschweig, 1902-03), hereafter cited as Königsberger, plus volume and page numbers; an abridged English translation was produced by Welby (London, 1906). Königsberger reproduced a significant number of letters and unpublished writings in his chronological account of Helmholtz's professional and personal life.

5. It is clear from Helmholtz's "Autobiographical Sketch" of 1881 that he hoped his philosophical writings could be reckoned alongside what he had accomplished using experimental and mathematical methods: VR 1:13–14, 16; SW, 473, 475.

6. Helmholtz and his contemporaries did not initiate the back-to-Kant movement. Beneke, who called for a return to the "true" epistemological Kant in 1832, taught philosophy and psychology at the University of Berlin from 1827–54; Helmholtz took courses at the University from 1838–43 in conjunction with his medical studies, and he had coursework in logic and psychology from Beneke (Galaty, 46, n. 22).

7. Helmholtz, *Die Thatsachen in der Wahrnehmung* (Berlin, 1879); trans. in EW, 115–63.

8. Königsberger, 2:72; the "epistemological" portion of the work was especially to blame (1865). See also his weary references to the last part of the *Optiks* in his correspondence with Emil du-Bois Reymond, in *Dokumente einer Freundschaft*, ed. Kirsten (Berlin, 1986), letters 96, 97, and 101, in which he calls this part of the work "wicked" and "thankless."

9. PO 3:26–31; lectures of 1862, 1868, and 1878, in SW, 131, 215, 222 and EW, 132, 136, 138. Erdmann, "Philosophischen Grundlagen," stresses this connection.

10. Helmholtz did not use the term "unbewusster Schluss" in the 1855 lecture *Ueber das Sehen des Menschen* (Leipzig; also in VR 1:85–117), but he used equivalent language. The 1894 paper was "Origin."

11. On the use of the term "empirist" to translate Helmholtz's "*empiristich*," see Appendix A.

12. Recent scholarship has tended to emphasize the Kantian influence on Helmholtz, in opposition to Boring's treatment of Helmholtz as an empiricist (*History*, ch. 15). Those who have emphasized the Kantian influence on Helmholtz include Galaty, "Emergence of Biological Reductionism," 159–66, Pastore, "Helmholtz on the Projection or Transfer of Sensation," 355–76, and Leary, "Immanuel Kant and the Development of Modern Psychology," 36. I agree with these authors in maintaining that Helmholtz subscribed to certain Kantian doctrines for a time, but I depart from

their position in emphasizing the extent to which Helmholtz's relation to Kant developed and changed.

13. Helmholtz, "Facts in Perception," EW, 117.

14. The statement is Helmholtz's own summary of his life-long position, in his "Autobiographical Sketch" of 1891. The first statement of the doctrine, in which Helmholtz spoke of color sensations as "symbols" (*Symbole*) rather than "signs" (*Zeichen*), occurred at the end of a lecture from 1852 entitled "Ueber die Natur der menschlichen Sinnesempfindungen," in WA 2:591–609.

15. "On the Conservation of Force," SW, 3–4, 49 (WA 1:12–13, 68); the higher page numbers refer to an appendix dated 1881 and published in 1882.

16. The Dutch Kantian J. P. Land published a scathing attack on Helmholtz's theory of geometry in "Kant's Space and Modern Mathematics," *Mind*, original series, 2 (1877), 38–46; he was replying to Helmholtz's "The Origin and Meaning of Geometrical Axioms," Mind, 1 (1876), 301–21. Some years later Ludwig Goldschmidt published an impassioned defense of Kant, *Kant und Helmholtz* (Hamburg and Leipzig, 1898). Among the more widely known neo-Kantians, Hermann Cohen appreciated Helmholtz's role in the "back-to-Kant" movement while at the same time rejecting any view that equated the a priori with the innate: *Kants Theorie der Erfahrung* (Berlin, 1871), iii and ch. 7. In a much expanded second edition (Berlin, 1885) Cohen defended Kant's treatment of geometry (222–37) against the objections raised by Riemann and Helmholtz regarding the application of geometry to physical space, contending that both of them had misunderstood the precise status Kant ascribed to space as the form of intuition and that neither of them had successfully shown that it is contingent which geometrical axioms describe physical space.

17. The juxtaposition is especially striking in PO, sec. 26, and in "Facts in Perception" (see sec. 4 of this chapter). Erdmann, "Grundlagen," notes this tension.

18. In his lecture on vision of 1868, entitled "Recent Progress in the Theory of Vision," Helmholtz explained the relation between retinal elements and sensations as follows: "We may assume that a single nervous fibril runs from each of these cones through the trunk of the optic nerve to the brain, without touching its neighbors, thereby transmitting the sensory impression, so that the excitation of each individual cone can produce a sensation distinct from the others" (VR 1:278; modified from SW, 153; see also SW, 165–8).

19. The use of the term "psychophysics" serves a useful function in this context, although it is anachronistic. Helmholtz described the primary aim of the third part of the *Optics* as that of discovering "the conditions which give rise to visual perceptions," among which he included the "physical stimulus and the physiological stimulation" that are responsible for various representations of external objects, as well as the psychological laws by which such representations are formed (PO 3:3). He conceived his project as one of discovering the relation between stimulus conditions and perceptual representation, which makes his project "psychophysical" in the broad sense in which that term is used today.

20. The bulk of Helmholtz's work was not theory-independent *simpliciter* but rather against the background of shared assumptions from the 1860s and 1870s. Later visual theorists have not, for the most part, questioned the validity of the facts Helmholtz gathered under the conditions he gathered them; they have questioned the theoretical motivations in accordance with which he gathered and interpreted those facts. In particular, the assumption of punctiform sensations that underlay much of Helmholtz's theorizing, and which he shared with his nativist rivals, was questioned by succeeding generations and most notably by the Gestalt psychologists: Köhler, *Gestalt Psychology* (New York, 1929), chs. 3–5, and Koffka, *Principles of Gestalt*

Psychology (London and New York, 1935), ch. 3. More recently, Gibson has questioned the approach shared by Helmholtz and his contemporaries in an even more fundamental way, beginning with the traditional analysis of optical stimulation and proceeding to the central role allotted to "sensations" postulated to arise from the stimulation of individual sensory receptors: *Perception of the Visual World* (Boston, 1950) and *The Senses Considered as Perceptual Systems* (Boston, 1966).

21. Helmholtz observed that "the rule given above has been stated in a form that does not anticipate the decision of" the question of the extent to which perception depends on prior experience; "it can be accepted even by those who have entirely different opinions as to the way representations of objects in the external world come into being" (PO 3:7).

22. The relation of visual direction to the terrain would require taking body posture and head rotation into account; Helmholtz explicitly abstracts from these conditions in his discussion of vision (PO 3:236).

23. Helmholtz did not take an unequivocal stand on the question of whether the genesis of visual spatial abilities depends upon the sense of touch. At PO 3:135, he describes the learning process in purely visual terms; at p. 139 he describes the process as dependent upon touch; at p. 433 he says that vision is not the only spatial sense, for touch is adequate for the sense of space in those born blind. He described prism experiments in which touch captures vision (206–7) without concluding that touch is the primary spatial sense. In practice, he helped himself to tactual spatial abilities freely in explaining the development of visual spatial abilities.

24. This question was not addressed in PO; it is implicitly raised in the 1868 lecture ("Recent Progress") when Helmholtz excuses himself from answering it (VR 1:356–7, SW, 216); in the "Facts in Perception" he maintains that space is a "transcendental" form of intuition, by which he simply means that it depends upon motor impulses of the will, the capacity for which we have an innate endowment—an answer that perhaps comes as close as possible to saying that the spatial form of intuition is both innate and acquired (EW, 122–5).

25. Wundt, *Beiträge zur Theorie der Sinneswahrnehmung* (Leipzig and Heidelberg, 1862), 151, 166–7. Wundt considered his theory to incorporate aspects of both the "motor theory" of Lotze and the "sensory fusion" theory of Waitz, which he saw as simply two sides of Herbart's theory of the development of the "construction" of the field of sight (100–104).

26. PO 3:47–9, 54, 192, 204–5. Helmholtz was willing to allow (but did not consider it necessary to do so) that local signs form a qualitatively well-ordered series, and that their qualitative similarity aids in the development of direction within the field of sight of a stationary eye (3:189–90, 436); but that is a different point.

27. This point is obscured in Southall's translation by the rendering of *Sehfeld* as "visual globe" (PO 3:137). The translation of "feld" as "globe" led to an implied reification of the two-dimensional representation: "in the field of sight" (*im Sehfelde*) became "on the visual globe" throughout Part Three (3:138, etc.).

28. For a recent discussion of the similarity between motion parallax and stereopsis, see Rogers and Graham, "Similarities between Motion Parallax and Stereopsis" (1982). As is still common in vision research, they cite Helmholtz's PO.

29. Helmholtz characterizes the cue to motion parallax as a difference between the "apparent" (*scheinbar*) motions of objects at different distances (PO 3:247).

30. PO 3:248–60, 270–99.

31. PO 3:5–7, 23–5; "Recent Progress," SW, 217, 220 (VR 1:358, 361–2).

32. One may get a sense of how various authors conceived the fundamental differences among theories of perception by considering how they portrayed the history of

theories of vision. Müller (the inventor of the law of specific nerve energies) organized his history around the question of the subjectivity of sensory representations (*Vergleichenden Physiologie des Gesichtssinnes*, v-xvii). Tourtual used the same framework (*Sinne des Menschen*, xxxiii-lx), even though he recognized that the innateness question served to separate his own position from that of his admired predecessor, Steinbuch. Wundt introduced two themes into his historical discussion: subjectivity and innateness (*Beiträge*, Second Part). Helmholtz organized the historical sketches at the end of the opening and closing sections of Part Three (PO, secs. 26 and 33) around the dispute between empirism and nativism.

33. Waitz, *Lehrbuch der Psychologie als Naturwissenschaft* (Braunschweig, 1849), argued that simultaneously obtained sensations, in forcing themselves into consciousness, must be represented as next to one another and that they become "fused" into a spatial representation (166–79). Cornelius, in his *Die Theorie des Sehen und räumlichen Vorstellens* (Halle, 1861), argued that spatial representation arises with the development of "reversible series" of representations (591–3); see also his *Zur Theorie des Sehens*. In the Herbartian tradition, both these authors emphasized the need for an autonomous science of psychology founded upon a dynamical view of the associative and inhibitive interactions of simple representations.

34. Classen, "Beitrag zur physiologischen Optik" (1862); Wundt, *Beiträge*, 28, 145–69.

35. Nagel, *Das Sehen mit zwei Augen* (Leipzig and Heidelberg, 1861), held that we project sensations externally according to the law of cause, which he accepted, along with the spatial and temporal forms of intuition, as an a priori given (174–7; he cited Schopenhauer in this regard). Such projection apparently takes place without learning, yielding a two-dimensional surface; experience is required to fine-tune the direction of the projection and also for the perception of distance (12–16, 177–80). Classen came to argue that the retina immediately, independently of muscle action, senses the two-dimensional pattern of light that falls upon it. While he thus denied the radical empirism of Steinbuch and Wundt in favor of Müller's view on the innateness of spatial representation, he agreed with Wundt that perception of a three-dimensional world is the result of a logical process of inference: Classen, *Das Schlussverfahren des Sehactes* (Rostock, 1863), v, 2–3. Helmholtz recognized that Classen was a nativist respecting the perception of two-dimensions but classified him as an empirist anyway; he seems not to have read Nagel as a nativist on any score: see PO 3:33, 454. Later, Classen criticized the empirism of both Wundt and Helmholtz and sought to base his theory of perception on thoroughly "Kantian" foundations: *Physiologie des Gesichtssinnes zum ersten Mal begründet auf Kants Theorie der Erfahrung* (Braunschweig, 1876).

36. Nagel, *Sehen*, 93–4; Classen, *Schlussverfahren*, 65–9, 75. See PO 3:33, 454.

37. Hering (1834–1918), *Beiträge zur Physiologie* (Leipzig, 1861–1864), 132–6, 146–7, and foreword to the fifth part (1864), iv; see also 161–2, 290.

38. According to Hering, there are two factors involved in space perception:

> Visual space, that is the embodiment of all of the intuitions that we have at any instant, is a product of our sensorium and arises from the cooperation of two factors, first, the light sensations and spatial sensations that are elicited directly, on the basis of an inborn mechanism, by the two retinal images [das Doppelnetzhautbild], and, second, the property of the sensorium that is conditioned by the countless experiences, judgments, and inferences through which the sensorium is, in the course of life (so to speak), transformed. [ibid., 323]

From his point of view, the distinction between nativism and empirism was not particularly significant and merely a matter of degree; all investigators must accord some role to both innate mechanisms and subsequent learning. As he saw it, the distinction did not mark a division between physiology and psychology, since learning must be given a physiological basis; see especially his *Zur Lehre vom Lichtsinne*, 2d ed. (Vienna, 1878), 1–5.

39. Schleiden, *Zur Theorie des Erkennens durch den Gesichtssinn* (Leipzig, 1861), spoke of "productive imagination" requiring an immaterial soul (8, 15, 73–5, 103–4). Ulrici, *Gott und der Mensch*, Vol. 1, *Leib und Seele* (Leipzig, 1866), spoke of unconscious discrimination and judgment (166–71).

40. PO 3:17: "I am aware that in the present state of knowledge it is impossible to refute the nativistic theory"; cf. 3:1, 558. In PO2 Helmholtz reduced the amount of material criticizing the nativism of sensory physiologists and introduced new material, from "Facts in Perception," questioning the a priori status of the axioms of geometry (PO2, sec. 26).

41. Helmholtz handled the evidence that young animals are perceptually sophisticated at birth by questioning whether the experience of the animals could be known, and by suggesting that animals are more controlled by instincts than human infants; "Recent Progress," SW, 221. He recognized that studies of the visual capacities of very young infants could in principle bear on the issue, but he considered the research impossible to carry out: how could one decide what an infant experiences? (PO 3:7). With respect to the newly-sighted blind, he reproduced portions of the Cheselden and Wardrop cases, admitting that the rapid development of Wardrop's spatial abilities indicated some innate facilitation of spatial learning that might be accounted for by postulating some regularity in the relation between the qualitative character of local signs and their retinal locations. He denied that the evidence entailed the acceptance of a complete nativism (3:182–190).

42. For a survey of work on single vision and stereoscopy, see Boring, *Sensation and Perception*, 226–30, 282–8, and Helmholtz, PO, secs. 30 and 31.

43. Isaac Newton, *Optics* (London, 1704), Query 15, hypothesized anatomical connections between the two retinas as the basis for single vision; his hypothesized anatomical connections were not accepted by all those who attributed single vision to an innate disposition of the nervous system. Thomas Reid believed that corresponding retinal elements (the elements that produce single vision when simultaneously stimulated) are innately given but eschewed any hypothesis about their anatomical basis; he mentioned reports of clinical observations that seemed to speak against Newton's views: Reid, *Inquiry*, in *Works*, ed. Hamilton, 1:163–182.

44. Müller, *Handbuch*, 2:381.

45. Valentin, *Lehrbuch*, 2:489; Volkmann, *Physiologische Untersuchungen*, chs. 5–6; Hering, *Beiträge*, 334. Hering did not speak of an anatomical connection, but he explained single vision through the hypothesis that stimulation from the two eyes combines to form a single physiological result that enters consciousness as a simple sensation. Helmholtz characterized Volkmann as an empirist (PO 3:454), presumably because he allowed a role for experience (and psychological operations: Volkmann, ibid., pp. 259–66) in accounting for deviations from the predictions of the identity theory. (As Helmholtz himself observed, 3:446, Hering also appealed to experience; however, Hering eschewed autonomously psychological explanations.) This classification of Volkmann confirms the interpretation of Helmholtz's classification scheme according to which allowing psychological—in opposition to physiological or "organic"—processes into one's explanation of binocular single vision is a sufficient condition for being classified an empirist. In this connection, it is interesting to note that Nagel, also an empirist by Helmholtz's lights, considered Volkmann to be essentially an adherent of the identity theory who appealed in an ad hoc manner to psychological processes to explain apparent deviations from the theory (*Sehen*, 85–88).

46. Robert Smith set forth one version of this empiristic position in his well-known *Compleat System of Optics* (Cambridge, 1738), 48–9:

    Now if it be asked why in seeing with both eyes we do not always see double, because of a double sensation; I think it is sufficient to say that in the ordinary use of our eyes, in which the pictures of an object are constantly painted upon corresponding places of the retinas, the predominant sense of feeling [i.e., touch] has originally and constantly informed us that the object is single . . . the judgments we make of the number and places of external objects are entirely the effect of experience.

47. Steinbuch, *Beytrag*, 246–50. A generically empiristic position was to be found in many reference works: Gehler, *Physikalisches Wörterbuch*, 2:652–3, 4:20–22; Fischer, *Physikalisches Wörterbuch* (Göttingen, 1798–1827), 4:586–7; Muncke, "Gesicht," in Gehler's *Physikalisches Wörterbuch*, revised ed. (Leipzig, 1825–1845), 4:1471–85. On the later literature, see the references in Helmholtz, PO, secs. 31 and 32; the importance that Helmholtz assigned binocular single vision in assessing the nativism-empirism controversy is apparent from his overview of the dispute in sec. 33.

48. Reid, *Inquiry*, ch. 6, secs. 15–17, in *Works*, 1:169–76; Helmholtz, PO 3:334–5.

49. Koffka discusses and criticizes the constancy hypothesis: *Principles*, 84–98.

50. Helmholtz, PO 3:438; Reid accepted the rule but rejected the empirical finding: *Works*, 1:175a.

51. Reid, *Works*, 1:174–176; Müller, *Gesichtssinne*, 76, and *Handbuch*, 2:380.

52. Hering, *Beiträge*, 323; *Lichtsinne*, 2–3; see also von Kries, PO 3:462–3.

53. "Recent Progress," SW, 212 (VR 1:351); see PO 3:417–20.

54. Helmholtz first argued from an empiristic interpretation of binocular rivalry to a psychological as opposed to anatomical explanation in his 1856 article, "Ueber die Erklärung des Glanzes" (WA 3:4–5); Wundt, *Beiträge*, part five, developed the argument at length, citing Helmholtz's article (303).

55. Müller, *Handbuch*, 2:387–8.

56. PO 3:17, 432, 453; SW, 213; EW, 135.

57. Hering, *Beiträge*, 42–3, 287–8.

58. Königsberger, 2:84; Welby trans., 238.

59. This is not a statement about the reception of Helmholtz's theory but an evaluative point, criticizing his argument as a response to Hering. On the the reception and ultimate demise of Helmholtz's position in Germany, see von Kries, PO 3:497–511. Woodward discusses the reception of empirism in "From Association to Gestalt." Turner examines the reception of Helmholtz's empirism in terms of institutional discipline-building: "Helmholtz, Sensory Physiology, and the Disciplinary Development of German Psychology."

60. PO 3:17, 134, 154, 440–1; SW, 196; EW, 135.

61. PO 3:441; "Recent Progress," SW, 196: The nativistic theory "obviously cuts short all further inquiry into the origin of these intuitions, since it requires them as original, inborn, and incapable of further explanation." It once was common for Helmholtz's interpreters to agree with this charge: e.g., Boring, *Sensation and Perception*, 33.

62. Helmholtz associated the nativists with German Idealism in the following passage from the *Optics*, which pertains to the "agreement" between sensations and external bodies (PO 3:17–18):

    Some have asserted that there is such an agreement, and others have denied it. In favor of it, a preestablished harmony between nature and mind was assumed. Or it was maintained that there was an identity of nature and mind, by regarding nature as the product of the activity of a general mind; the human mind being supposed to be an emanation from it. The nativistic theory of spatial intuition is connected with these views to the extent that, by some innate mechanism and a certain preestablished harmony, it admits of the origin of perceptual images that are supposed to correspond with reality, although in a rather imperfect fashion.

This passage represents an attempt to align nativism with the views of Schelling and Hegel, which were held in contempt by growing numbers of mid-nineteenth-century German scientists. Helmholtz himself went on to posit a merely pragmatic correspondence between spatial intuitions and the external world.

63. Helmholtz acknowledged that the new theory was controversial but characterized the uncertainty as pertaining to its scope rather than its correctness (SW, 239–240). I agree with Lenoir, *Strategy of Life*, 236, that Helmholtz did not reveal in print his appreciation of Darwin's work prior to 1869. However, I do not agree that Helmholtz sought to eliminate the notion of *Zweckmässigkeit* from his view of nature (he thought Darwin allowed him to accommodate that notion), nor that he considered the Darwinian notion of variation to violate Newtonian determinism (ibid., 237, 238–40).

64. Hering suggested this possibility in speaking of the genesis of spatial abilities during "the life of the species," as opposed to the life of the individual (*Beiträge*, 273–4)—the very language Helmholtz later used to contrast natural selection with learning; see also *Die Lehre vom binocularen Sehen* (Leipzig, 1868), 102, and *Lehre vom Lichtsinne*, 3. Although Hering does not invoke Darwin by name on the cited pages, he read Darwin while a student and cultivated a Darwinian outlook: see Metzner, *Die biologischen Ansichten Ewald Herings* (Heidelberg, 1945), 9–14, 21–5, 27.

65. Hering, *Lichtsinne*, 1–2.

66. PO 3:445: "Instead of postulating an organic basis for the case of single vision with disparate retinal locations, as Mr. Panum did, Mr. Hering accepts a psychical basis, in that he relies upon the claim that practice and precise training is necessary for the separation of compound sensations."

67. See also PO 3:453, where Helmholtz wrote that "as I learned to control my eye movements and my attention with free will, I became more and more convinced that the essential phenomena in this area could not be explained through a previously given nervous mechanism." If "previously given" simply means "innate," then this passage is consistent with free will's having as its basis a nervous mechanism that undergoes change and development through learning. But only in the one offhand remark quoted above does Helmholtz suggest this possibility; otherwise, his remarks are remarkably dualistic in spirit.

68. Helmholtz's strong antivitalistic stance and his program to reduce physiological processes to chemistry and physics have sometimes led historians to classify him, erroneously, as a materialist—see, e.g., Cranefield, "The Philosophical and Cultural Interests of the Biophysics Movement of 1847," and Galaty, "Emergence of Biological Reductionism," who allows the vulgar materialist Karl Vogt to speak for Helmholtz and friends (p. 11). Like his co-reductionist du Bois-Reymond, Helmholtz attacked materialism as a metaphysical hypothesis (as we shall see, he also attacked spiritualism, a radical form of dualism in which it was argued that mental processes are unfathomable). Only recently have historians begun to appreciate the gulf between materialism and the limited reductionism of Helmholtz, Ludwig, du Bois-Reymond and others; see Gregory, *Scientific Materialism*.

69. Helmholtz's wording is similar to a passage by the avowed dualist Ulrici (*Gott und der Mensch*, 1:1) in a work cited by Helmholtz (PO 3:33), though the "positive" conclusion is different.

70. Helmholtz expressed this point repeatedly; here is a typical statement: "The same great importance which experiment has for the certainty of our scientific convic-

tions, it has also for the unconscious inductions of the perceptions of our senses. It is only by voluntarily bringing our organs of sense in various relations to the objects that we learn to be sure as to our judgments of the causes of our sensations" (PO 3:28). See also "Recent Progress" (SW, 215, 222) and "Facts in Perception" (EW, 132, 136).

71. The evidence that Helmholtz wrote the bulk of sec. 26 in 1865 comes from his complaint of frequent headaches in composing the "epistemological" portion of Part Three (Königsberger, 2:72). Helmholtz's first published use of the term *unbewusster Schluss*—along with a characterization of such inferences as conclusions from analogy (*Analogieschlüsse*)—appeared in several pages of sec. 26 published along with Part Two in 1860 (the signature containing this portion of sec. 26 ended on p. 432 of the first edition, which corresponds to the middle of the third full paragraph on PO 3:8). Wundt had already used the term in 1858. A priority dispute arose, on which see Richards, "Wundt's Early Theories of Unconscious Inference," 47–50.

72. VR 1:112, 115; PO 2:101–2, 243–4, 250.

73. Helmholtz went on to claim that psychologists "previously have for the most part assigned the mental acts spoken of here immediately to sensory perception, and have not sought to attain a finer explanation of them" (VR 1:111), a characterization that clearly does not apply to one large group of German psychologists prior to Helmholtz, the Herbartians.

74. The letter to his father was dated 4 March 1857; here is the relevant portion:

> *I feel the crying want of a special treatment of certain questions, which have not, so far as I know, been attacked by any modern philosopher, and which lie wholly within the field of a priori concepts which Kant investigated, e.g., the derivation of the principles of geometry and mechanics, the reason why we are logically bound to reduce reality to two abstractions— matter and energy, etc., or again, the laws of the unconscious arguments from analogy* (unbewussten Analogieschlüsse), *by which we pass from sensations to sense perceptions. I see plainly that these can only be solved by philosophical investigation, and are resolvable by it, so that I feel the need of more profound philosophical knowledge* . [Königsberger, 2:292–3; trans. modified from Welby, 160–1]

The description of perception as resting upon "unconscious arguments from analogy" does not go beyond the lecture of 1855.

75. PO 3:6; in Part Two he had spoken of judgments that pertain to sensations of which we are not conscious: PO 2:243–4.

76. "The Relation of the Natural Sciences to Science in General," delivered at Heidelberg on 22 November 1862 (SW, 1302). In the same lecture Helmholtz mentioned John Stuart Mill, but he did not credit him with the foregoing analysis of inductive reasoning.

77. The question of priority is a difficult one to decide. Wundt was Helmholtz's junior colleague at Heidelberg and taught a laboratory section of Helmholtz's course in physiology, for which he attended at least some of the lectures. Yet these lectures were in the introductory physiology course, so it is not clear how deeply Helmholtz would have gone into his views on the psychology of perception, if they had in fact crystallized prior to 1862. Moreover, assuming little or no influence of Helmholtz on Wundt, some of the similarities can be accounted for by the fact that the two drew on a common set of sources. Any remaining unaccounted for similarity must be explained by Helmholtz's familiarity with Wundt's work, which he cited often and with praise in the *Optics*, where he credited Wundt with making an important contribution to the theory of the effect of eye movements on the perception of bidimensional spatial configurations (PO 3:192, 453). On the relation between

Wundt and Helmholtz during the period in question, see Richards, "Wundt's Early Theories."

78. Wundt, *Beiträge*, iv-v, xi-xxxi, 441–5. Wundt maintained that this psychology would borrow methods from physiology, but that it would apply these methods to its own distinct subject matter. On Wundt's self-conscious efforts to make the discipline of experimental psychology exist more than "in name only" (ibid., vi), see Mischel, "Wundt and the Conceptual Foundations of Psychology," Danziger, "Wundt and the Two Traditions of Psychology," and Woodward, "Wundt's Program for the New Psychology."

79. Wundt, *Beiträge*, 148–64, 28.

80. Ibid., 65, 149, 169, 297, 438.

81. The preface to the reprinted series of articles is dated January, 1862 (ibid., vi).

82. Ibid., 441–2.

83. Ibid., 442–5.

84. Wundt preferred the term "elemental sensations" or "retinal sensations" to "local signs," for he believed that such signs result from the properties of the nervous fibers in various retinal locations and do not depend on muscle activity as in Lotze's account of local signs (ibid., 54, 150–4, 444).

85. Ibid., 443–4, 28. The details of the example are very sketchy, for Wundt could expect his reader to recall his previous extensive discussion of the theories of Weber, Lotze, the Herbartians, and others (pts. 1–3).

86. Recent Wundt commentators have tended to downplay the influence of Mill on Wundt and to highlight what they see as Leibnizian and Kantian influences, perhaps mediated by Wolff, Herbart, or Whewell (Danziger, "Wundt and the Two Traditions," 75–82; Richards, "Wundt's Early Theories," pp. 45–6, 50–1). Although I agree that the German intellectual tradition had a more direct influence on Wundt than the British, I question the positing of strong Leibnizian or Kantian influence. In the *Beiträge* the most evident influences on his psychology (apart from those of sensory physiologists proper) stem from Herbart and Lotze, who were themselves at least indirectly influenced by the associationist tradition. Wundt mentions Mill and Whewell in direct connection with his account of unconscious inference, and even though his use of the term *Verschmelzung* smacks of Herbart (Richards, 51, n. 12), it is nonetheless associationist (and non-Kantian).

87. Reprinted in PO2, 601; trans. modified from SW, 507. See also PO 3:28–9.

88. *We may rightly claim that the idea of the stereometric form of a material object plays the role of a concept, formed on the basis of the combination of an extended series of sensuous intuitional images. It is a concept, however, which, unlike a geometric construct, is not necessarily expressible in a verbal definition. It is held together or unified only by the clear idea of the laws in accordance with which its perspective images follow one another.* [ibid.]

89. "Facts in Perception," EW, 143; see also PO2, 598–602, which largely reproduces material from "Origin and Interpretation."

90. John Stuart Mill, *System of Logic*, in his *Collected Works*, 7:664. Mill expressed his unwillingness to endorse outright an associational account of the psychology of inference in a note added to this passage in 1851; he contended that the ultimate basis of associative laws is uncertain. He took up the question at some length in his edition of James Mill's *Analysis of the Phenomena of the Human Mind* (London, 1869); in a long note appended to ch. 11 he sought to remove objections to his father's associational account of belief, but he left open the question of whether he would give the account his own unqualified approval.

91. See also "Origin and Interpretation," SW, 484. The point had been implicit at PO 3:23–4.

92. At the end of his account of spatial vision, he summed up the status of sensations as follows: "The sensations of the senses are signs for our consciousness, it being left to the understanding to learn how to comprehend their meaning" (PO 3:433).
93. The quoted questions are from "Facts in Perception" (EW, 117), but the conception of the task of epistemology is consonant with earlier statements in lectures of 1852, 1853 and 1863 (WA 2:608–9; SW, 70, 126). Schlick's observation occurred in his commentary to "Facts in Perception," note 2 (EW, 163). Helmholtz had discussed his use of "true" at PO 3:22.
94. "Recent Progress," SW, 187, where Helmholtz also mentioned Herbart and Kant.
95. Helmholtz, "Aim and Progress," SW, 241–2 (VR 1:393).
96. "Recent Progress," SW, 222: *In the region of qualities, we can prove conclusively that in some instances [viz., color sensations] there is no correspondence at all between sensations and their objects. Only the relations of time, space, and equality and those which are derived from them (number, size, and regularity of coexistence and of sequence)—mathematical relations, in short—are common to the outer and the inner worlds. Here we may indeed anticipate a complete correspondence between our conceptions and the objects which excite them.*
97. See also SW, 243. Helmholtz remarked that the faithfulness of the agreement for temporal relations is reduced by the different transmission times for the various sense organs, since they are located at different distances from the brain (PO 3:21). Helmholtz discovered the finite speed of neural transmission (WA 3:1–3), which Müller had believed to be infinite. See also Erdmann, "Grundlagen."
98. Here is the relevant passage (PO 3:21–2):

    The idea of an individual table which I carry in my mind is correct and exact, provided I can derive from it correctly the precise sensations I shall have when my eye and my hand are brought into this or that definite relation with respect to the table. I do not know how to conceive any other sort of similarity between such an idea and the body it represents. One is the mental symbol of the other. The kind of symbol was not chosen by me arbitrarily, but was forced on me by the nature of my sense-organ and of my mind. This is what distinguishes this sign-language of our ideas from the arbitrary phonetic signs and alphabetical characters that we use in speaking and writing. A writing is correct when he who knows how to read it forms correct representations from it; the idea of a thing is correct for him who knows how to determine correctly from it in advance what sensory impressions he will get from the thing when he places himself in definite external relations to it.

    See also a passage from the *Nachlässe*, quoted in Königsberger, 2:159: "Representations are signs, which are re-translated into reality through movement. The only real similarity is the temporal relation." Königsberger juxtaposed this passage with "Facts in Perception"; if this was intended to imply that the passage was contemporary with the latter work, it may have been incorrect, for by then Helmholtz had dropped the contention that any similarity could be established between representations and the external world by admitting that idealism could be true.
99. See Elkana, *Discovery of the Conservation of Energy*, on the development of Helmholtz's terminology and thinking from the conservation of *Kraft* to the conservation of *Energie*.
100. He expressed this as temporal priority ("vor aller Erfahrung") and equated it with Kant's position, thereby missing Kant's point about the law's transcendental status.
101. Helmholtz characterized the law of cause as a "purely logical law" (PO 3:30) and cited his 1855 discussion, in which he equated his point with Kant's. That he should so characterize the law suggests that he did not fully grasp, or was uncomfortable

with, Kant's description of the law as synthetic a priori. This impression is reinforced by Helmholtz's description of Kant's philosophy in a lecture from 1862: "Kant's philosophy rested upon exactly the same foundations as the physical sciences, as is evident from his own scientific works and especially from his cosmological hypothesis. . . . According to his teaching, a principle established a priori by pure thought was a rule applicable only to the methods of thought and to nothing else; it had no real, positive content" (VR 1:162–3; SW, 124–5).

102. Here is the passage: "But if it is found that the natural phenomena can be subsumed under a definite causal connection, this is certainly an objectively valid fact, and corresponds to objective specific relations among natural phenomena, which we express in our thinking as their causal connection, and indeed do not know how else to express it" (PO 3:31). Helmholtz has in mind here the discovery, through empirical investigation, that certain laws are always observed to hold.

103. WA 1:68; trans. modified from SW, 49. Juxtaposition of the statement just quoted with point (3) in the quotation from 1878 might lead one to conclude that Helmholtz had changed his mind in the intervening three or four years. That is not likely. I think he realized the position he had expressed in 1878 was not truly Kantian.

104. Königsberger, 1:248. Here the numbering is Helmholtz's.

105. Helmholtz, "Ueber den Ursprung und die Bedeutung der geometrischen Axiome," lecture presented in Heidelberg in 1869 and first published in volume 2 of his *Populäre wissenschaftliche Vorträge*, (Braunschweig, 1870); reprinted in VR 2:1–31 and translated as "On the Origin and Significance of the Axioms of Geometry," EW, 1–26. Helmholtz (EW, 8, 12) cited Riemann, "Ueber die Hypothesen, welche der Geometrie zugrunde liegen" (1867): Riemann's paper is reprinted in his *Gesammelte mathematischen Werke*, ed. Weber, 2d edition, 272–87 (this paper was given as Riemann's *Habilitationschrift* at Göttingen in 1854). Helmholtz also cited Beltrami, *Saggio di Interpretazione della Geometrie Non-Euclidea* (Naples, 1868; misprinted as 1848 in Helmholtz's paper). See also Helmholtz's "Ueber die Tatsachen, die der Geometrie zugrunde liegen," (1868), reprinted in WA 2:618–39 and translated as "On the Facts Underlying Geometry," EW, 39–58, which cites only Riemann. On the relation of Helmholtz's work in geometry to his other work, see Joan Richards, "The Evolution of Empiricism."

106. On the relation of Herbart's and Grassmann's work on manifolds to that of Riemann, see Torretti, *Philosophy of Geometry*, 107–9. Although Torretti doubts that Riemann was aware of Grassmann's work, Helmholtz certainly was; he cited Grassmann's application of the theory of manifolds to color mixtures (PO 2:111–23, 137) in connection with his own discussion of the same problem.

107. In the 1868 paper, Helmholtz failed to appreciate the possibility of negative curvature; he thus implied at the end of that paper that if space is taken to have three dimensions and to be infinite, it must (necessarily) be Euclidean (EW, 58). He corrected this oversight in the lecture from 1869 ("Origin and Significance"), and so came to recognize that measurements inconsistent with Euclid's geometry conceivably could be obtained (assuming the "fixity" or "rigidity" of one's instruments). For a discussion of Helmholtz's work on geometry from a contemporary point of view, see Torretti, *Philosophy of Geometry*, 155–71.

108. Helmholtz, EW, 4–5, 39, 130.

109. Helmholtz treated continuity as implying differentiability (EW, 42, 46–7); Torretti infers that he was aware of Weierstrauss's discovery of nowhere differentiable continuous functions, and chose to narrow the definition of continuity (*Philosophy of Geometry*, 159).

110. "Facts Underlying Geometry," EW, 41; "Origin and Significance," ibid., 4–5, 18–19. For comparison of Helmholtz and Riemann on this point, see Richards, "Evolution of Empiricism," sec. 2.

111. The assumption of the existence of rigid bodies is by no means trivial, as Helmholtz well knew (EW, 19); it invites the charge of circularity on the grounds that it is only by having a measuring system in place that one could determine which bodies are rigid (further discussion to come).

112. Beltrami, *Geometrie Non-Euclidea*, as described in EW, 20–3.

113. "Facts in Perception," Appendix 2 (EW, 151).

114. "Origin and Significance," EW, 25. Cohen, *Kants Theorie*, 2d. ed., 232, complained that Helmholtz here followed the letter of Kant's analytic/synthetic distinction while violating its spirit; Cohen proceeded to show that Helmholtz had misunderstood the notion of intuition in Kant. However, it is fair to say that in light of the work of Riemann and Helmholtz, the categorization of neo-Kantian claims about the necessity of Euclid's axioms as descriptions of physical space became problematic. Helmholtz's proposal—that the Kantian would effectively be appealing to definition in stipulating that, in the face of empirical findings of the sort envisioned by Helmholtz, space should be considered flat and mechanics should be altered—had merit.

115. As Helmholtz observed (EW, 19–20), assuming that one's body underwent the same systematic spatial alterations as one's instruments, it would be impossible to detect the alteration by the senses.

116. EW, p. 153 and Schlick's note 75. Schlick merely observed that the reading of a clock involves spatial perception; he might also have appealed to the fact that when Helmholtz wrote, the exact calibration of time pieces depended upon astronomical (hence, spatial) observation. Helmholtz thus was unable to rule out the position that would later be known as "conventionalism," a position implied in his discussions of 1869 and 1878 (EW, 19–20, 156). On conventionalism, see Friedman, *Foundations of Space-Time Theories*, who concludes that the position can be rejected through appeal to simplicity as a criterion of hypothesis choice.

117. See Torretti, *Philosophy of Geometry*, ch. 2, sec. 1, on the development of non-Euclidean geometry. As he makes clear, although Kant may well have been aware of the problematic relation between the "parallel postulate" and the other postulates, the work of his contemporaries did not establish, nor even envision, the possibility of non-Euclidean geometries.

118. "Facts in Perception," VR 2:230; trans. modified from EW, 129. Helmholtz went on to argue that the question of the transcendental status of geometry should be separated from the question of nativism with respect to visual perception; apparently he believed that nativism was not essential to Kant's position but merely was his own personal opinion (EW, 150).

119. In terms of the example cited above, this would be a system in which parallel lines in space with negative curvature did not appear to converge more rapidly than do parallel lines in Euclidean space for the observer with Euclidean habits.

120. "Facts in Perception," EW, 122–4, 128–9.

121. "Relation of Optics to Painting," SW, 297–329 (VR 2:93–135).

122. Helmholtz, *Tonempfindungen*, translated by Ellis, *On the Sensations of Tone as a Physiological Basis for the Theory of Music*, 4th ed. (London, 1912), chs. 4, 10, and pp. 226–33; "The Physiological Causes of Harmony in Music," SW, 75–108 (VR 1:119–55).

123. *Tonempfindungen*, 357 (modified from Ellis, 234).

124. *Tonempfindungen*, 358 (modified from Ellis, 234).

125. *Tonempfindungen*, 358 (in later editions Helmholtz qualified the passage to read "does not rest solely upon . . . but is, at least in part" etc.: Ellis, 235).

126. "Recent Progress," SW, 220.
127. "Relation of Optics to Painting," SW, 328–9.
128. "Scientific Researches of Goethe," SW, 56–74 (VR 1:23–45).
129. "Goethe's Anticipation of Subsequent Ideas," SW, 481–5. Helmholtz contended that his account of artistic imagination in terms of "the laws of memory" was much preferable to the Romantic emphasis on "the free play of fancy" (SW, 485; VR 2:343–4).
130. Gregory, *Scientific Materialism*, ch. 7; Erdmann, "Grundlagen," 13–14.
131. See the end of section 4.

Chapter 6, Summary

1. Diderot was a materialist who developed an articulated theory of mind; see Anderson, "Encyclopedic Topologies." But much of his work was unpublished, and the published portion had no significant influence in the development of German philosophical and psychological theories of mind.
2. Hartley, *Observations on Man*, 1:33–4; Lossius, *Physische Ursachen*, 41–2.
3. Hartley, *Observations on Man*, 1:65–7; Lossius, *Physische Ursachen*, 149.
4. D'Holbach, *System of Nature*, Pt. 1, chs. 8, 10.
5. John Dewey, "Psychologism," in the *Dictionary of Philosophy and Psychology* (New York, 1901-05), 2:382; Abbagnano, "Psychologism," in the *Encyclopedia of Philosophy* (New York, 1967), 7:520–1.

Chapter 7, Conclusions

1. See Danziger, "Wundt and the Two Traditions of Psychology," Woodward, "Wundt's Program for the New Psychology," and Koch, "Wundt's Creature at Age Zero—and as Centenarian."
2. See Arens, *Structures of Knowing*, chs. 4–6.
3. Dewey was member and chair of the department of philosophy, psychology, and education at the University of Chicago from 1894 to 1904; for statements of his conception that the method of psychology goes beyond experimental study to include what might be termed historical and cultural studies, see his 1884 essay "The New Psychology," in his *Early Works*, 1:48–60, and his *Psychology* (New York, 1887), ch. 1, sec. 2. On Munsterberg, see Daston, "Theory of Will," 98–100. Stout was influential in British psychology; for his conception of the subject-matter and methods of psychology, see his *Groundwork of Psychology* (New York, 1903), chs. 1–2, esp. 15–17.
4. Klemm, *Geschichte der Psychologie* (Leipzig, 1911); as translated by Wilm and Pintner, *History of Psychology* (New York, 1914), 7, 147–55.
5. Flugel, *Hundred Years of Psychology* (New York, 1933), 15–16 and Pt. III, chs. 6–8.
6. Boring, *History of Experimental Psychology* (New York, 1929), chs. 12–15.
7. On the process of institutionalization in Germany, see the articles in Ash and Geuter, eds., *Geschichte der deutschen Psychologie im 20. Jahrhundert*, and Geuter's book, *Professionalisierung*, which argues that German psychology was not fully institutionalized until the time of World War II.
8. For a philosophical endorsement of the division, see Kyburg, "Rational Belief"; for psychological uses of normative theories that respect this division, see Kahneman and Tversky, "On the Psychology of Prediction," and Lanning, "Some Reasons for Distinguishing."

9. This remark is more commonly uttered than printed. Notable published discussions include Ryle's suggestion in *The Concept of Mind* (London, 1949) that there is nothing for scientific psychology to explain about normal achievements in thinking and perceiving, including "our correct estimations of shape, size, illumination, and speed"; he suggests that the role of psychology is to explain perceptual illusion, not veridical perception, and parapraxis, not "normal" behavior (325–7). Hamlyn developed a similar argument in great detail with respect to the scientific study of perception: *Psychology of Perception* (London, 1959).

10. This attitude is present in conversation and in the tone of published works but is rarely made explicit. For an example of the tone, see Johnson-Laird, *Computer*, ch. 1; for something closer to an explicit statement, see Nisbett and Thagard, "Psychology, Statistics, and Analytical Epistemology," and Falmagne, "Normative Theory and the Human Mind."

11. Patricia Churchland, *Neurophilosophy* (Cambridge: The MIT Press, 1986).

12. Examples of natural scientific achievement in the study of perception are replete in Gibson, *Senses Considered as Perceptual Systems*, Rock, *Logic of Perception*, and Marr, *Vision*.

13. Marr, *Vision*, ch. 5. The psychology of classification itself has been invigorated by works such as Rosch and Lloyd, *Cognition and Categorization*, and Smith and Medin, *Categories and Concepts*.

14. For a guide to this literature, see Massaro, *Experimental Psychology*.

15. On the distinction between the "logical space of reasons" and the "logical space of causes," see Rorty, *Philosophy and the Mirror*, ch. 3, esp. 141, and Sellars, "Empiricism and the Philosophy of Mind," 131, 144–5, 166–9, 179.

16. In treating Descartes' position as "psychological," one might be imposing a principle of charity—that is, discounting the metaphysically normative talk of mental processes, and reading it "in the only way in which it is coherent" (as one might say), that is, as naturalistic causal talk. But this "act of charity" will do little to make sense of the "compulsion of the will" Descartes describes in the Fourth Meditation. Of course, a wholesale causal account of "causal conditions in which assent is elicited" might be substituted for the Fourth Meditation account of judgment, but that would hardly be a charitable reading of the Fourth Meditation: it would not be a *reading* of it at all.

17. See Dewey, Hook, and Nagel, "Are Naturalists Materialists?" *Journal of Philosophy*, 42 (1945), 515–30.

18. See the essays in *Naturalism and the Human Spirit*, ed. Krikorian (New York, 1944), especially Dewey, "Antinaturalism in extremis" (1–16), Krikorian, "A Naturalistic View of Mind" (242–69), and Dennes, "The Categories of Naturalism" (270–94).

19. Dewey, "Antinaturalism," 1–11.

20. Ibid., 12.

21. See Krikorian, "Naturalistic View of Mind," 242–3, 252, and Dennes, "Categories of Naturalism," 288–9.

22. Quine's physicalist program informs *Word and Object*, secs. 1, 5, 54 (1–4, 20, 264–5); see also *Roots of Reference*, 6, 10–14, 33–4.

23. *Word and Object*, sec. 45 (218–21); *Roots of Reference*, 33. Solomon, "Quine's Point of View," ably defends an interpretation of Quine that makes method primary (the method of "natural empiricism"), and she also reviews other interpretations (which make Quine's physicalism, or his behaviorism, primary).

24. Quine is by no means the first to call for a naturalized epistemology, but in recent years his work has been viewed as providing the paradigmatic statement of the project: Quine, "Epistemology Naturalized," esp. 82–3. The project of naturalization is carried out most fully by Quine in *Roots of Reference*.

25. "Epistemology Naturalized," pp. 72–4; *Roots of Reference*, p. 3.
26. *Word and Object*, secs. 5–6 (17–20, 22–3).
27. *Roots of Reference*, 33–4.
28. Paul Churchland, "Eliminative Materialism"; Patricia Churchland, *Neurophilosophy*, ch. 9.
29. P. F. Strawson, *Skepticism and Naturalism*, 68.
30. Ibid., 24.
31. Ibid., 37–42.
32. Ibid., 39, 42–5.
33. Paul Churchland, *Scientific Realism and the Plasticity of Mind*, ch. 5, sec. 21; Patricia Churchland, *Neurophilosophy*, 450–8.
34. Putnam, "Why Reason Can't be Naturalized," in his *Realism and Reason*, 229–47.
35. Baker, *Saving Belief*, ch. 7.
36. Fodor advances this program in his *Language of Thought* (New York, 1975). Earlier "naturalists" such as Arthur Lovejoy also attempted to bring the mental within the domain of the natural without reduction: see his *Revolt against Dualism* (La Salle, 1929), ch. 9.
37. Dretske, *Knowledge and the Flow of Information*; Robert Nozick, *Philosophical Explanations*, ch. 3; Alvin Goldman, *Epistemology and Cognition*, Part I.
38. See Millikan, *Language, Thought, and Other Biological Categories*.
39. Quine thinks evolutionary theory may be "almost explanation enough" for inductive success: "Natural Kinds," 125–8.
40. Putnam, "Computational Psychology and Interpretation Theory"; Taylor, "Interpretation and the Sciences of Man."
41. Geertz, *Interpretation of Cultures*, chs. 2–3.
42. Gould, *Ever Since Darwin*, selections 5, 7, 8.
43. Gould's position seemingly might be explained through his political position within the "nature" versus "nurture" debate, evident in ibid., selections 7–8. However, the issues of nature and nurture that are central to the political debates (questions of hereditary differentials in the "ability to learn") are irrelevant to assessing the force of neotony for the naturalistic approach to cognition. Only on the assumption that the standards of scientific judgment are "preprogrammed" will nativism lead to a hard naturalist reduction of the normative.
44. This argument should be distinguished from one that may appear similar to it. Suppose one were to argue that our natural scientific theories are not objectively true because they are culturally contingent: not every culture developed them. But the text argument does not question the validity of natural science. Far from endorsing this "parallel" argument, it needs to reject it. The text argument pertains to the contingency of the object of a certain natural scientific investigation (the knowing mind), not to the results of that investigation. It claims that, from the point of view of natural science, the categories that provide for human self-description are culturally contingent. Since there is no single "correct" set of cultural practices and institutions, there is no single "correct" set of categories for descriptions of the human animal as a cultural creature. But there are no human beings that are not cultural creatures, for it is part of our biology to be such creatures.
45. An evolutionary epistemologist could respond that a truth detector adequate for converging on the true scientific theory has been programmed in, but it simply takes a long time to work. But if the standards of theoretical evaluation have changed over the course of the history of science, so that something approaching the "correct" standards arose only in historical time, then Putnam's general response to evolutionary epistemology is effective: "our concepts depend on our physical and social

environment in a way that evolution (which was completed, for our brains, about thirty thousand years ago) couldn't foresee" (*Representation and Reality*, 15).

Appendix A, Nativism-Empirism

1. Boring, *Sensation and Perception*, 28–9, used the terms "nativism" and "empiricism" to name the opposing positions Helmholtz made prominent; some recent historians of psychology have preferred to use the term "empirism." See Pastore, "Reevaluation of Boring," n. 3; Woodward, "From Association to Gestalt," n. 7; Wertheimer, *Brief History of Psychology*, rev. ed., 22; and O'Neil, *Beginnings of Modern Psychology*, 2d ed., 9. Pastore, Woodward, and O'Neil are among the psychologists who distinguish between empiricism and empirism, not only terminologically, but conceptually.

2. Hochberg, "Nativism and Empiricism in Perception," in *Psychology in the Making*, ed. Postman, 257, 265 (n. 4), 271, 293, 327; Piaget and Garcia, *Psychogenesis and the History of Science*, trans. Feider, viii, 23, 265. This attitude is often implied by tone rather than stated directly; see, for example, McCleary's "Where We Stand Today," in McCleary, ed., *Genetic and Experiential Factors in Perception*, 139.

3. Descartes considered intellectual perceptions and their objects (intellectual ideas, or pure conceivings) to be independent of sensory experience, but he admitted "intellectual experience," that is, acts of intellect that are phenomenally available to consciousness and have no sensory content. For further discussion, see Grimaldi, *L'Expérience de la Pensée dans la Philosophie de Descartes* (Paris, 1978), 103-08, and my "The Senses and the Fleshless Eye."

4. Hoffmeister, *Wörterbuch der philosophischen Begriffe* (Hamburg, 1955), 507.

5. Tenneman, *Geschichte der Philosophie* (Leipzig, 1798–1819), 11:iv.

6. Hegel, *Vorlesungen über die Geschichte der Philosophie*, ed. Michelet, 2d ed. (1840); translated as *Lectures on the History of Philosophy*, by Haldane and Simson (London, 1892). Hegel divided philosophy up to Kant into a metaphysical period (with rationalist and empiricist subdivisions), followed by idealism and skepticism, terminated by criticism (*Lectures*, 218–9). But his most basic division was between realist and idealist, or objective and subjective moments in the history of modern philosophy, which led to the third moment of the synthetic unity between subject and object—a synthesis Hegel claimed to have effected (ibid., 162–6, 359–60, 409, 545–54).

7. Rixner, *Handbuch der Geschichte der Philosophie* (Sulzbach, 1829), vol. 3, developed a highly speciated division of the history of philosophy into three epochs. The first was characterized by an opposition between "versuchenden" and "raisonnirenden Weltweisheit," respectively represented by the "Empirismus" of Bacon and the "dialektisch-raisonnirende Idealistik" of Descartes. The second epoch was characterized by the battle of "empirischen Realismus" (Locke) against the tendencies toward speculation (Leibniz), skepticism (Hume), and mysticism (Malebranche), but it also included the "empirisch Idealismus" of Berkeley and the "Dogmatismus" of Wolff. The third epoch was dominated by Kant's reform of philosophy through "Vernunft-Kritik." The Kantian Kuno Fischer, in his *Geschichte der neuern Philosophie* (Mannheim, 1854), used the classificatory scheme of empiricism, idealism, dogmatism, and criticism; Erdmann, in his *Grundriss der Geschichte der Philosophie* (Berlin, 1866), vol. 2, used the scheme of realism (with empiricism as a subheading), idealism, and criticism.

8. Ritter, *Geschichte der Philosophie* (Hamburg, 1829–53), applied the term *Sensualismus* to the philosophies of Bacon, Hobbes, Gassendi, and Locke (10:357, 474, 547; 11:455,

535), and mentioned it as a presupposition adopted by Hume (12:308); Fischer, *Geschichte der neuern Philosophie*, 3d ed. (Heidelberg, 1889–93), 3:ix, applied the term to Locke, as did Wundt, *Beiträge* (Leipzig, 1862), 81, 86, 90.

9. Ueberweg, *Grundriss der Geschichte der Philosophie*, 8th ed. (Berlin, 1894–7), vol. 3, pt. 1, 59–60. Here the term "origin" (*Ursprung*) pertains to the basis of justification of knowledge, as Ueberweg made clear. Ueberweg used *empiristisch* as the adjective for "Empirismus," understood in its epistemological sense (ibid., 61); however, in discussing Helmholtz, he mentioned a distinction between "nativistischen" and "empiristischen Raumtheorie" (vol. 3, pt. 2, 206).

10. Falckenberg, *Geschichte der neueren Philosophie*, 2d ed. (Leipzig, 1892), 66.

11. Hoffmeister, *Wörterbuch*, 421, credits Helmholtz with establishing the term *nativistisch* to indicate a theory that imputes a large measure of innate ability to an organism. He dates Helmholtz's introduction of the term from the 1855 address "Ueber das Sehen des Menschen"; however, the term is not found in that address, although Helmholtz distinguishes between theories emphasizing innate abilities and theories emphasizing learning.

12. Aguilon, *Opticorum libri sex* (Antwerp, 1613), 155–8; Chale, *Cursus seu mundus mathematicus* (Annison, 1674), vol. 2, Bk. 2, prop. 1. Descartes is of course the most notorious adherent of the doctrine of natural geometry, but he proposed a noncognitive, psychophysical brand of it, thus making his position different from the usual cognitive theory of "natural geometrical" inference, but still making him a nativist in Helmholtz's sense because he attributed the perception of distance to an innate physiological mechanism.

13. Condillac, *Essai sur l'Origine des Connoissances Humaines* (Amsterdam, 1746), Pt. 1, sec. 6, attacked Berkeley's theory and proposed that we see distance innately; in his *Traité des Sensations* (London, 1754), Pt. 3., ch. 3, he modified his position, allowing touch a role in teaching us to attend to visual space. Reid was a "nativist" about single vision: *Inquiry*, ch. 6, sec. 13–19, *Works*, ed. Hamilton, pp. 163–82. On Reid's account of single and double vision, see Morgan, *Molyneux's Question*, ch. 5. He also considered our tendency to believe in the external existence of the objects of perception a native or natural endowment (*Inquiry*, ch. 6, sec. 20; *Intellectual Powers*, Essay 2, ch. 5); his "common sense" reply to Hume rested on this foundation.

14. Stewart, *Outlines of Moral Philosophy* (Edinburgh, 1793), Pt. 1, sec.2, art. 1 (*Collected Works*, ed. Hamilton, 2:16): "Among the phenomena of vision, more immediately connected with the philosophy of the human mind, the most important are those which depend on the distinction between the *original* and the *acquired* perceptions of sight."

15. Helmholtz, *Handbuch* (Leipzig, 1867), 435; translation modified from Southall, 3:10. On the connection between nativism and an explanatory emphasis on sensation, see Appendix B.

16. For examples see O'Neil, *Beginnings*, 9, and Woodward, "From Association to Gestalt," 572, n. 7.

17. Both the philosophical meaning, contrasted with *Rationalismus*, and the psychological meaning, contrasted with *Nativismus*, are given by Hehlmann, *Wörterbuch der Psychologie* (Stuttgart, 1959), 95–6; Beigel, *Wörterbuch der Psychologie* (New York, 1971) gives only the psychological meaning.

18. The quoted phrase is from *Handbuch*, 3d ed., 3:32; his placement of Locke among the *Sensualisten* occurred at 3:190, and also in a lecture of 1869, "Die neueren Fortschritte in der Theorie des Sehens," in *Populäre wissenschaftliche Vorträge* (Braunschweig, 1865–76), pt. 2, 66. Wundt had earlier classified Locke as an adherent of *Sensualismus*, which he opposed to Bacon's *Empirismus* (*Beiträge*, 90). Helmholtz tended to use *empiristische*

Theorie and *nativistische Theorie* to name the doctrines that others would call *Empirismus* and *Nativismus*; perhaps he did so in order to avoid the ambiguity of the term *Empirismus*.

19. Ribot, *Psychologie Allemande, ch. 4*; Klemm, *Geschichte* (1911), ch. 11.

20. In addition to empirism and nativism with respect to visual space-perception, there has been empirism and nativism with respect to spatiality in general. The innateness of spatiality as a general form of experience common to vision and touch was at issue in late eighteenth and nineteenth century discussions.

21. There is some disagreement on whether Berkeley held that bidimensional visual form is innately given; Pastore, *Selective History*, 94, suggests that he did not, whereas I have argued that he did, in an article written with William Epstein, "Sensory Core," 380–1. In my view, Berkeley's frequent mention of "visual magnitude," "visible figure," and "visible extension" in the *Essay towards a New Theory of Vision* (Dublin, 1709) puts the matter beyond doubt (see secs. 43–5, 49–50, 54, etc.), as does his description of the proper object of vision in his *Theory of Vision Vindicated and Explained* (London, 1733), sec. 44: "The proper, immediate object of vision is light, in all its modes and variations, various colours in kind, in degree, in quantity; some lively, others faint; more of some and less of others; various in their bounds or limits; various in their order and situation." Passages from the *New Theory* seeming to deny that visible figure is perceived—such as the denial that the immediate objects of sight are "flat or plain figures" (sec. 157)—are presented as corollaries to Berkeley's denial that touch and sight share a common idea of extension. His point was not to deny visible extension but to demarcate two genera of extension, visible and tangible.

22. Among early modern authors, Locke certainly recognized this point, maintaining that all ideas are acquired through experience (*Essay*, I.ii.15; II.i.2) but nonetheless granting innate "capacities for understanding" (I.ii.22), such as *discerning, comparing, compounding*, and *abstracting*, that depend upon innate powers or capacities of the mind (II.xi). Leibniz, *New Essays*, trans. Remnant and Bennett, codified this point in the slogan, "Nothing is in the intellect that was not first in the senses, except the intellect itself" (p. 111). Although Leibniz surely meant this slogan to include truths known by reason alone, it may be read as a more minimal response to Locke, that even he must admit some innate learning ability.

23. See the careful discussion of innateness by Stich in his introduction to *Innate Ideas*, 1–22.

24. Locke, *Essay*, I.ii.5–18.

25. See the articles by Chomsky and Putnam in Stich, ed., *Innate Ideas*, 121–44.

26. Aristotle and the Aristotelians, who are not ususally counted among empiricists even though they share with Locke the slogan that all knowledge must come through the senses and the image of the mind as a blank tablet, thought that reason could grasp such essences as it worked on sense experience. Classical empiricists such as Locke or Hume were skeptical that such essences could be grasped, with or without the aid of the senses.

Appendix B, Sensation and Perception

1. See n. 15 and the related quotation in Appendix A.

2. Nicholas Malebranche, *Recherche de la Vérité*, 5th ed. (Paris, 1700), Bk. I, ch. 1, art. 1, in his *Oeuvres*, 21 vols. (Paris, 1958–70), 1:42. Descartes had earlier, with some consistency but without fanfare, used the Latin term "sensus," and the French "sentiment," and "apercevoir," in accordance with the above distinction; see his

Optics, Sixth Discourse, and *Principles of Philosophy*, Pt. I, art. 70–1. Michael Ayers drew my attention to Malebranche's distinction.

3. Reid, *Inquiry*. ch. 6, sec. 20, in his *Works*, ed. Hamilton, 1:183a.
4. Ibid.
5. Ibid., 185.
6. Descartes emphasizes the physiological aspect of "*sentiment*" in the opening paragraph of the Sixth Discourse of the *Optics*, and of "*sensus*" in the *Principles*, Pt. IV, art. 189–90, 195.
7. Wundt, *Beiträge*, 446.

References

Abbagnano, Nicola. "Psychologism," in *The Encyclopedia of Philosophy*, ed. by Paul Edwards, 8 vols., 7:520–521. New York: Macmillan, 1967.

Abicht, J. H., and F. G. Born, eds. *Neues philosophisches Magazin, Erläuterungen und Anwendungen des Kantischen Systems bestimmt*. Leipzig, 1789–91.

Aguilon, Frances. *Opticorum libri sex*. Antwerp, 1613.

Alexander, Peter. "Boyle and Locke on Primary and Secondary Qualities." *Ratio*, 16 (1974), 51–67.

Alexander, Peter. *Ideas, Qualities, and Corpuscles: Locke and Boyle on the External World*. Cambridge: Cambridge University Press, 1985.

Allison, Henry E. *Kant's Transcendental Idealism*. New Haven: Yale University Press, 1983.

Ameriks, Karl. *Kant's Theory of Mind: An Analysis of the Paralogisms of Pure Reason*. Oxford: Clarendon Press, 1982.

Anderson, Wilda. "Encyclopedic Topologies." *MLN*, 101 (1986), 912–929.

Aquila, Richard E. *Intentionality: A Study of Mental Acts*. University Park: Pennsylvania State University Press, 1977.

Aquinas, Thomas. *Summa Theologica*, 61 vols. Latin text and English translation. Cambridge: Blackfriars, 1964–81.

Arens, Katherine. *Structures of Knowing: Psychologies of the Nineteenth Century*. Boston: Kluwer Academic Publishers, 1989.

Ash, Mitchell G. "Reflections on Psychology in History." In *The Problematic Science*, Woodward and Ash, eds. 347–367.

Ash, Mitchell G., and Ulfried Geuter, eds. *Geschichte der deutschen Psychologie im 20. Jahrhundert*. Opladen: Westdeutscher Verlag, 1985.

Ayer, Alfred J. *The Foundations of Empirical Knowledge*. New York: Macmillan, 1940.

Baker, Lynne Rudder. *Saving Belief: A Critique of Physicalism*. Princeton: Princeton University Press, 1987.

Bartels, Carl M. N. *Beiträge zur Physiologie des Gesichtssinnes*. Berlin, 1834.

Baumgarten, Alexander. *Metaphysica*, 7th ed. Halle, 1779. Reprint. Hildesheim: Olms, 1982.

Beck, Lewis White. *Early German Philosophy: Kant and His Predecessors*. Cambridge, MA: Harvard University Press, 1969.

Beigel, Hugo. *Wörterbuch der Psychologie und verwandter Gebiete*. New York: Ungar, 1971.

Beiser, Frederick C. *The Fate of Reason: German Philosophy from Kant to Fichte*. Cambridge, MA: Harvard University Press, 1987.

Beltrami, Eugenio. *Saggio di Interpretazione della Geometrie Non-Euclidea*. Naples, 1868.

Beneke, Friedrich Eduard. *Kant und die philosophische Ausgabe unserer Zeit*. Berlin: Posen and Bromberg, 1832.

Beneke, Friedrich Eduard. *Die neue Psychologie*. Berlin: Posen and Bromberg, 1845.

Beneke, Friedrich Eduard. *Lehrbuch der Psychologie als Naturwissenschaft*, 2d ed. Berlin: Posen and Bromberg, 1845.

Bennett, Jonathan. *Kant's Analytic*. Cambridge: Cambridge University Press, 1966.

Bennett, Jonathan. *Locke, Berkeley, and Hume*. Oxford: Clarendon, 1971.

Berkeley, George. *An Essay towards a New Theory of Vision*. Dublin, 1790. As printed in his *Philosophical Works*, ed. Ayers.

Berkeley, George. *Treatise Concerning the Principles of Human Knowledge*. Dublin, 1710. As printed in his *Philosophical Works*, ed. Ayers.

Berkeley, George. *Three Dialogues between Hylas and Philonous*. London, 1713. As printed in his *Works*, eds. A. A. Luce and T. E. Jessop, 9 vols., vol. 2. London: Nelson, 1948–57.

Berkeley, George. *Theory of Vision Vindicated and Explained*. London, 1733. As printed in his *Philosophical Works*, ed. Ayers.

Berkeley, George. *Philosophical Works*, ed. Michael R. Ayers. London: Dent, 1975.

Berthold, Arnold Adolph. *Das Aufrechterscheinungen der Gesichtsobjecte*. Göttingen, 1830.

Bieri, Peter, Rolf-Peter Horstmann, and Lorenz Krüger, eds. *Transcendental Arguments and Science: Essays in Epistemology*. Boston: Reidel, 1979.

Bonnet, Charles. *Betrachtung über die Natur*, trans. Johann Daniel Titius. Leipzig, 1766.

Bonnet, Charles. *Analytischer Versuch über die Seelenkräfte*, trans. Christian Gottfried Schütz, 2 vols. Bremen, 1770–71.

Boring, Edwin G. *A History of Experimental Psychology*. New York: Century, 1929.

Boring, Edwin G. *Sensation and Perception in the History of Experimental Psychology*. New York: Appleton-Century-Crofts, 1942.

Boring, Edwin G. *A History of Experimental Psychology*, 2d ed. New York: Appleton-Century-Crofts, 1950.

Brandt, Francis Burke. "Friedrich Eduard Beneke: The Man and His Philosophy." Ph.D. Dissertation, Columbia University. New York, 1895.

Bréhier, Emile. "The Creation of the Eternal Truths in Descartes' System." In *Descartes*, Willis Doney, ed.,192–208. Notre Dame: University of Notre Dame Press, 1968.

Broad, C. D. *Scientific Thought*. London: Kegan Paul, 1923.

Brown, Harold I. "Normative Epistemology and Naturalized Epistemology." *Inquiry*, 31 (1988), 53–78.

Buroker, Jill Vance. *Space and Incongruence: The Origin of Kant's Idealism*. Boston: Reidel, 1981.

Cassirer, H. W. *Kant's First Critique*. New York: Macmillan, 1954.

Chale, Claude de. *Cursus seu mundus mathematicus*. Annison, 1674.

Chambers, Ephraim. *Cyclopaedia: or, An Universal Dictionary of Arts and Sciences*, 2d ed., 2 vols. London, 1738.

Chaplin, James P., and T. S. Krawiec. *Systems and Theories of Psychology*, 3d ed. New York: Holt, 1974.

Chomsky, Noam. *Cartesian Linguistics: A Chapter in the History of Rationalist Thought*. New York: Harper and Row, 1966.

Chomsky, Noam. "Recent Contributions to the Theory of Innate Ideas." In Stich, ed., *Innate Ideas*, 121–131.

Chomsky, Noam. *Knowledge of Language*. New York: Praeger, 1985.

Churchland, Patricia Smith. *Neurophilosophy: Toward a Unified Science of the Mind-Brain*. Cambridge, MA: The MIT Press/Bradford Books, 1986.

Churchland, Paul M. *Scientific Realism and the Plasticity of Mind*. Cambridge: Cambridge University Press, 1979.

Churchland, Paul M. "Eliminative Materialism and the Propositional Attitudes." *Journal of Philosophy*, 78 (1981), 67–90.

Classen, August. "Beitrag zur physiologischen Optik." *Archiv für pathologische Anatomie und Physiologie* , 25 (1862), 1–38.

Classen, August. *Ueber das Schlussverfahren des Sehactes.* Rostock, 1863.

Classen, August. *Physiologie des Gesichtssinnes zum ersten Mal begründet auf Kants Theorie der Erfahrung.* Braunschweig, 1876.

Cohen, Hermann. *Kants Theorie der Erfahrung.* Berlin, 1871; 2d ed., Berlin, 1885.

Condillac, Etienne Bonnot de. *Essai sur l'Origine des Connoisances Humaines.* Amsterdam, 1746.

Condillac, Etienne Bonnot de. *Traité des Sensations.* London, 1754.

Cornelius, Carl Sebastian. *Die Theorie des Sehen und räumlichen Vorstellens.* Halle, 1861.

Cornelius, Carl Sebastian. *Zur Theorie des Sehen mit Rücksicht auf die neuesten Arbeiten in diesem Gebiete.* Halle, 1864.

Cranefield, Paul. "The Philosophical and Cultural Interests of the Biophysics Movement of 1847." *Journal of the History of Medicine,* 21 (1966), 1–7.

Cronin, Timothy J. *Objective Reality in Descartes and Suarez.* Rome: Gregorian University Press, 1966.

Crusius, Christian August. *Entwurf der nothwendigen Vernunft-Wahrheiten wiefern sie den zufälligen entgegen gesetzet werden.* Leipzig, 1745. Reprinted as vol. 2 of his *Philosophische Hauptwerke,* ed. G. Tonelli, 4 vols. Hildesheim: Olms, 1964–87.

Curley, E. M. "Locke, Boyle, and the Distinction between Primary and Secondary Qualities." *Philosophical Review,* 81 (1972), 438–464.

Czermak, Johann. "Physiologische Studien." *Sitzungsberichte der kaiserliche Akademie der Wissenschaften zu Wien. Mathematisch-naturwissenschaftliche Klasse,* 15 (1855), 425–521, and 17 (1855), 563–600.

Dancy, Jonathan. *Introduction to Contemporary Epistemology.* Oxford: Blackwell, 1985.

Danziger, Kurt. "Wundt and the Two Traditions of Psychology." In R. W. Rieber, ed., *Wilhelm Wundt and the Making of a Scientific Psychology,* 73–87. New York: Plenum, 1980.

Darwin, Charles. *On the Origin of Species by Means of Natural Selection, or, The Preservation of Favoured Races in the Struggle for Life.* London, 1859. Reprint. Cambridge, MA: Harvard University Press, 1964.

Daston, Lorraine J. "The Theory of Will versus the Science of the Mind." In *The Problematic Science,* Woodward and Ash, eds., 88–115.

Dennes, William. "The Categories of Naturalism." In *Naturalism and the Human Spirit,* Krikorian, ed., 270–294.

Descartes, René. *Discourse on Method, Optics, Geometry, and Meteorology,* trans. Paul J. Olscamp. Indianapolis: Bobbs-Merrill, 1965.

Descartes, René. *Treatise on Man,* trans. Thomas S. Hall. Cambridge, MA: Harvard University Press, 1972.

Descartes, René. *Philosophical Writings of Descartes,* trans. John Cottingham, Robert Stoothoff, and Dugald Murdoch, 2 vols. Cambridge: Cambridge University Press, 1984. Cited as CSM.

Dessoir, Max. *Geschichte der neueren deutschen Psychologie,* 2nd ed. Berlin, 1902.

Dewey, John. "The New Psychology." As reprinted in his *Early Works,* 1:48–60.

Dewey, John. *Psychology.* New York, 1887. As reprinted in his *Early Works,* vol 3.

Dewey, John. "Psychologism." In *Dictionary of Philosophy and Psychology,* James Mark Baldwin, ed., 3 vols, 2:382. New York: Macmillan, 1901-05.

Dewey, John. *Early Works,* 1882–98, 5 vols. Carbondale: Southern Illinois University Press, 1967–72.

Dewey, John. "Antinaturalism in extremis." In *Naturalism and the Human Spirit,* Krikorian, ed., 1–16.

Dewey, John, Sidney Hook, and Ernst Nagel. "Are Naturalists Materialists?" *Journal of Philosophy*, 42 (1945), 515–530.

d'Holbach, Paul Henri Thiry, baron. *System of Nature*, trans. H. D. Robinson. New York: Franklin, 1970.

di Giovanni, George, and H. S. Harris, trans. *Between Kant and Hegel: Texts in the Development of Post-Kantian Idealism*. Albany: State University of New York Press, 1985.

Donagan, Alan. "Berkeley's Theory of the Immediate Objects of Vision." In Machamer and Turnbull, eds., *Studies in Perception*, 312–335.

Dretske, Fred. *Knowledge and the Flow of Information*. Cambridge: The MIT Press/Bradford Books, 1981.

Ebbinghaus, Hermann. *Abriss der Psychologie*. Leipzig: Veit, 1908.

Eberhard, Johann August. *Allgemeine Theorie des Denkens und Empfindens*. Berlin, 1776.

Eberhard, Johann August, ed. *Philosophisches Magazin*. Halle, 1788–95.

Elkana, Yehuda. *The Discovery of the Conservation of Energy*. Cambridge: Harvard University Press, 1974.

Erdmann, Benno. "Die philosophischen Grundlagen von Helmholtz' Wahrnehmungstheorie." *Abhandlungen der Preussischen Akademie der Wissenschaften. Philosophisch-historische Klasse* (1921), Nr. 1, 1–45.

Erdmann, Johann Eduard. *Grundriss der Geschichte der Philosophie*, 2 vols. Berlin, 1866.

Erdmann, Johann Eduard. *Grundriss der Geschichte der Philosophie*, 2d ed., 2 vols. Berlin, 1870.

Erpenbeck, Heinrich. *Das Recht-Sehen trotz des umgekehrten Sehbildes auf der Retina*. Würzburg, 1833.

Falckenberg, Richard. *Geschichte der neueren Philosophie*, 2d ed. Leipzig, 1892. Translated by A. C. Armstrong as *History of Modern Philosophy*, 3d ed. New York, 1897.

Falmagne, Rachel Joffe. "Normative Theory and the Human Mind." *Behavioral and Brain Sciences*, 8 (1985), 750–751.

Fichte, Johann Gottlieb. *Sämtliche Werke*, ed. I. H. Fichte, 8 vols. Berlin, 1834–46.

Fichte, Johann Gottlieb. "Zweite Einleitung in die Wissenschaftslehre," in his *Sämtliche Werke*, ed. I. H. Fichte, 1:471–9. Translated by P. Heath and J. Lachs as "Second Introduction to the *Science of Knowledge*," in Fichte, *Science of Knowledge*. Cambridge: Cambridge University Press, 1982.

Fischer, Johann Karl. *Physikalisches Wörterbuch*, 10 vols. Göttingen, 1798–1827.

Fischer, Kuno. *Geschichte der neuern Philosophie*, 10 vols. Mannheim: Bassermann & Mathy, 1854.

Fischer, Kuno. *Geschichte der neuern Philosophie*, 3rd ed., 8 vols. Heidelberg: Winter, 1889–93.

Flugel, J. C. *A Hundred Years of Psychology*. New York, 1933.

Fodor, Jerry. *Language of Thought*. New York: Crowell, 1975.

Fodor, Jerry. *Representations*. Cambridge, MA: The MIT Press, 1981.

Fodor, Jerry. *Psychosemantics*. Cambridge, MA: The MIT Press, 1987.

Förster, Eckart, ed. *Kant's Transcendental Deductions: The Three Critiques and the Opus Postumum*. Stanford: Stanford University Press, 1989.

Friedman, Michael. *Foundations of Space-Time Theories: Relativistic Physics and Philosophy of Science*. Princeton: Princeton University Press, 1983.

Friedman, Michael. "Kant's Theory of Geometry." *Philosophical Review*, 94 (1985), 455–506.

Fries, Jakob Friedrich. *Neue oder anthropologische Kritik der Vernunft*, 1st ed., 3 vols. Heidelberg, 1807. 2d ed., 3 vols. Heidelberg, 1828–31.

Fries, Jakob Friedrich. *Sämtliche Schriften*, 25 vols. Aalen: Scientia Verlag, 1967-.

Galaty, David H. "The Emergence of Biological Reductionism." Ph.D. dissertation, Johns Hopkins University. Baltimore, 1971.

Gardner, Howard. *The Mind's New Science: A History of the Cognitive Revolution.* New York: Basic Books, 1985.

Garnett, Christopher Browne. *Kantian Philosophy of Space.* New York: Columbia University Press, 1939.

Geertz, Clifford. *The Interpretation of Cultures: Selected Essays.* New York: Basic Books, 1973.

Gehler, Johann Samuel Traugott. *Physikalisches Wörterbuch, oder Versuch einer Erklärung der vornehmsten Begriffe und Kunstwörter der Naturlehre,* 6 vols. Leipzig, 1787–96.

Geuter, Ulfried. *Die Professionalisierung der deutschen Psychologie im Nationalsozialismus.* Frankfurt: Suhrkamp, 1984.

Gibson, James J. *Perception of the Visual World.* Boston: Houghton-Mifflin, 1950.

Gibson, James J. *The Senses Considered as Perceptual Systems.* Boston: Houghton-Mifflin, 1966.

Giere, Ronald. *Explaining Science: A Cognitive Approach.* Chicago: University of Chicago Press, 1988.

Gleitman, Henry. *Basic Psychology,* 2d ed. New York: Norton, 1987.

Goldman, Alvin. "Innate Knowledge." In Stich, ed., *Innate Ideas,* 111–120.

Goldman, Alvin. *Epistemology and Cognition.* Cambridge: Harvard University Press, 1986.

Goldschmidt, Ludwig. *Kant und Helmholtz: Populärwissenschaftliche Studie.* Hamburg and Leipzig, 1898.

Goodman, Nelson. *Languages of Art,* 2d ed. Indianapolis: Hackett, 1976.

Gould, Stephen J. *Ever Since Darwin.* New York: Norton, 1977.

Green, T. H. *Works,* ed. R. L. Nettleship, 3 vols. London, 1885.

Greenlee, Douglas. "Locke's Ideas of 'Idea'." In *Locke on Human Understanding,* I. C. Tipton, ed. Oxford: Oxford University Press, 1977.

Gregory, Frederick. *Scientific Materialism in Nineteenth Century Germany.* Boston and Dordrecht: Reidel, 1978.

Gregory, Frederick. "Neo-Kantian Foundations of Geometry in the German Romantic Period." *Historia Mathematica,* 10 (1983), 184–201.

Grimaldi, Nicolas. *L'Expérience de la pensée dans la philosophie de Descartes.* Paris: Vrin, 1978.

Guyer, Paul. *Kant and the Claims of Knowledge.* Cambridge: Cambridge University Press, 1987.

Hall, Thomas Steele. *History of General Physiology.* Chicago: University of Chicago Press, 1975.

Hamlyn, D. W. *The Psychology of Perception: A Philosophical Examination.* London: Routledge & Kegan Paul, 1959.

Hamlyn, D. W. *Sensation and Perception: A History of the Philosophy of Perception.* London: Routledge & Kegan Paul, 1961.

Harman, Gilbert. "Epistemology." In *Historical and Philosophical Roots of Perception,* Edward C. Carterette and Morton P. Friedman, eds. New York: Academic Press, 1974.

Hartley, David. *Various Conjectures on the Perception, Motion, and Generation of Ideas.* Bath, 1746.

Hartley, David. *Observations on Man: His Frame, His Duty, and His Expectations,* 2 vols. London, 1749.

Hartley, David. *Betrachtungen über den Menschen,* 2 vols. Rostock, 1772–73.

Hasenclever, Richard. *Die Raumvorstellung aus dem Gesichtssinne.* Berlin, 1842.

Hatfield, Gary. "Force (God) in Descartes' Physics." *Studies in the History and Philosophy of Science,* 10 (1979), 113–140.

Hatfield, Gary. "The Senses and the Fleshless Eye: The *Meditations* as Cognitive Exercises." In *Essays on Descartes' Meditations*, Amélie O. Rorty, ed., 45–79. Berkeley: University of California Press, 1976.

Hatfield, Gary. "Metaphysics and the New Science." In *Reappraisals of the Scientific Revolution*, David C. Lindberg and Robert Westman, eds., 93–166. Cambridge and New York: Cambridge University Press, 1990.

Hatfield, Gary, and William Epstein. "The Sensory Core and the Medieval Foundations of Early Modern Perceptual Theory." *Isis*, 70 (1979), 363–384.

Hegel, Georg W. F. *Vorlesungen über die Geschichte der Philosophie*. In his *Sämtliche Werke*, 26 vols. Stuttgart, 1927–38. Translated by E. S. Haldane and F. H. Simson as *Lectures on the History of Philosophy*, 3 vols. London: Humanities, 1892.

Hehlmann, Wilhelm. *Wörterbuch der Psychologie*. Stuttgart: Kröner, 1959.

Helmholtz, Hermann. *Ueber die Erhaltung der Kraft: Eine physikalische Abhandlung*. Berlin, 1847. As printed in his *Wissenschaftliche Abhandlungen*, 1:12–75. Translated by John Tyndall as "On the Conservation of Force: A Physical Memoir"; printed in Helmholtz's *Selected Writings*, ed. Kahl, 3–55, with editorial revisions.

Helmholtz, Hermann. *Die Lehre den Tonempfindungen als physiologische Grundlage für die Theorie der Musik*. Braunschweig: Vieweg, 1863. Translated by A. J. Ellis as *On the Sensations of Tone as a Physiological Basis for the Theory of Music*. 4th ed. London: Longman's Green, 1912.

Helmholtz, Hermann. *Handbuch der physiologischen Optik*. Leipzig: Voss, 1867; 2d ed., 1886–96; 3d ed., 1909–11. Translated by James P. C. Southall as *Treatise on Physiological Optics*, 3 vols. Milwaukee: Optical Society of America, 1924–25.

Helmholtz, Hermann. *Populäre wissenschaftliche Vorträge*, 2 vols. Braunschweig: Vieweg, 1870.

Helmholtz, Hermann. *Wissenschaftliche Abhandlungen*, 3 vols. Leipzig: Barth, 1882–95. Cited as WA.

Helmholtz, Hermann. *Vorträge und Reden*, 4th ed., 2 vols. Braunschweig, 1896. Cited as VR.

Helmholtz, Hermann. *Selected Writings*, ed. Russell Kahl. Middletown, Conn.: Wesleyan University Press, 1971. Cited as SW.

Helmholtz, Hermann. *Epistemological Writings*, eds. Robert S. Cohen and Yehuda Elkana. Boston: Reidel, 1977. Cited as EW.

Helmholtz, Hermann. *Dokumente einer Freundschaft: Briefwechsel zwischen Hermann von Helmholtz und Emil du-Bois-Reymond, 1846–1894*, ed. Christa Kirsten. Berlin: Adademie-Verlag, 1986.

Henrich, Dieter. "Kant's Notion of a Deduction and the Methodological Background of the First *Critique*." In *Kant's Deductions*, Förster, ed., 29–46.

Herbart, Johann Friedrich. *Hauptpuncte der Metaphysik*. Göttingen, 1808.

Herbart, Johann Friedrich. *Lehrbuch zur Einleitung in die Philosophie*. Königsberg, 1813; 4th ed., 1837.

Herbart, Johann Friedrich. *Lehrbuch der Psychologie*. Königsberg and Leipzig, 1816; 2d ed., 1834.

Herbart, Johann Friedrich. *Allgemeine Metaphysik nebst den Anfängen der philosophischen Naturlehre*. Königsberg, 1828–29.

Herbart, Johann Friedrich. *Psychologie als Wissenschaft neu gegründet auf Erfahrung, Metaphysik und Mathematik*. Königsberg, 1824–25.

Herbart, Johann Friedrich. *Sämtliche Werke*, ed. G. Hartenstein, 12 vols. Leipzig: Voss, 1850–51.

Hering, Ewald. *Beiträge zur Physiologie*. Leipzig: Engelmann, 1861–64.

Hering, Ewald. *Die Lehre vom binocularen Sehen*. Leipzig: Engelmann, 1868.

Hering, Ewald. *Zur Lehre vom Lichtsinne*, 2d ed. Wien: C. Gerolds Sohn, 1878.

Hilts, Victor. "Statistics and Social Science." In *Foundations of Scientific Method: The Nineteenth Century*, Ronald Giere and Richard Westfall, eds., 206–233. Bloomington: Indiana University Press, 1973.

Hobbes, Thomas. *De Corpore*. London, 1655. Published in English as *Elements of Philosophy*, Part 1, *De Corpore*. London, 1656. As printed in his *English Works*, ed. William Molesworth, 11 vols., vol. 1. London, 1839–45.

Hobbes, Thomas. *Leviathan*. London, 1651. Reprint. Baltimore: Penguin, 1968.

Hochberg, Julian. "Nativism and Empiricism in Perception." In *Psychology in the Making*, Leo Postman, ed., 255–330. New York: Knopf, 1964.

Hoffmeister, Johannes. *Wörterbuch der philosophischen Begriffe*. Hamburg: Felix Meiner, 1955.

Hothersall, David. *History of Psychology*. New York: Random House, 1984.

Hueck, Alexander. *Das Sehen seinem äussern Process nach entwickelt*. Riga and Göttingen, 1830.

Hume, David. *Treatise of Human Nature*. London, 1739–40. Reprint. London, 1888.

Hume, David. *An Inquiry Concerning Human Understanding*. London, 1748.

Hume, David. *Philosophische Versuche über die menschliche Erkenntniss*, 1755. Vol. 2 of his *Vermischter Schriften*, ed. Johann Georg Sulzer, 2 vols. Hamburg and Leipzig, 1754–55.

Jardine, Nicholas. "Epistemology of the Sciences." In Schmitt, ed., *Cambridge History of Renaissance Philosophy*, ch. 19.

Johnson-Laird, Phillip. *The Computer and the Mind: An Introduction to Cognitive Science*. Cambridge, MA: Harvard University Press, 1988.

Kahneman, D., and A. Tversky. "On the Psychology of Prediction." *Psychological Review*, 80 (1973), 237–251.

Kant, Immanuel. *On the Form and Principles of the Sensible and Intelligible World*, Innaugural Dissertation of 1770. In his *Selected Pre-Critical Writings*, trans. G. B. Kerferd and D. E. Walford. Manchester: Manchester University Press, 1968.

Kant, Immanuel. *Critique of Pure Reason*, trans. Norman Kemp Smith. New York: St. Martin's, 1964. Original German edition published in 1781; 2d ed., 1787.

Kant, Immanuel. *Prolegomena to Any Future Metaphysics*, trans. James W. Ellington. Indianapolis: Hackett, 1977. Original German edition published in 1783.

Kant, Immanuel. *Metaphysical Foundations of Natural Science*, trans. James Ellington. Indianapolis: Bobbs, 1970. Original German edition published in 1786.

Kant, Immanuel. *Critique of Practical Reason*, trans. Lewis White Beck. Indianapolis: Bobbs-Merrill, 1956. Original German edition published in 1788.

Kant, Immanuel. *Critique of Judgment*, trans. James C. Meredith. Oxford: Oxford University Press, 1952. Original German edition published in 1790; 2d ed., 1793.

Kant, Immanuel. *Anthropology from a Pragmatic Point of View*, trans. Mary Gregor. The Hague: Martinus Nijhoff, 1974. Original German edition published in 1798.

Kant, Immanuel. *Die philosophischen Hauptvorlesungen Immanuel Kants*, ed. Arnold Kowalewski. Munich and Leipzig, 1924.

Kant, Immanuel. *Gesammelte Schriften*, Akademie Ausgabe. Berlin, 1902-. Cited as Ak.

Kant, Immanual. *Philosophical Correspondence*, ed. and trans. Arnulf Zweig. Chicago: University of Chicago Press, 1967.

Katz, J. J. *Cogitations*. New York: Oxford University Press, 1986.

Kenny, Anthony. *Descartes*. New York: Random House, 1968.

Kessler, Eckhard. "The Intellective Soul." In *Cambridge History of Renaissance Philosophy*, Schmitt, ed., ch. 15.

Kitcher, Patricia. "Discovering the Forms of Intuition." *Philosophical Review*, 96 (1987), 205–248.

Kitcher, Philip. *The Nature of Mathematical Knowledge*. New York: Oxford University Press, 1983.

Klemm, Otto. *Geschichte der Psychologie* (Leipzig: Teubner, 1911). Translated by E. C. Wilm and R. Pintner as *A History of Psychology*. New York: Scribner's, 1914.

Koch, Sigmund. "Wundt's Creature at Age Zero—and as Centenarian: Some Aspects of the Institutionalization of the 'New Psychology.'" In Sigmund Koch and David E. Leary, eds., *A Century of Psychology as a Science*, 7–35. New York: McGraw-Hill, 1985.

Koffka, Kurt. *Principles of Gestalt Psychology*. London and New York, 1935.

Köhler, Wolfgang. *Gestalt Psychology*. New York, 1929.

Köhnke, Klaus Christian. *Entstehung und Aufstieg des Neukantianismus*. Frankfurt: Suhrkamp, 1986.

Königsberger, Leo. *Hermann von Helmholtz*, 3 vols. Braunschweig: Vieweg, 1902–03.

Königsberger, Leo. *Hermann von Helmholtz*, abridged trans., F. A. Welby. Oxford: Clarendon Press, 1906.

Kornblith, Hilary, ed. *Naturalizing Epistemology*. Cambridge: The MIT Press/Bradford Books, 1985.

Krikorian, Yervant, ed. *Naturalism and the Human Spirit*. New York: Columbia University Press, 1944.

Krikorian, Yervant. "A Naturalistic View of Mind." In Krikorian, ed., *Naturalism and the Human Spirit*, 242–269.

Kuehn, Manfred. *Scottish Common Sense in Germany, 1768–1800: A Contribution to the History of Critical Philosophy*. Kingston and Montreal: McGill-Queen's University Press, 1987.

Kyburg, Henry E. "Rational Belief." *Behavioral and Brain Sciences*, 6 (1983), 231–245, 263–273.

La Mettrie, Julien Offray de. *Man a Machine*. La Salle: Open Court, 1912.

Lambert, Johann Heinrich. *Neues Organon*, 2 vols. Leipzig, 1764. As printed in his *Philosophische Schriften*, ed. H. W. Arndt, 6 vols., vols. 1–2. Hildesheim: Olms, 1965–69.

Land, J. P. "Kant's Space and Modern Mathematics." *Mind*, original series, 2 (1877), 38–46.

Lanning, Kevin. "Some Reasons for Distinguishing between 'Non-Normative Response' and 'Irrational Decision'." *Journal of Psychology*, 121 (1987), 109–117.

Leary, David E. "The Philosophical Development of the Concept of Psychology in Germany." *Journal of the History of the Behavioral Sciences*, 14 (1978), 113–121.

Leary, David E. "Herbart's Mathematization of Psychology." *Journal of the History of the Behavioral Sciences*, 16 (1980), 150–163.

Leary, David E. "Immanuel Kant and the Development of Modern Psychology." In *The Problematic Science*, Woodward and Ash, eds., 17–42.

Leibniz, Gottfried Wilhelm. *New Essays on Human Understanding*, trans. Peter Remnant and Jonathan Bennett. Cambridge: Cambridge University Press, 1981.

Lenoir, Timothy. *The Strategy of Life: Teleology and Mechanics in Nineteenth Century German Biology*. Boston: Reidel, 1982.

Lewis, C. I. *Mind and the World Order*. New York: Scribner's, 1929.

Lindberg, David C. *Theories of Vision from Al-Kindi to Kepler*. Chicago: University of Chicago Press, 1976.

Locke, John. *Essay Concerning Human Understanding*, ed. Peter Nidditch. Oxford: Oxford University Press, 1975.

Lossius, Johann Christian. *Physische Ursachen des Wahren*. Gotha, 1775.

Lotze, Rudolph Hermann. *Medizinische Psychologie; oder, Physiologie der Seele*. Leipzig: Weidmann, 1852.

Lotze, Rudolph Hermann. *Mikrokosmus: Ideen zur Naturgeschichte und Geschichte der Menschheit. Versuch einer Anthropologie,* 3 vols. Leipzig, 1856–64. Translated by E. Hamilton and E. E. Constance Jones as *Microcosmus: An Essay Concerning Man and His Relation to the World,* 2 vols. Edinburgh, 1885.

Lotze, Rudolph Hermann. *Logik.* Leipzig: Hirzel, 1874. Translated by Bernard Bosanquet as *Logic,* 2d ed., 2 vols. Oxford, 1888.

Lovejoy, Arthur O. *The Revolt against Dualism: An Inquiry Concerning the Existence of Ideas.* Chicago: Open Court, 1930.

Löw, Reinhard. *Philosophie des Lebendigen: Der Begriff des Organischen bei Kant, sein Grund and seine Aktualität.* Frankfurt: Suhrkamp, 1980.

Lowry, Richard. *Evolution of Psychological Theory,* 2d ed. New York: Aldine, 1982.

Ludwig, Carl F. *Lehrbuch der Physiologie der Menschen,* 2d ed., 2 vols. Leipzig and Heidelberg, 1858–61.

Machamer, Peter K., and Robert G. Turnbull, eds. *Studies in Perception.* Columbus: Ohio State University Press, 1978.

Mackie, J. L. *Problems from Locke.* Oxford: Clarendon, 1976.

Malebranche, Nicolas. *Oeuvres,* 21 vols. Paris: Vrin, 1958–70.

Mandelbaum, Maurice. *History, Man, and Reason: A Study in Nineteenth Century Thought.* Baltimore: Johns Hopkins University Press, 1971.

Marr, David. *Vision: A Computational Investigation into the Human Representation and Processing of Visual Information.* San Francisco: Freeman, 1982.

Marshall, Marilyn E. "Physics, Metaphysics, and Fechner's Psychophysics." In *The Problematic Science,* Woodward and Ash, eds., 65–87.

Massaro, Dominic. *Experimental Psychology: An Information Processing Approach.* San Diego: Harcourt, Brace, Jovanovich, 1989.

Mayne, Zachary. *Two Dissertations Concerning Sense, and the Imagination; with an Essay on Consciousness.* London, 1728.

McCleary, Robert A., ed. *Genetic and Experiential Factors in Perception.* Glenview: Scott, Foresman and Co., 1970.

Meissner, Georg. *Physiologie des Sehorgans.* Leipzig, 1854.

Metzner, Karl Hans. *Die biologischen Ansichten Ewald Herings.* Heidelberg, 1945.

Mill, James. *Analysis of the Phenomena of the Human Mind.,* ed. with notes by A. Bain, A. Findlater, and G. Grote; ed. with additional notes by J. S. Mill, 2 vols. London, 1869.

Mill, John Stuart. *A System of Logic,* in his *Collected Works,* vol. 7. Toronto: University of Toronto Press, 1963-.

Millikan, Ruth Garret. *Language, Thought, and Other Biological Categories.* Cambridge: The MIT Press/Bradford Books, 1984.

Mintz, Samuel I. *Hunting of Leviathan.* Cambridge: Cambridge University Press, 1962.

Mischel, Theodore. "Wundt and the Conceptual Foundations of Psychology." *Philosophy and Phenomenological Research,* 31 (1970–71), 1–26.

Molyneux, William. *Treatise of Dioptricks.* London, 1692.

Moore, G. E. "The Status of Sensa." In his *Philosophical Studies,* ch. 5. London: Kegan Paul, 1922.

Morgan, M. J. *Molyneux's Question: Vision, Touch and the Philosophy of Perception.* Cambridge: Cambridge University Press, 1977.

Müller, Johannes. *Zur vergleichenden Physiologie des Gesichtssinnes des Menschen und der Thiere.* Leipzig: Cnobloch, 1826.

Müller, Johannes. *Handbuch der Physiologie des Menschen,* 2 vols. Coblenz: Holscher, 1833–37. Translated by William Baly as *Elements of Physiology,* 2 vols. London, 1838–42.

Muncke, Georg Wilhelm. "Gesicht." In Gehler's *PhysikalischesWörterbuch,* rev. ed., 11 vols. Leipzig, 1825–45.

Nagel, Albrecht E. *Das Sehen mit zwei Augen und die Lehre von dem identischen Netzhautstellen.* Leipzig and Heidelberg, 1861.

Newton, Isaac. *Opticks; or, A treatise on the Reflections, Refractions, Inflections and Colours of Light,* 4th ed. London, 1730. Reprint. New York: Dover, 1952.

Nisbett and Thagard, "Psychology, Statistics, and Analytical Epistemology." *Behavioral and Brain Sciences,* 6 (1983), 257–258.

Nozick, Robert. *Philosophical Explanations.* Cambridge: Harvard University Press, 1982.

O'Neil, W. M. *The Beginnings of Modern Psychology,* 2nd ed. Sussex: Harvester, 1982.

O'Neill, Brian. *Epistemological Direct Realism in Descartes' Philosophy.* Albuquerque: University of New Mexico Press, 1974.

Pappas, George S., and Marshall Swain, eds. *Essays on Knowledge and Justification.* Ithaca: Cornell University Press, 1978.

Park, Katharine, and Eckhard Kessler. "The Concept of Psychology." In *Cambridge History of Renaissance Philosophy,* Schmitt, ed., ch. 13.

Pastore, Nicholas. *Selective History of Theories of Visual Perception,* 1650–1950. New York: Oxford University Press, 1971.

Pastore, Nicholas. "Reevaluation of Boring on Kantian Influence, Nineteenth Century Nativism, Gestalt Psychology and Helmholtz." *Journal of the History of the Behavioral Sciences,* 10 (1974), 375–390.

Pastore, Nicholas. "Helmholtz on the Projection or Transfer of Sensation." In *Studies in Perception,* Machamer and Turnbull, eds., 355–376.

Pecham, John. *Perspectiva communis,* ed. and trans. David Lindberg. Madison: University of Wisconsin Press, 1970.

Piaget, Jean, and Rolando Garcia. *Psychogenesis and the History of Science,* trans. by Helga Feider. New York: Columbia University Press, 1989.

Pitcher, George. *Berkeley.* London: Routledge & Kegan Paul, 1977.

Porterfield, William. *A Treatise on the Eye.* Edinburgh and London, 1759.

Price, H. H. *Perception.* London: Methuen, 1932.

Priestley, Joseph. *Disquisitions Relating to Matter and Spirit.* London, 1777.

Putnam, Hilary. "The 'Innateness Hypothesis' and Explanatory Models in Linguistics." In *Innate Ideas,* Stich, ed., 133–144.

Putnam, Hilary. *Realism and Reason.* Cambridge: Cambridge University Press, 1983.

Putnam, Hilary. "Computational Psychology and Interpretation Theory." In his *Realism and Reason,* 139–154.

Putnam, Hilary. "Why Reason Can't be Naturalized." In his *Realism and Reason,* 229–247.

Putnam, Hilary. *Representation and Reality.* Cambridge: The MIT Press / Bradford Books, 1988.

Quine, W. V. *Word and Object.* Cambridge: The MIT Press, 1960.

Quine, W. V. "Epistemology Naturalized." In his *Ontological Relativity and Other Essays,* 69–90. New York: Columbia University Press, 1969.

Quine, W. V. *The Roots of Reference.* La Salle: Open Court, 1973.

Reid, Thomas. *An Inquiry into the Human Mind.* London and Edinburgh, 1764. As printed in his *Works,* ed. Hamilton, vol. 1.

Reid, Thomas. *Essays on the Intellectual Powers of Man.* Edinburgh, 1785. As printed in his *Works,* ed. Hamilton, vol. 1.

Reid, Thomas. *Works,* ed. William Hamilton, 8th ed., 2 vols. Edinburgh, 1895.

Reidel, Rupert. *Biology of Knowledge: The Evolutionary Basis of Reason,* trans. Paul Foulkes. New York: Wiley, 1984.

Reil, Johann Christian. "Von der Lebenskraft." *Archiv für die Physiologie,* 1 (1796), 8–162. Reissued as *Von der Lebenskraft,* ed. Karl Sudhoff. Leipzig, 1910.

Ribot, Theodule. *La Psychologie Allemande Contemporaine*. Paris, 1879. Translated by J. M. Baldwin as *German Psychology of Today*. New York, 1886.

Richards, Joan L. "The Evolution of Empiricism: Hermann von Helmholtz and the Foundations of Geometry." *British Journal for the Philosophy of Science*, 28 (1977), 235–253.

Richards, Robert J. "Wundt's Early Theories of Unconscious Inference and Cognitive Evolution in their Relation to Darwinian Biopsychology." In *Wundt Studies*, W. Bringmann and R. Tweney, eds., 42–70. Toronto and Göttingen: Hogrefe, 1980.

Rickert, Heinrich. *Gegenstand der Erkenntnis*. Freiburg, 1892.

Riehl, Alois. *Philosophische Kriticismus und seine Bedeutung für die positive Wissenschaft*, 2 vols. Leipzig, 1876–79.

Riemann, Bernard. "Ueber die Hypothesen, welche der Geometrie zugrunde liegen," *Abhandlungen der königlichen Gesellschaft der Wissenschaften zu Göttingen*, 13 (1867). As printed in his *Gesammelte mathematischen Werke und wissenschaftlicher Nachlass*, ed. Heinrich Weber, 2d edition, 272–287. New York: Dover, 1953.

Ritter, Heinrich. *Geschichte der Philosophie*, 12 vols. Hamburg, 1829–53.

Rixner, Thaddä Anselm. *Handbuch der Geschichte der Philosophie*, 3 vols. Sulzbach, 1829.

Robinson, Daniel N. *Intellectual History of Psychology*, rev. ed. New York: Macmillan, 1981.

Rock, Irvin. *The Logic of Perception*. Cambridge, MA: The MIT Press/Bradford Books, 1983.

Roe, Shirley. *Matter, Life, and Generation*. Cambridge: Cambridge University Press, 1981.

Rogers, Brian J., and Maureen Graham. "Similarities between Motion Parallax and Stereopsis in Human Depth Perception." *Vision Research*, 22 (1982), 261–270.

Rorty, Richard. *Philosophy and the Mirror of Nature*. Princeton: Princeton University Press, 1979.

Rosch, Eleanor, and Barbara Lloyd, eds. *Cognition and Categorization*. New York: Wiley, 1978.

Russell, Bertrand. "The Relation of Sense-Data to Physics." *Scientia*, 16 (1914), 1–27. As printed in his *Mysticism, Logic, and Other Essays*, ch. 8. New York: Norton, 1929.

Ryle, Gilbert. *The Concept of Mind*. London: Hutchinson, 1949.

Sabra, A. I. "Sensation and Inference in Alhazen's Theory of Visual Perception." In *Studies in Perception*, Machamer and Turnbull, eds., 160–185.

Satura, Vladimir. *Kants Erkenntnispsychologie*. Kantstudien Ergänzungshefte no. 101. Bonn: Bonvier, 1971.

Schleiden, Matthias. *Zur Theorie des Erkennens durch den Gesichtssinn*. Leipzig, 1861.

Schmitt, Charles B., ed. *Cambridge History of Renaissance Philosophy*. Cambridge: Cambridge University Press, 1988.

Schnädelbach, Herbert. *Philosophy in Germany, 1831–1933*, trans. Eric Matthews. Cambridge: Cambridge University Press, 1984.

Schultz, Duane P., and Sydney Ellen Schultz. *A History of Modern Psychology*, 4th ed. San Diego: Harcourt, Brace, Jovanovich, 1987.

Sellars, Wilfrid. "Empiricism and the Philosophy of Mind." In his *Science, Perception, and Reality*, 127–196. London: Routledge & Kegan Paul, 1963.

Sellars, Wilfrid. "Berkeley and Descartes: Reflections on the Theory of Ideas." In Machamer and Turnbull, eds., *Studies in Perception*, 259–311.

Smith, A. Mark. "Getting the Big Picture in Perspectivist Optics." *Isis*, 72 (1981), 568–589.

Smith, Edward E., and Douglas Medin. *Categories and Concepts*. Cambridge: Harvard University Press, 1981.

Smith, Norman Kemp. *New Studies in the Philosophy of Descartes: Descartes as Pioneer*. London: Macmillan, 1953.

Smith, Robert. *A Compleat System of Optics.* Cambridge, 1738.

Solomon, Miriam. "Quine's Point of View." *Journal of Philosophy,* 86 (1989), 113–136.

Specht, Rainer. *Commercium mentis et corporis: Ueber Kausalvorstellungen im Cartesianismus.* Stuttgart: Frommann, 1966.

Steinbuch, Johann Georg. *Beytrag zur Physiologie der Sinne.* Nurnberg: Johann Leonard Schragg, 1811.

Stewart, Dugald. *Outlines of Moral Philosophy.* Edinburgh, 1793. As printed in his *Collected Works,* ed. William Hamilton, 11 vols. Edinburgh, 1854.

Stich, Stephen, ed. *Innate Ideas.* Berkeley: University of California Press, 1975.

Stout, G. F. "The Herbartian Psychology." *Mind,* original series, 13 (1888), 321–338, 473–498.

Stout, G. F. *Groundwork of Psychology.* New York, 1903.

Strawson, P. F. *The Bounds of Sense.* London: Methuen, 1966.

Strawson, P. F. *Skepticism and Naturalism: Some Varieties.* New York: Columbia University Press, 1985.

Suarez, Francisco. *De anima.* Madrid: Sociedad de Estudios, 1978–81.

Summers, David. *Judgment of Sense: Renaissance Naturalism and the Rise of Aesthetics.* Cambridge: Cambridge University Press, 1987.

Taylor, Charles. "Interpretation and the Sciences of Man." In his *Philosophy and the Human Sciences,* 15–57. Cambridge: Cambridge University Press, 1985.

Tennemann, Wilhelm Gottlieb. *Geschichte der Philosophie,* 11 vols. Leipzig: Barth, 1798–1819. Reprint. Brussells: Culture and Civilisation, 1969–74.

Tetens, Johann Nicolas. *Philosophische Versuche über die menschliche Natur und ihre Entwicklung,* 2 vols. Leipzig, 1777. Reprint. Hildesheim: Olms, 1979.

Torretti, Roberto. *Philosophy of Geometry from Riemann to Poincaré.* Boston: Reidel, 1978.

Tourtual, Caspar Theobald. *Die Sinne des Menschen in den wechselseitigen Beziehungen ihres psychischen und organischen Lebens: Ein Beitrag zur physiologischen Aesthetik.* Münster: Friedrich Regensberg, 1827.

Tourtual, Caspar Theobald. *Die Dimension der Tiefe im freien Sehen und im stereoskopischen Bilde.* Münster: Coppenrath, 1842.

Träger, Franz. *Herbarts realistisches Denken: Ein Abriss.* Amsterdam: Rodopi, 1982.

Turner, R. Steven. "Helmholtz, Sensory Physiology, and the Disciplinary Development of German Psychology." In *The Problematic Science,* Woodward and Ash, eds., 147–166.

Ueberweg, Friedrich. *Grundriss der Geschichte der Philosophie,* 8th ed., ed. by Max Heinze, 3 vols. Berlin, 1894–97.

Ulrici, Hermann. *Gott und der Mensch. I. Leib und Seele. Grundzuge einer Psychologie der Menschen.* Leipzig, 1866.

Valentin, Gabriel. *Lehrbuch der Physiologie des Menschen,* 2 vols. Braunschweig, 1844.

Volkmann, Alfred Wilhelm. "Sehen." In Wagner, ed., *Handwörterbuch,* vol. 3, pt. 1, 265–351.

Volkmann, Alfred Wilhelm. *Physiologische Untersuchungen im Gebiete der Optik.* Leipzig, 1863–64.

Wagner, Rudolph, ed. *Handwörterbuch der Physiologie,* 4 vols. Braunschweig: Vieweg, 1842–53.

Waitz, Theodor. *Lehrbuch der Psychologie als Naturwissenschaft.* Braunschweig, 1849.

Ward, James. "Psychological Principles." *Mind,* original series, 8 (1883), 153–169.

Warnock, G. J. *Berkeley,* rev. ed. Baltimore: Penguin, 1969.

Warren, Howard. *A History of the Association Psychology.* New York, 1921.

Watson, Robert I. *The Great Psychologists: From Aristotle to Freud.* Philadelphia: Lippincott, 1963.

Weber, Ernst Heinrich. "Der Tastsinn und das Gemeingefühl." In Wagner, ed., *Handwörterbuch*, vol. 3, pt. 2, 481–543.

Weber, Ernst Heinrich. "Ueber den Raumsinn und die Empfindungskreise in der Haut und im Auge." *Berichte über die Verhandlungen der königlich sächsischen Gessellschaft der Wissenschaften zu Leipzig. Mathematisch-physisiche Classe* (1852), 85–164.

Wertheimer, Michael. *Brief History of Psychology*, rev. ed. New York: Holt, 1979.

Wertheimer, Michael. "Historical Research—Why?" In *Historiography of Modern Psychology*, Josef Brozek and Ludwig J. Pongratz, eds., 3–23. Toronto: Hogrefe, 1980.

Wilson, Margaret. *Descartes*. London: Routledge, 1978.

Wolff, Christian. *Psychologia empirica methodo scientifica pertractata, qua ea, quae de anima humana indubia experientiae fide constant, continentur*, new ed. Frankfurt and Leipzig, 1738. Reprint. Hildesheim: Olms, 1980.

Wolff, Christian. *Psychologia rationalis methodo scientifica pertractata, qua ea, quae de anima humana indubia experientiae fide innotescunt, per essentiam et naturam animae explicantur*, new ed. (Frankfurt and Leipzig, 1740). Reprint. Hildesheim: Olms, 1980.

Wolff, Christian. "Prolegomena to Empirical and Rational Psychology." Translation and commentary by Robert J. Richards. *Proceedings of the American Philosophical Society*, 124 (1980), 227–239.

Wolff, Robert Paul. *Kant's Theory of Mental Activity*. Gloucester: Peter Smith, 1973.

Woodward, William R. "From Association to Gestalt: The Fate of Hermann Lotze's Theory of Spatial Perception, 1846–1920." *Isis*, 69 (1978), 572–582.

Woodward, William R. "Wundt's Program for the New Psychology: Vicissitudes of Experiment, Theory, and System." In *The Problematic Science*, Woodward and Ash, eds., 167–197.

Woodward, William R., and Mitchell G. Ash, eds. *The Problematic Science: Psychology in Nineteenth-Century Thought*. New York: Praeger, 1982.

Wundt, Wilhelm. *Beiträge zur Theorie der Sinneswahrnehmung*. Leipzig and Heidelberg: Winter, 1862.

Yolton, John W. *Perceptual Acquaintance from Descartes to Reid*. Minneapolis: University

Index